SCIENCE, HISTORY AND SOCIAL ACTIVISM

A TRIBUTE TO EVERETT MENDELSOHN

SCIENCE, HISTORY AND SOCIAL ACTIVISM

A TRIBUTE TO EVERETT MENDELSOHN

Edited by

GARLAND E. ALLEN
Washington University, St. Louis, U.S.A.

and

ROY M. MacLEOD
University of Sydney, Sydney, Australia

KLUWER ACADEMIC PUBLISHERS
DORDRECHT / BOSTON / LONDON

Library of Congress Cataloging-in-Publication Data is available.

ISBN 1-4020-0495-8

Published by Kluwer Academic Publishers BV,
PO Box 17, 3300 AA Dordrecht, The Netherlands.

Sold and distributed in North, Central and South America
by Kluwer Academic Publishers,
101 Philip Drive, Norwell, MA 02061, USA

In all other countries, sold and distributed
by Kluwer Academic Publishers, Distribution Center,
PO Box 322, 3300 AH Dordrecht, The Netherlands

Printed on acid-free paper

Printed and bound in Great Britain by MPG Books Limited.

Everett Mendelsohn, 1997. Photograph by Martha Stewart

TABLE OF CONTENTS

PREFACE

"To earn a degree, every doctoral candidate should go out to Harvard Square, find an audience, and explain his [or her] dissertation". Everett Mendelsohn's worldly advice to successive generations of students, whether apocryphal or real, has for over forty years spoken both to the essence of his scholarship, and to the role of the scholar. Possibly no one has done more to establish the history of the life sciences as a recognized university discipline in the United States, and to inspire a critical concern for the ways in which science and technology operate as central features of Western society. This book is both an act of homage and of commemoration to Professor Mendelsohn on his 70th birthday. As befits its subject, the work it presents is original, comparative, wide-ranging, and new.

Since 1960, Everett Mendelsohn has been identified with Harvard University, and with its Department of the History of Science. Those that know him as a teacher, will also know him as a scholar. In 1968, he began – and after 30 years, has just bequeathed to others – the editorship of the *Journal of the History of Biology*, among the earliest and one of the most important publications in its field. At the same time, he has been a pioneer in the social history and sociology of science. He has formed particularly close working relationships with colleagues in Sweden and Germany – as witnessed by his editorial presence in the *Sociology of Science Yearbook*. Less visible, but even more in keeping with the man, has been his contribution – both in person and in print – to the pursuit of peace in the Middle East. Whether in his own university, or at a broader level, Everett Mendelsohn has always acted on a global stage. He has contributed to the work of UNESCO and other international organizations, and has been honored by scholars throughout the world.

This volume offers the editors – both, former students – as well as other students and colleagues who number among the leading scholars in their fields – an opportunity to give thanks to a respected teacher and friend. The volume is divided into four sections, each representing an aspect of Everett Mendelsohn's diverse career, reflecting some of the many areas in which his influence has been felt. Part I focuses upon the history of the life sciences, with essays on

G.E. Allen and R.M. MacLeod (eds.), Science, History and Social Activism: A Tribute to Everett Mendelsohn, ix–x.

subjects ranging from radical views of nature in the enlightenment to Mendelism and the rise of molecular biology. Part II continues with contributions to the social studies of science, ranging from the concept of science as social practice to the social movements in science of the 1960s and 1970s. It concludes with a reflection by one who, having taken his courses, and was inspired by his example, left academia for other walks of life. For Everett Mendelsohn, knowing the social roles that scientists play is central to our understanding of science. Part III therefore focuses upon the social responsibility of scientists, a topic of considerable interest to Everett from early on in his career. In Part IV, which concludes the book, a colleague reviews Everett's role and work in the Middle East peace process with the American Friends Service Committee – an important aspect of his life that has rarely found mention in print, but is well remembered by many who have never known him as a university professor.

For their assistance in the preparation of this volume, we are indebted to Mrs. Irma Morose of Washington University, St. Louis, Missouri, who mastered so magnificently the challenging process of preparing camera-ready copy; to Ms Jill Barnes of the University of Sydney and to Everett's assistant Yuri Hospadar, who helped assemble the complete bibliography, published at the end of this volume. We also thank the staff at Kluwer Academic Publishers, especially Charles Erkelens, who saw the manuscript through the final stages of production. The index of proper names was expertly prepared by Kim Kleinman. The entire project was initially encouraged by Annie Kuipers, now retired from Kluwer. We are also indebted to Everett Mendelsohn himself, for agreeing to be interviewed in Oxford and Boston during the summer of 2000. It is a measure of the man that he would wish to be involved in the preparation of such a work, and actively so. We look forward to seeing more from his pen, as formal retirement beckons, and as new opportunities arise, for one to whom the very idea of scholarly inactivity is assuredly absurd.

Garland E. Allen and Roy M. MacLeod
St. Louis, Missouri, and Sydney, Australia

GARLAND E. ALLEN AND ROY MacLEOD

INTRODUCTION

"All new developments in the history of knowledge have been due to those scientists who did more in their social roles than their critics wanted and expected them to do".[1] So Florian Znaniecki concluded his landmark contribution to the sociology of knowledge. Alongside the technologist and the sage, the scholar who systematizes knowledge and the scholar who fights for truth, we see from time to time emerge those whom we call "explorers, those who create new knowledge", and who stand at the apogée of human achievement. Their achievement, however, lies not only in the domain of ideas. They are above all those who see and understand cultural realities, and who "accept as normal in the domain of knowledge ceaseless and unexpected change".[2]

Znaniecki's description fits well the character and life of Everett Mendelsohn, mentor and friend, to whom the following essays are dedicated. For over forty years, Everett Mendelsohn's career has rested squarely on his service to Harvard as classroom teacher and mentor to thousands of graduate and undergraduate students; to the history of science through his writings and penetrating questions at meetings far and wide, and to the international community through his wide-ranging peace efforts in conjunction with the American Friends Service Committee (AFSC). During this time, he has seen massive changes in all three areas – indeed, he has been an agent of change, both in his profession, and in the way in which his profession is seen by others.

From fledgling beginnings in Widener Library, in the shadow of George Sarton, the history of science has become an established feature not only at Harvard, but many other universities around the world. Today, over 600 undergraduates at Harvard each year take courses in the History of Science Department, and over 400 take the courses that Mendelsohn offers. His Social Sciences 119 (currently Historical Study A 18), "Science and Society in the Twentieth Century", is among Harvard's most popular undergraduate offerings. Few who attend his lectures fail to be impressed by his easy familiarity with complex issues, and his ability to motivate the least motivated. For some, it becomes an experience in learning how to think about professional issues as matters of public interest. For others, it becomes an introduction to a world in

G.E. Allen and R.M. MacLeod (eds.), Science, History and Social Activism: A Tribute to Everett Mendelsohn, 1–20.

which the factors that have shaped modern science are to be challenged, understood, and made accountable.

Blessed with apparently limitless energy (he once said, that to get things done you have to "stay up late, get up early, and read fast"), Everett has called into existence two generations of scholars devoted to the social studies of science and the history of biology. Today, his roll call of graduate students numbers over fifty. Students gravitate to his recent work on Lewis Mumford, technology, and the challenges of modernism, and to his work in the history of nineteenth century abiogenesis, physiology, and human behavior. Then, there are those who know him best through his work in peace studies. Among very few others in his profession, certainly in the United States, Mendelsohn has shown the way a public intellectual can speak of justice and provide a challenge to traditional power structures.

Beginnings

Everett Mendelsohn was born on 28 October 1931 in Yonkers, New York, the only son and second child of first-generation Jewish immigrants from Romania and Russia. As he recalls,

> I was born and grew up during the intense years of the depression, so it was a family which was constrained in terms of resources, but there were members of the family who had independent intellectual interests ... [professionally] they were small business people, a lot of school teachers and people of that sort. I was raised in a family in which my mother was from a highly secular Jewish family – the type of background which valued social justice. My mother as a child was sent off to socialist Sunday school, and I'm sure that framed the background against which I was then to look at the world. We were, however, also in that period [when] Jewish families were haunted by Nazism. My father's older sisters had remained in Europe, in Romania, and I'm sure that there was continuing discussion in the household as relatives came into the States – and if not our relatives, relatives of friends – so that these were parts of the world that I grew up in, although I was not necessarily conscious of all the implications at the time.*

An enterprising business family in the Bronx, the Mendelsohns educated themselves and their children, and young Everett went – ironically, as he put it, "as far away as I could get on the New York subway" – to the Brooklyn Technical High School, where he enjoyed and excelled in science. After his sister established the precedent of going West – to the University of Wisconsin – he went to Antioch, a good place for a "hyperactive child", as he describes himself, a place that would 'let me move around.' Indeed, the Antioch system – study for twelve weeks, then working at a job for twelve weeks – was perfect for his needs.

Antioch proved a turning point, for it was there that one of his science teachers, Oliver Loud, a Harvard graduate in organic chemistry (and an Ohio

State Ph.D.), brought him to reflect upon history, the history of science and the political issues of the day. Loud was a knowledgeable Marxist, with a strong social conscience:

> ... [he was] truly a socialist in outlook, and he introduced me to the works of J.D. Bernal and the range of left intellectual commentaries on the sciences. Personally, he was very influential in my life. He was very, very thoughtful, a kind of moderated voice, but very principled, and he introduced me to ways of thinking about science – to the ways of framing questions and ways of pursuing issues – which matched my own general progressive outlook on the world. To a very significant extent, it was his influence that brought me into the history of science, via an interest in science and society. He was a Marxist, no question. He gave me a number of his books, he was the kind of person who read and underlined copiously, and you could, of course, watch the development of his own thought in the books that he read and underlined ... I became interested in those sorts of issues [science and society] under Loud's influence, but my own sense from early on was that a fundamental Marxist interpretation of the nature of society, the role of different elements of that society, its science, its technology, its medicine, its agriculture, this made a lot of sense. I found it a good analytical tool.

However, international affairs soon intervened. Everett came of age during the Korean War, and like so many other progressives, found himself confronting questions driven by the ambiguities of a world at war. It was then that he came into contract with the pacifist movement. His closest counselors had instilled in him a sense, enriched by his early religious beliefs and training, that pacifism was the better path; and in facing the draft, he contemplated registering as a conscientious objector. But pacifism raised problems for Marxism:

> Having early on as a college student become a pacifist, I also found some of the doctrinaire attitudes of some Marxists not very useful, they just didn't help, they got you trapped places. You didn't seem to be able to move with the nature of issues that came towards you and wanted everything to fit in neat compartments – at least from where I looked at the world, the world was much too complicated to fit in simple compartments, and perhaps it was the pacifism that I had become involved with and developed that left me much more sensitive to elements of oppression and coercion from all sides. It made my approach to the sciences recognize in part that the old Marxist view that science was a terribly powerful tool for changing society. But I did not want to slip into what was for many the easily anti-scientific attitudes of some critics on the left who had no use at all for organized scientific activity or society.

It was at this point, as Everett recalls, that he chose to adopt a "commitment to spend some proportion of my life ... making the world a better place to live", a self-commitment that guided him in his future scholarship and practice. He became involved with the Society of Friends (Quakers), and began what would become a life-long affiliation with the American Friends Service Committee (AFSC). As an undergraduate, he contemplated a career in the labor movement. However, prospects were not promising:

> Organizing within the labor movement – labor education – seemed to be the way to go. One or two of my close friends did it. It turned out not to be a particularly possible choice because it was the beginning of the Cold War and the enactment of the Taft-Hartley Act [anti-labor legislation forbidding the closed shop], which made legal movements very, very sensitive politically.

Eventually, scholarship and family made the larger claims. Everett met Mary Maule Leeds, a deeply-committed Quaker from Philadelphia, who had graduated from Antioch a year before him. Planning to marry, they looked for graduate schools that could offer places to them both. The choice of Harvard was circumstantial, as Everett recalls:

> My wife-to-be [a biologist] and I were both looking for places to apply and we had both applied to Cornell – they had a history of science department [headed by Henry Guerlac] – and to Harvard [where the department was headed by I. Bernard Cohen]. The Biology Department at Cornell lost my wife's application, and since we had both been accepted at Harvard, that's where we went.

Everett and Mary entered the Graduate School of Arts and Sciences at Harvard in the autumn of 1953 – he in the History of Science and she in Biology. They were married in 1954.

Harvard

To the present generation, Harvard in 1953 must seem like a different country. Experiences of wartime mobilization and military service were still close to many of the staff, both who had stayed in Cambridge, and those who had returned from government or military service. The troubled beginnings of the Cold War, followed by the outbreak of the Korean War in 1950, underlined the central role of the universities in what President Eisenhower, in his Farewell Address to the Nation in 1962, would later call the "military-industrial complex". Loyalty oaths were stipulated by the National Science Foundation for all new grantees, and in the spring of 1954 Harvard required oaths from all new Faculty appointments. Fear stalked the campus, in Cambridge as elsewhere, as radicals of many persuasions were confronted by the threat of expulsion, or worse.[3] That semester, Everett, a newly appointed teaching fellow in biology, and thus at the bottom of the academic ladder, was called in to see the Dean, McGeorge Bundy. Apparently, his name had appeared on one of the lists circulated by right wing groups. Bundy asked Everett if he was a communist. Everett said that he was not, but that he *was* a socialist. Bundy did not pursue the matter, and later apologized for the incident. In retrospect, Everett was lucky; others were less so.[4] Certainly, in such matters he enjoyed the warm support of I. Bernard Cohen, who equally supported Mendelsohn's contemporary, Owen Gingerich, who was a member of a strongly pacifist Mennonite family.

Inevitably, in such a small department, Cohen exerted an important influence on the young graduate assistant in biology,[5] with whom he soon began to collaborate on several scholarly projects as well as in teaching.[6] In the survey courses, the two divided history between them (Cohen took the 17th and 18th centuries; Everett, the 19th and 20th). Everett took the biological sciences, and Cohen, the physical sciences. In 1957, in his fifth year at Harvard, Everett was elected to the distinguished Society of Fellows. In 1960, he completed his Ph.D. and was appointed an Instructor in the History of Science. He soon began to attract honors and graduate students. Although his research and teaching efforts eventually took him into contemporary issues, the nineteenth century held a special place among his interests. His first love was the history of nineteenth century physiology, which formed the basis of his doctoral dissertation and his first book.[7]

The first year of his formal teaching career, 1960-61, saw another side emerge. That year, he first offered what would become his near-legendary, upper level course, Social Sciences 119, entitled "The Social Context of Science". This course – unprecedented at Harvard, and indeed, in the United States – was listed in the General Education Program, the remarkable creation of James Bryant Conant, who had become Harvard's president in 1933.[8] Open to all undergraduates, "Soc Sci 119" was an immediate success, and established Everett's reputation as a lecturer. The course also attracted new friends and colleagues, as Everett recounts:

> I did have one of those funny experiences the first day I showed up in front of those forty students. There in the first row was David Reisman, whom I had met through some of my anti-war interests; and he announced to me that he had the tradition of always sitting in on the course of some young scholar and would I mind if he sat in on mine. He is very expressive: if you say something he likes, he beams, or [if you say something he doesn't like he] scowls – it was quite an education. What he did after that, he sent me carbons of letters he had sent to other people which in any way referred to one of the topics I had dealt with, with his own little hand-scribbles of what I might find interesting.

By the early 1960s, Harvard, respected for having given sanctuary, if not whole-hearted endorsement, to George Sarton's fledgling program in the history of science a generation earlier, was now ready to support the history of science in more substantial ways. As a discipline, the history of science fitted well into Conant's vision of a liberal education suited for an age of science.[9] Equally, it seemed just the right approach for those who, in the metaphor of the day, sought to "bridge" the "two cultures" of the arts and the sciences. With Bernard Cohen, Everett pioneered in 1962 a new undergraduate Field of Concentration (i.e. "major") in "History and Science", which soon began to attract outstanding students, many destined for careers in the professions. Asked what most shaped his thinking about the history of science in those early days, Everett is quick to reply:

> Dirk Struik, the mathematician at MIT [was an important influence]; he was the father of my classmate at Antioch, Gwen Struik. We saw a lot of him when I was an undergraduate, and then when I came [to Cambridge] as a graduate student. He was very generous and had me over often with several other students and other people. He engaged in very thoughtful, very well-schooled, not doctrinaire, Marxism. The introduction he did to Engels' book [*The Dialectics of Nature*] was a very thoughtful historical piece. He was a Marxist who absolutely enjoyed taking a group of young people off to see the Saugus Iron Works in Saugus, Massachusetts. He was like a kid when he was explaining how they got the iron or how it was smeltered in a certain kind of way. That's a good kind of model to have, a combination of Oliver Laud and Dirk Struik ... .

Much stimulation also came from Thomas Kuhn, Marie Boas (later Marie Boas Hall), Robert K. Merton, and, of course, the work of J.D. Bernal:

> As a young instructor or assistant professor I did find Tom Kuhn's *Structure of Scientific Revolutions* provoking. I had heard a lot of it before, he had been [at Harvard] when I was a graduate student, and was on my graduate doctoral committee, one of my examiners, but I found he had a fairly systematic way in which he attempted to unpack what it is that goes into the development of scientific concepts ... In that way I would say it was a very useful study for me. ... I found some of Marie Boas' work on the early history of atomism extremely interesting, looking at an intellectual historian's approach. I also liked the kinds of studies that Robert Merton did: I read *Science, Technology and Society in the Seventeenth Century* early on in my career, and I liked the approach. ... I liked Bernal's little book on *Science and Industry in the Nineteenth Century.* I found his way of analyzing useful and, not surprisingly, congenial.

Everett's timely appearance on this busy stage, shared with other teachers in the General Education program, including Thomas Kuhn, enabled him to consolidate a position of leadership in both history and contemporary affairs, and to make a mark for himself in the "social history of science".[10] Harvard in the 1960s was at the center of what became known later as the "golden age" of science policy studies, a subject that quickly seized Everett's interest. He soon became a Research Associate in Don Price's program in Science and Public

Policy at the Graduate School of Public Administration, and later served as Director of the Research Group on the Biomedical Sciences at Harvard (1966–68), and as a Research Associate of the IBM-funded (but ultimately short-lived) Program on Technology and Society (1966–71) at Harvard and MIT.

In the same period, he rose in the ranks at the University, becoming an Associate Professor in 1965 and a full Professor in 1969. Not unlike many Harvard colleagues, Everett spent time in these years giving short courses to public servants in Washington, and attending conferences where policy issues were being discussed. Between 1968 and 1973, he served on the Committee on Life Sciences and Social Policy of the National Academy of Sciences/National Research Council, and contributed to a number of reports on issues facing the future of the biomedical sciences. This period also saw him working with Harvard Medical School on the study and definition of "brain death".

By the late 1960s, Everett had become an outspoken critic of the Vietnam war, and of the deployment of nuclear weapons. On behalf of the American Friends Service Committee he visited Vietnam between December 1967 and January 1968, and so experienced the Tet Offensive. The violence of war became a first-hand reality for him. On returning to the States, his experience was canvassed by Senator Everett Dirkson (R-IL), then Senate minority leader, and by others who were critical of the war. By then, he had become active in the anti-war movement. Supporting petitions, getting Harvard staff to take up positions on the war – including weapons research and the use of herbicides in Southeast Asia – he also helped organize one of the largest campus moratoria on the war, which brought nearly 200,000 people together on Boston Common in October, 1969.

Activism came at a price, however. In the years 1968-70, political tensions were high on college campuses. Colleagues and associates who happened to disagree with Everett's position on political issues, or who questioned the methods of protest in which he was involved, were often hostile, and the criticism hurt. Sometimes, his reputation preceded him. He recalls, with mild amusement, when

> ... in the fall of 1968 and spring of 1969, I was a fellow at Churchill College [Cambridge]. I remember coming back to Harvard at the beginning of the [next] term and being at a cocktail party, and someone coming up to me and grabbing me by both lapels [saying] 'Everett I think what you did last spring was absolutely despicable, yelling at me. ... I explained I wasn't there, but away on leave. That did not suffice. His assailant replied, 'But if you had been there you would have ... .

There were also less amusing moments. In the 1970s, an ungenerous reporter for the *Wall Street Journal* portrayed Everett in a rather negative light.[11] Recalling the incident, Everett remarked,

> A young woman reporter asked if she could follow me around, she wanted to do a piece on activist professors. It was a funny piece. She picked up from

> one student that I had been traveling a lot. ... The article was fairly accurate, assessing the things I was doing, and then the spin was 'activist professors'. They were shooting at the category and I happened to be the person they had come to. ... I was not overly wounded [by the tone of the article] but some of my colleagues at Harvard treated me as though I had been hit by a truck; but I shoved it off. My guess is that the reporter wrote one thing, and then her editors took some of the piece even further.

A highly visible political profile inevitably brought strong reactions from university administrators. Everett tells the following story, with a touch of humor, from June 1971:

> I had given a talk earlier that year at the Radcliffe Institute ... about science and social action. ... In the audience was the wife of MIT President Jerome Wiesner, whom I knew at the time. I know the talk created some controversy, but then at the MIT Commencement, Wiesner was giving his President's address, and he singled out three critics of science who he thought were really detrimental: Lewis Mumford, Herbert Marcuse, and Everett Mendelsohn. That was my moment in the sun. After that it was [all] downhill! I heard about Wiesner's remarks because I had just had a ruptured appendix, and I was lying in the recovery section in Holyoke Center, and someone brought me a copy of the commencement address.

Such notoriety was annoying, but not damaging. If Everett's public posture cost him friends in some quarters, it brought him admiration and respect in others. Over the years, he has represented the "loyal opposition" at Harvard. His role, as he puts it, has been "consciously modest":

> I know [that] institutions have lives of their own which roll right over the rest of us. I would say there were a number of us who [have] remained visible critics, who made sure that Harvard didn't edge away from their policy of, for example, not accepting any classified research.

In many ways, Harvard gave him both the scope and resources he needed to speak critically on issues affecting the nation and the world:

> Their [the administration's] displeasure never silenced my voice. There were times when I'm sure there was kickback. I know that at the time I was up for tenure a number of the people who were supporting me were very worried that politics may have come back. Fortunately, it didn't.

In the last analysis, Harvard protected Everett. Tenure helped him "make a difference" not only in the fields in which he writes, but also in the larger community in which he moves. He has shown how a public intellectual, by taking a stand, can initiate or encourage change. Change was a characteristic feature of his term on the Committee on Science, Arms Control and National

Security of the American Association for the Advancement of Science (1980-82), and the Committee on International Security Studies, of the American Academy of Arts and Sciences.[12] To change thinking was also a principal intention, and a major result, of his work for the American Friends Service Committee.

From time to time, friends have observed that time devoted to political issues meant time diverted from research and writing. To the extent that this is true, Everett recalls a trade-off that he made early in his career:

> There have been several academic costs [to activism] and I have been asked periodically to confront them. I can remember quite a few years ago several of my former graduate students got me at a meeting ... and sat me down and told me they were worried about what I was doing because of the articles and things that were not coming out in the history of science. ... They were good people, these were not people who disagreed with me but they said I was wasting time and they were absolutely right. I was very conscious of that. Somewhere along the line I came to terms with the fact that I might write a book or two less. That was a commitment I had made as an individual. That was one cost. I would have been more productive [in academia].

True enough. But there are many measures of academic "productivity" apart from numbers of students and publications. Everett's influence upon his undergraduate and graduate students, his ideals of service to academia and the world, together with his writing, have made him a "role model" many have found inspiring. There can be no question that Everett has had a "productive" career.

Historical Studies in the Life Sciences

When Everett Mendelsohn entered the infant history of science program at Harvard, there was no sub-discipline called the history of biology. As a disciplinary activity, the history of science was principally seen as the history of the physical sciences, especially physics, and especially the physics associated with the Scientific Revolution of the seventeenth century. Although there had been a flurry of activity surrounding the centennial of Darwin's *On the Origin of Species*, even Darwinism was regarded more as the business of Victorian social and intellectual history, than as an aspect of the history of scientific ideas *per se*. Everett was one of the first Americans to insist that the history of biology was a legitimate area to be pursued within the larger field being claimed by the history of science.

Everett's early work in the history of science focused upon the development of eighteenth- and nineteenth-century German cell theory and physiology. What he found particularly intriguing was the change in explanatory modes in the work of people such as Theodor Schwann, Maurice Francois Xavier Bichat,

Richard Mayow, and Antoine Laurent Lavoisier. Explanations of cell structure and function were shifting from the descriptive and comparative to the functional, and from *biological* (meaning usually *vitalistic*), to the materialistic and physico-chemical. It was a shift in criteria of theory formation whose importance was as profound as that in the physical sciences in the sixteenth and seventeenth centuries, and in chemistry at the end of the eighteenth century. At a broader level, Everett went into this field to explore whether such vast theoretical changes could be taken as models for the biological sciences,[13] a question that Kuhn had raised in *The Structure of Scientific Revolutions*, but not explored.

As a graduate student, Everett's dissertation advisor was Bernard Cohen. Although the history of biology was not Cohen's field, he knew a great deal about seventeenth and eighteenth century science, and far from trying to woo Everett into working on the physical sciences, encouraged his interest in biology:

> In a very interesting way he was very happy and supportive of my continuing work in biology. His image of an historian of science was someone who had a field in science and he liked the fact that they gave me lab space in bio labs, that I was a teaching assistant [in biology] that I regularly gave seminars there. ...My relationship was close with Bernard, but my work was actually on my own, it wasn't his field.

Fortunately, Everett found support from biologists, including the insect physiologist Carroll Williams (who later proposed him for the Society of Fellows). But the topic of animal respiration, which formed the subject of his dissertation, came from his own reading and laboratory experience.

Reflecting on the role of science in the training of an historian of science, Everett recalls his own experience:

> My own sense is that in the training of the historian of any of the sciences there is an enormous amount to be gained by spending time in the lab or working theories among [scientists], hearing them talk in the labs, feeling the stuff, the uncertainties, going to the colloquia, and taking courses to give you a greater confidence in being able to understand the material. ... In that context, if I were to take any historian of biology coming along, I would say go spend a summer at Woods Hole in [one of the] courses, to get immersion with a group of people who are making their career in biology. It's exciting to interact and see and feel how they are thinking and then, of course, have them turn to you and say, 'You are an historian ... what does that mean? Here you are doing these messy experiments, what good is that going to do you?" On the other hand, you know that, in retrospect, all those hours you spent isolating neuro-secretory hormones and doing bioassays pay off in terms of your feel for what you can and can't do in an experimental set-up.

What is evident even from an early stage, however, was Everett's keen interest in the larger intellectual and social issues that form part of the history of any science. In looking at the development of the theory of animal heat, for example, he was first and foremost motivated by the "desire to find out what elements went into concept formation in the biological sciences and what the relation of this process was to the knowledge and techniques of the physical sciences".[14] More specifically, he was particularly intrigued by the use of analogies to physical and chemical processes, what work they did for the physiologist, and how their use changed over time.

In *Heat and Life,* the book that grew from his dissertation, he wrote:

> The history of biology gives evidence of the growing reliance of the biologist on mechanical and physical analogy. ... What is of particular interest is the manner in which physical and mechanical analogies are utilized in the genesis of biological theories.[15]

Heat and Life traced the changing analogies of respiration and the "inner fire of life", from their characterization as strictly biological, or "vital" processes – comparable to fire in the physical sense only by analogy – to their characterization as a true combustion process physically and chemically *identical* to that found in a flame (although occurring at a slower rate). Through the nineteenth century, physiology came more and more to incorporate the physical sciences not merely as analogy, but as process. As such, this work took Everett deeply into German philosophy, particularly the philosophy of mechanism and materialism, a field that intrigued him, and one in which he might have stayed had not other interests and activities taken precedence. Asked what areas he would have liked to explore more thoroughly in the history of biology, he replies:

> I would have loved to have gone more deeply into that nineteenth century biological material. ... [I would have liked to look] in greater depth at the nature of the physiologists' activity, the physiologists' role in the nineteenth century – in a way to add the degree of contextual analysis that might have been there. ... I knew the task ahead of me was going to be big and probably by default, rather than absolutely saying, 'No, I'm not going to do that,' I diverted ... .

In fact, his research in the history of physiology lead directly to his pioneering work on the factors that shaped the professionalization of science in the nineteenth century.[16] It also led him later, in the 1980s, to take a brief look at eighteenth- and nineteenth-century theories of the origin of life, work that he found fascinating but did not pursue.

When Everett went on to more sociological and policy-related issues, his many graduate students and others working with him continued to mine an enormous range of topics within the history of biology. Over the last forty years, a good percentage of American – and not an insignificant number of

European – historians of biology have passed through the Harvard program under Everett's generous guidance. As so many have been highly productive, Everett's influence has spread widely within the field. Perhaps more significantly, many of his students have incorporated – sometimes explicitly – Everett's wider sociological perspective.

Social Studies of Science

From about 1965, Everett's interests turned increasingly towards what was coming to be called, notably in Britain, the "social studies of science".[17] Several factors played into this change of direction. One was his own earlier work in the history of physiology, which raised a number of questions about the structure of scientific communities.[18] Another was his political activism – focused especially upon nuclear weapons and the alleged misuses of science and technology – which led logically to an interest in science policy. Indeed, this was one way to integrate his activism and his scholarship:

> Part of what I did do ... was to go off into work on social studies of science – policy things. Some of the activism carried over successfully, into the scholarly world.

A third, even more influential development, came with his course, Social Sciences 119. It became clear that this field was one in which extremely important questions were being posed, and about which there was much ill-informed debate. Everett found himself pulled into debates where his expertise, and perspective, were needed, and welcomed. Similar commitments also came about with his advice to government, and his work with Don Price on science and public policy, especially on the issue of classified research. Recognizing this new shift in his research, he was invited (by one of us) in 1970 to join the inaugural editorial board of *Science Studies* (now *Social Studies of Science*), and to share in the development of what was rapidly becoming, in Britain and elsewhere, a recognizably new field of research and teaching. At the same time, Everett played an active role in the History of Science Society, and also served as chair of what had metamorphosed from a Committee to a Department of History of Science at Harvard (1971–78), with a growing graduate program, in which the sociology of science was an increasingly important component.

The social studies of science gave a new dimension to Everett's early transatlantic turn. Rather "by accident", he recalls, and in consequence of his work on 19th century physiology, he became involved in the activities of contemporary German philosophers, sociologists, and political scientists. In the early 1970s, he was introduced to Jürgen Habermas and Carl Friedrich von Weizsäcker, then working at a newly-established Max Planck Institute in Starnberg, near Munich. This association led to a series of appointments in Germany – at the Zentrum für Interdisziplinäre Forschung in Bielefeld (1978–79), and the Wissenschaftskolleg in Berlin (1983–84) – and from 1976, to a long and productive association with Peter Weingart and Helga Nowotny, among others, in the conduct of the *Sociology of Science Yearbook,* which began in 1977.[19] Similar interests brought him many friends overseas – and visiting appointments at Churchill College, Cambridge (1968–69), the Van Leer Foundation in Jerusalem (1978), the Conservatoire National des Arts et Métiers (CNAM) in Paris (1989), and the Swedish Center for Advanced Studies in the Social Sciences (SCASSS) in Uppsala (1994). In 1974, he formed collaborations with prominent scholars in the United States, including Arnold Thackray (on the volume, *Science and Values*) and Marjorie Grene (on *Topics in the Philosophy of Biology*).[20]

During the 1980's, Everett was particularly sympathetic to the work of those who would see "science in context", a subject perhaps better understood in Britain and Europe than in the United States, where the roots of enquiry in the history of science remained damaged by Cold War paranoia. As the discipline began to mature, however, and easy political categorizations became more difficult. Everett grew impatient with those who, like Joseph Ben-David, "vigorously denied the notion that social activities, social forums, social processes in any way affected knowledge, that the knowledge system was immune, separate".[21] On the other hand, he sympathized with the intellectual position associated with Robert Merton,[22] and faulted (in his words) those (especially in Britain) who were "forgetting the realm of the social beyond the micro-social". It seemed, he says, "that they lost [sight of] the relationship of society as a whole, [that is] what [role] in society science [plays], not only in the micro-sociological world of the scientific laboratory or department". His relationships with sociologists were to remain cordial, if cool; as he recalls, "it may be because of my regularly wanting to deal with society as an element [in the practice of science] that I was not seen as a threat by [them], albeit I was pushing some of them places they didn't want to go".

Today, he believes the "internalist-externalist debate", which in the 1970s sowed such discord,[23] is now passé. Still, he well recalls that:

> I was chided for taking people in those directions; I think some of my students were chided for being engaged in some of the same, but what became funny is [that] the colleagues who picked up the crusade against the social in the processes of science [did so] almost as an ideological bludgeon, and I think they were just bypassed by the new generation. ... I think ... that the way you dealt with that was not through the creation of manifestos, but through the nature of the work you did in your historical studies; and by and large I would say that today's generation of historians of science totally ignore that debate ... and just go ahead with doing the empirical historical study and the interpretations of what they find, regardless of that old distinction ... .

Whether or not one believes, as Everett does, that the "science wars" have also had their day, it is tempting to share his view that "there is no argument any more" that science is "influenced in all sorts of ways by social, economic and philosophical currents". Perhaps scientists will read more of the history of science, and find it questioning; and historians will ask more questions, without being considered subversive of science. What is clear is that Everett has encouraged productive conversation between historians and scientists. His method, he says, is to distinguish between people and the positions they hold – in his words, making a "political argument without making it antagonistic to the individual". His method succeeds. Everett has never been "anti-science" – far from it. Science, he recognizes, is a practice of vital significance to human progress. On the other hand, as he demonstrated during the recombinant DNA controversy in Cambridge in the 1980s.[24] He has been highly critical of the way in which the scientists and scientific institutions can operate. There is, and remains, an important difference.

The Wider World: Academics and Social Responsibility

Outside academic life, Everett is best known for his contributions to public debate on global issues of war and peace. Among the issues that have preoccupied him, perhaps the most important have been nuclear disarmament, peace in the Middle East, and public responsibility in genetics research. In 1967, working with his first wife, Mary – a leading member of the Quaker community in Cambridge – he became a member of the Executive Committee of the American Friends Service Committee (AFSC). Although he has never become a member of the Society of Friends, he became chairman of the New England Regional Office in 1972, and continued as a member of the International Division.

At about the same time, he began to bring together – or "facilitate" – the formation of small groups of Soviet and American scholars and former government people, for "intense discussions of issues of American-Soviet relations". These gatherings were held alternatively in the US and the USSR, and continued for several years. To Everett, as to Alfred North Whitehead, the process was the reality:

> Getting people together discussing the issues, not necessarily agreeing, but discussing the issues in such ways that you can develop a common discourse [was the idea]; and what we watched over the years was a broadening circle in the Soviet Union and in the United States, a broadening circle of people who knew how to talk to each other. I was not the only one doing that; Pugwash started doing a number of those things, and we had some interactions. Anti-nuclear weapons, anti-testing, this overlapped with the Soviet-American [meetings] because very often they were interested in the weapons as the Americans wanted to talk to Soviets about it. That involved a kind of activism, petitions, campaigns to stop nuclear weapons testing; issue by issue, they were focused and they were operational.

Everett went on to develop a similar "focused and operational" approach towards the Middle East.[25] Oddly, as it became a subject on which he has published, his first involvement in the Middle East was, he recalls, "an accident". In 1968, the AFSC Executive Committee was attempting to establish a program in Gaza, but was having trouble getting Israeli permission for their staff to enter the area. Everett, who was on the Middle East panel, was asked to intervene:

> I was in Europe, they [AFSC] called me at the airport, and said could I divert and go to see the Israeli authority ... to see if I could work it out. I went down and visited in Gaza and visited back in Tel Aviv and the Defense Ministry in Jerusalem and the Foreign Ministry, and wrote a report. I had been there only once before ... but ... have been going back at least once or twice every year since.

This introduction led to structured dialogues between Israelis, Arabs and Palestinians, similar to those Everett had facilitated in Soviet-American relations, and contributed to a process that surmounted distrust, built confidence, and even led to co-authored papers. Begun as an activity of the AFSC, the idea was passed in the 1980s to the American Academy of Arts and Sciences, and its Committee on International Security Studies, of which Everett was (and is) a member. Following the Oslo Accords, and with help from the MacArthur Foundation, a regular series of meetings took place. An "invisible college" was created, and a conversation continues – even by mobile phones – today. Once again, Everett demonstrated how a single individual, with a willingness to hear both sides, can "make a difference".

How much difference? On that no one, especially Everett himself, can be definite. Optimistic in his belief that change is possible, Everett is also realistic. In referring to his work on the Presidential Commission on Research in the Life Sciences, which dealt with the recombinant DNA debates of the 1970s, he wrote:

> We came out with a report with recommendations on how to deal with recombinant DNA. I think it added a nice element to the discussion at the time. Was it world-shaking? No, but it was one of those several things which were adding to the reshaping of how you think about governance of science.

Individuals do not often by themselves, or by any one act, significantly alter large social patterns. But organized group efforts, such as those sponsored by the AFSC, have had an influence. For Everett, every small step – individual or group – counts, and his activism has helped fulfill his early promise to himself that he would devote a part of his life to making the world "a better place".

Looking to the Future

Looking ahead, Everett sees science and the academy facing undiminished challenges.[26] Where once the Cold War drove the universities into an embrace with the military, now globalism drives them into the arms of the transnationals. The consequences for science are enormous. "If I finish up a paper that

I was going to do for the latest *Yearbook* in our series," he recalls,

> it was [to be] on the transformations of biology which at so many levels of the field has changed, and with it, science. The 19th century, turn of the century image is gone. Science is now part of a much more integrated system, so the argument that I made about there being a social context and there being a tension between the role of the university and the outside were accurate; but university scientists never stood wholly outside that broader context. Some of their day-to-day work may be more or less relevant, but they were becoming part of the system, and at the moment it is the knowledge system itself and the practice of these fields in molecular biology, which are part of the voice; so whether you are working in the university or outside, your research agenda largely looks the same ... .

What is apparent, is that Everett is not only a scholar, but also someone who appreciates the necessity of bringing the intellectual and social aspects of life together in an effective way. Everett could not be an ivory tower scholar. Neither could he be a full-time social or political organizer. For him, it is the Weberian synthesis of the two – the man of knowledge in the midst of the tensions of the real world – that matters. He does not "bridge" the two cultures, because for him there are no two cultures to begin with. This belief has not only informed Everett's ethics and social conscience, but has also inspired many who have come to know and work with him. Always optimistic and positive, despite adversities, both personal and professional, he has pushed ahead with a secular faith far more resolute than many with more conventional religious convictions. It is in this spirit, and to recognize his generous and principled commitment, that we dedicate this volume, to Everett, a citizen of the world, on the occasion of his seventieth birthday.

Looking Ahead: The Essays

In the following essays, Everett's students and colleagues tackle a range of historical and historiographical issues that have proved central to the study of science as an intellectual and social activity. Part I is devoted to Everett's first love, the History of Biology. Jane Maienschein opens the section by asking a question that is central both to Everett's life, and to the present book. 'Why study the history of biology?" she asks, and then offers persuasive answers, shaped by the period in which she developed as a scholar. Historians can and must 'speak out' on issues of the day, she argues, and can 'make a difference.' In the next essay, devoted to Diderot's *Encyclopédie*, Shirley Roe shows compellingly how ideological preferences have structured intellectual debate, obliging those who wish to reorder knowledge and society to reflect upon the 'subversive' tendency in all serious scholarship. Fred Churchill continues with a characteristically 'modest' claim, but finishes by showing how different models of explanation have been forged by deep debate between competing views of nature, and of knowledge. Raphael Falk also shows how, in the history of biology, understandings of nature are guided by deep convictions, arising from influences outside science itself. Lily Kay, in what was to prove one of her last essays, reflects on contemporary developments in molecular biology as seen from "Mt. Everett", as she puts it; a view that challenges simplistic reductionism, and strives towards the development of a wide range of explanations that reflect a deeper respect for human self-consciousness, experience, and diversity.

In Part II, devoted to the Social Studies of Science, Anne Harrington, stimulated by Everett's example, considers the *qigong* system in China as an opportunity to reflect upon characteristic features of human 'diversity' that impart different cultural explanations to observed natural phenomena, and the necessity to 'engage' with them, without prejudice. Helga Nowotny continues with a closely observed account of her own attempts to study explanatory diversity in different media, showing how the 'telling of a story' can in many ways affect its understanding, and perhaps even its outcome. It is a story familiar in literature since Scheherazade, but no less relevant to science. Peter Buck tells a different story of 'fact-finding', and shows how institutional preferences – in this case, in public health – can use so-called 'objective knowledge' to map what may be very different, not to say 'acceptable', ways of viewing the world. Jean-Jacques Salomon continues in a similarly cautionary vein, questioning (not for the first nor surely for the last) the contradictions of Western science and its belief in the "idea of progress". Peter Weingart moves behind the scenes, to examine how certified knowledge comes into existence, and how it has sought to remain unchallenged, despite the pressures of an increasingly informed democracy. Garland Allen considers the way in which Marxism can provide considerable insights into the history of science (in this case genetics and agriculture). The contexts within which people generate knowledge – the "social" as well as the "cognitive" – are clearly relevant to all human institutions, and, as Gary Werskey reminds us, the act of questioning their origins and consequences formed a central feature of Everett's 'SocSci 119'. As Werskey's own experience demonstrates, the course had far-reaching implications for many who took it, and he, for one, has taken its lessons to the wider world.

In Part III, seven authors consider different ways of viewing the nebulous concept of 'social responsibility' in science. Mark Barrow examines the American experience of nature conservation, and Robert Proctor, the practice of human experimentation – issues with a long history, but brought to new prominence in the 1930s and the Second World War. Mark Adams offers an appreciation of the changing sensibilities of Soviet scientists during the Cold War, while Gilbert Whittemore, who combines legal and historical training, shows how the facts of plutonium were constructed by government, well before they were debated by scientists. A different perspective on the nuclear issue is offered by Elizabeth Watkins, who shows how, even during the Cold War, the raising of public consciousness on environmental issues could, and did occur, sometimes forcing scientists to take up adversarial positions. In his study of the Smithsonian and the Cold War, Roy MacLeod, however, shows how even leading institutions could be altogether complacent – and complicit – in accepting government funds for science, without first establishing what consequences might ensue. To the extent that Western society now accepts the need to debate such issues, and to set clear requirements of public accountability – a subject on which Everett has argued passionately – it owes much to historians' efforts in revealing and describing such cases. If governments have a 'duty of care' to their citizens, so, too, must trans-national corporations and others in corporate science who stand to profit from the outcomes of new discoveries related to the human genome. Such is one inference to be drawn from the essay by Michael Fortun, which canvasses the Human Genome Project, and foresees dangers in the structured ways in which historians and scientists are accounting for its past, and changing its future.

In the concluding Part IV, which we have called a 'Postscript', we have but one essay, by Yaron Ezrahi, which highlights the large volume of work and substantial contribution that Everett has made to improving the climate of cooperation among intellectuals and scholars in the Middle East. Everett's experience, in many ways, has given a new sense of purpose and definition to the idea of the "public intellectual", in approaching one of the most intractable problems confronting modern society. As he demonstrated, and the Oslo Declaration confirmed, there is a vital role to be played by disinterested 'third parties', particularly where basic human rights are at issue, and where governments prove irresolute. In this case, as in many others, it may well be the individual David, rather than the corporate Goliath, that holds the key – and it is pleasing to see and report that the ideas that Everett inspired have retained a powerful hold on the imagination of negotiators and scholars alike. If wise counsels prevail, even in these dark hours of internecine struggle, we may yet see the light of humanity dawn in this and other troubled parts of our fast-shrinking world. For the process to succeed – and for scholarship to be truly engaged with social change – we need people like Everett, and those whom he inspires.

Washington University and University of Sydney

Notes

* Unless otherwise indicated, all quotations from Everett are taken from two interviews conducted by the editors in the summer of 2000 in England and in the United States.

1 Florian Znaniecki, *The Social Role of the Man of Knowledge* (New York: Columbia University Press, 1940), p. 164.
2 Znaniecki, *The Social Role* (cit. n. 1), p. 194.
3 See Michael P. Rogin, *McCarthy and the Intellectuals* (Cambridge, Mass: MIT Press, 1967); Sigmund Diamond, "Veritas at Harvard', *New York Review of Books,* 28 April 1977, pp. 13–17; David Caute, *The Great Fear: Anti-Communist Purge under Truman and Eisenhower* (New York: Simon and Schuster, 1978); *Regulating the Intellectuals: Perspectives on Academic Freedom in the 1980s,* Craig Caplan and Ellen W. Schrecker, eds. (New York: Praeger, 1983); Ellen W. Schrecker, *No Ivory Tower: McCarthyism and the Universities* (New York: Oxford University Press, 1986); Lionel S. Lewis, *Cold War on Campus: A Study of the Politics of Organisational Control* (New Brunswith: Transaction Books, 1988); Richard M. Fried, *Nightmare in Red: The McCarthy Era in*

Perspective (New York: Oxford University Press, 1990); Sigmund Diamond, *Compromised Campus: The Collaboration of Universities with the Intelligence Community, 1945–1955* (New York: Oxford University Press, 1992).

[4] Anna Marie Cox, "Harvard Acknowledges 'Regret' for Dismissal of Professor in 1954", *Chronicle of Higher Education* 42(31) (April 13, 2001): p. A-19.

[5] An influence that Everett has frequently and aptly acknowledged. See Everett Mendelsohn, *Transformation and Tradition in the Sciences* (Cambridge: Cambridge University Press, 1984).

[6] *A Treasury of Scientific Prose: A Nineteenth Century Anthology*, Everett Mendelsohn, I. Bernard Cohen and Howard Mumford Jones, eds. (Boston: Little Brown, 1963); and "Science in America: The Twentieth Century", in *The Evolution of American Thought*, A.M. Schlesinger, Jr. and Morton White eds. (Boston: Houghton Mifflin, 1963), pp. 432–445.

[7] Everett Mendelsohn, *Heat and Life: The Development of the Theory of Animal Heat* (Cambridge, Mass: Harvard University Press, 1964).

[8] See James Hershberg, *James B. Conant: Harvard to Hiroshima and the Making of the Nuclear Age* (New York: Alfred A. Knopf, 1993), and more recently (and controversially) Steve Fuller, *Thomas Kuhn: A Philosophical History for Our Time* (Chicago: University of Chicago Press, 2000), especially Chapter 3, "The Politics of the Scientific Image in the Age of Conant".

[9] Herschberg, *Conant* (cit. n. 8); Fuler, *Thomas Kuhn* (cit. n. 8); see also Joy Harvey, "History of Science, History and Science, and Natural Science: Undergraduate Teaching of the History of Science at Harvard, 1938–1970", *Isis* 90 (1999), pp. S270–S294.

[10] From his course came material that informed several works, including "Science Has A Social Context: Comments on Papers by Dr. Hanss Bahrdt and Dr. Jacob Schmookler, in *Economic and Social Factors in Technological Research and Development* (Columbus: Ohio State University Press: 1965): pp. 51–58; and also his influential essay, "The Emergence of Science as a Profession in Nineteenth-Century Europe", *The Management of Scientists*, in Karl Hill ed. (Boston: Beacon Press, 1964), pp. 3–48; and "Three Scientific Revolutions", in *Science and Policy* Issue, Paul J. Piccard, ed. (Itasca, IL: F.E. Peacock Publishers, 1969), pp. 19–35.

[11] Liz Roman Galese, "The Good Life: A Harvard Professor Can Teach, Research and Travel the Globe", *The Wall Street Journal* (December 14, 1976), p. 1.

[12] These topics were ones to which he has returned on several occasions. See, for example, Everett Mendelsohn, "The Historian Confronts the Bomb", *Proceedings of the Symposium on the Role of the Academy in Addressing the Issues of Nuclear War* (Washington, DC: Hobart and William Smith College, 1982), pp. 44–57; "Knowledge and Power in the Sciences", in *Science under Scrutiny*, R.W. Home, ed. (Dordrecht, Reidel, 1983), pp. 31–47; "Science, Technology and the Military: Patterns of Interaction", in *Science, War and Peace*, Jean-Jacques Salomon, ed. (Paris; Economica, 1990), pp. 49–70; "Science and the Military", in *Science in the Twentieth Century*, John Krige and Dominique Pestre, eds. (Reading: Harwood Academic Publishers, 1997), pp. 175–202.

[13] See, for example, Everett Mendelsohn, "Physical Models and Physiological Concepts: Explanation in Nineteenth-Century Biology". *British Journal for the History of Science* 2 (1965), pp. 201–219; "Cell Theory and the Development of General Physiology", *Archives Internationals d'Histoire des Sciences* 65 (1963), pp. 419–429.

[14] Mendelsohn, *Heat and Life* (cit. n. 7), p. ix.

[15] Mendelsohn, *Heat and Life* (cit. n. 7), p. 3.

[16] Everett Mendelsohn, "The Emergence of Science as a Profession ..." (cit. n. 10), pp. 3–48.

[17] This development was initiated on a European basis by Roy MacLeod, Bernard Lecuyer and Gerard Lemaine, with the support of Clemens Heller, through Project PAREX, forerunner of the European Association for the Social Studies of Science and Technology.

[18] Everett Mendeloshn, "Revolution and Reduction: The Sociology of Methodological and Philosophical Concerns in Nineteenth Century Biology", in *The Interaction between Science and Philosophy*, Yehuda Elkana, ed. (Atlantic Highlands, N.J.: Humanities Press, 1974), pp. 407–426

[19] *The Social Production of Scientific Knowledge: Sociology of the Sciences Yearbook*, Everett Mendelsohn, Perter Weingart and R. Whitley, eds. (Dordrecht: Reidel, Vol. 1, 1977). For subsequent volumes, see Mendelsohn's bibliography, at the close of this Volume; see also Everett Mendelsohn, "Thinking Like a Mountain: The Epistemological Puzzle of Environmentalism", in *Grenzbeschreitungen in der Wissenschaft*, Peter Weingart, ed. (Baden-Baden: Nomos Verlagsgesellschaft, 1995), pp. 152–167.

[20] *Science and Values*, Everett Mendelsohn and Arnold Thackray, eds. (New York, Humanities Press, 1974); and *Topics in the Philosophy of Biology*, Everett Mendelsohn, Arnold Thackray and Marjorie Grene, eds. (Dordrecht: Reidel, 1976).

[21] See, for example, Joseph Ben-David, *The Scientist's Role in Society: A Comparative Study* (Englewood Cliffs, N.J., Prentic-Hall, 1971).

[22] Everett Mendelsohn, "Robert K. Merton: The Celebration and Defense of Science", *Science and Context* 3 (1989), pp. 269–289.
[23] Roy MacLeod, "Changing Perspectives in the Social History of Science", in *Science, Technology and Society: A Cross-Disciplinary Perspective*, Ina Spiegel-Rösing and Derek de Solla Price, eds. (London: Sage, 1977), pp. 149–195.
[24] Everett Mendelsohn, James Sorenson and Judith Swayze, "Recombinant DNA: Science as a Social Problem", Appendix, *Special Study* (Washington, DC: National Commission for the Protection of Human Subjects of Biomedical and Behavioral Research, 1976); also, Everett Mendelsohn, "Frankenstein at Harvard: The Public Politics of Recombinant DNA Research", in *Transformation and Tradition in the Sciences: Essays in Honor of I. Bernard Cohen*, Everett Mendelsohn, ed. (Cambridge: Cambridge University Press, 1984), pp. 317–335.
[25] For an introduction to his thinking in this area, see Everett Mendelsohn, "Grasping the Elusive Peace in the Middle East", *Harvard Middle Eastern and Islamic Review* 1 (1994), pp. 1–16; and *A Compassionate Peace: A Future for the Middle East* (New York: Hill and Wang; revised edition, London: Penguin, 1989). Everett is also a Faculty Associate in the Center for Middle Eastern Studies at Harvard.
[26] Everett Mendelsohn, "Prophet of Our Discontent: Lewis Mumford Confronts the Bomb", in *Lewis Mumford, Public Intellectual*, Thomas and Agatha Hughes, eds. (Oxford: Oxford University Press, 1990), pp. 343–360.

I

HISTORICAL STUDIES IN THE LIFE SCIENCES AND SOCIETY

JANE MAIENSCHEIN

ADVOCATING THE HISTORY OF SCIENCE

ABSTRACT

Everett Mendelsohn has advised candidates for the Ph.D. degree to go out to Harvard Square and explain their dissertation. That it is important to communicate with an audience is one message. That we, in the academy, should take our knowledge to society, is another. That is one use of scholarship, and has been a driving force in Everett's career. Everett founded the *Journal of the History of Biology* in 1968 and edited it for 31 years, and he trained several generations of historians of biology. Indeed, he was instrumental in founding the field. It is therefore appropriate to ask: what do we gain by studying the history of biology and the history of science generally? This paper explores answers to that question. History of science helps make science better and helps make us better citizens. In effect, this is an argument for the promotion of history of science in the public interest.

When I was growing up in Oak Ridge, Tennessee, it seemed clear what social responsibility the scientist had. He (usually "he") should go to the lab in the morning wearing a suit, work all day, then come home and leave his work behind. After all, it was privileged knowledge (not that we called it that) with a top security clearance. Certainly it was not something to be shared with families, as the billboards at the edges of town reminded everyone with their exhortations to "sssh ..." and "protect American security". This all seemed perfectly normal, and soon the atom really would produce energy "too cheap to meter". Atomic bombs were effective deterrents to war, and besides, my dad was head of the peaceful and friendly neutron physics division. Pure science, right? Promoting the public good, right? Was I naïve? Yes. Was this comfortable and reassuring? Yes. Surrounded by all this lovely science, there seemed to be no question that my brother and I would become scientists too.

Then my freshman year at MIT in 1968–1969 brought a tumult of challenges. Against the backdrop of the Vietnam War, leading scientists offered a day-long "teach-in" to discuss "science in the public interest" and "the social responsibility of the scientist". One after another, these eminent scientists

G.E. Allen and R.M. MacLeod (eds.), Science, History and Social Activism: A Tribute to Everett Mendelsohn, 23–36.

challenged the easy assumptions that scientists have a license simply to pursue their "pure science", leaving it to a presumably responsible public to decide on appropriate uses. They all accepted J. Robert Oppenheimer's lament that "scientists have known sin" and set out to debate what followed from that realization. What should a socially responsible scientist do? What did it even mean to be "socially responsible", much less a "socially responsible scientist"? Heady stuff for a college freshman, and I don't pretend to have gone very far at the time in thinking it all through. But up Massachusetts Avenue at Harvard, Professor Everett Mendelsohn had already embarked on a long career of trying to make sense of the history of science, and of what that history tells us about the meaning of social responsibility.

Everett's brand of social responsibility has focused on promoting world peace, especially through his decades of work with the American Friends Service Committee. Fortunately, Harvard has afforded him the support necessary to carry on that service while sustaining his academic life. He has managed to bring these interests together through courses that explore the social context and the social nature of science – even before it was trendy to do so. Through his editorial ventures, he has encouraged scholars to publish social analyses of science. In the History of Science Department, Everett has nurtured interest in social responsibility at many levels.

Cambridge served as a visible center for Vietnam War protests and advocacy for political change, with some scientists playing important roles in accepting and even leading the challenges. Cambridge also became a center for lively debates about the social responsibilities of biologists through Matthew Meselson's widely-acknowledged concern with biological and chemical weapons; and again, as recombinant DNA research provoked new concerns about safety and science. Issues of racism and IQ, and the need to temper an excessive genetic determinism that can serve as justification for racism and other social inequities, have likewise placed Cambridge at the center of debate. Everett has played a part in many of these discussions.

It is easy among academic liberals to point to one side of these various issues and to applaud advocates' courage in taking a stand for what they believe is right. Indeed, most discussion of the "social responsibility of the scientist" has assumed that we know what is right, and has focused on how active and social a role the scientist should play in effecting the presumed proper and just position. In the Cambridge context, the Vietnam War was bad, most war is bad, biological and chemical warfare is bad, genetic engineering is at least potentially bad, racism is bad, sexism is bad. While these views seemed defensible and widely shared in academia, they are not all universally held and are not often held simply by virtue of being a scientist.

This brings us to questions about the appropriate social role of the scientist: is it to do good science, or also to be a good scientist in some wider sense and to raise social issues insofar as they derive from scientific work? In this case, the developmental biologist might say "please notice that I have made it possible to clone humans, and you people should figure out the ethical issues". Or is it, even further, to serve as a public intellectual and to take positions of presumed social good more generally: "I have invented cloning techniques and urge you not to use them for cloning people"? What social responsibilities arise by virtue of being a scientist, and what responsibilities arise by virtue of being an educated citizen? These are complex questions with a multitude of conflicting interpretations. While traditional liberalism would have us focus on rights and associated responsibilities, other political doctrines emphasize the obligations of contracts, or the needs of the presumed existence of community or absolute normative truths. Thus, people can – and do – disagree about what ideas and responsibilities particular players have in the social order. This is often difficult for academic liberals to accept. In their often-isolated realm, they believe that they know what is good and right, all the while calling for inclusion of a babble of diverse voices and "stories". Indeed, some attempts to call for "tolerance" in the name of political correctness have foundered upon the reality that some views and some behaviors are intolerable.

My intention is not to discuss social and political theory, however, but rather to acknowledge the verifiable fact that social responsibility of scientists has been an important category of interest for many, including Everett Mendelsohn. Everett's career pushes us to think about this more carefully. I want to consider a special aspect of that discussion, asking about the social role of the *historian* of science. Not the social responsibility, because I do not have compelling grounds for claiming any moral imperative or clearly defined responsibility. Rather, I wish to look at the social opportunity offered the historian of science. And for those who accept that there is responsibility, realizing this opportunity will be a moral good. For others, it may be pragmatically and politically, as well as socially, expedient for reasons that I will discuss. I contend that the historian (and the philosopher) of science has valuable, socially useful knowledge. This is knowledge about what science is, how it works, what forces change science over time, and its past significance and value. The historian of science has insight, perspective, and information in the form of examples and generalizations that can shape the way society thinks about science and its products. Historians of science can help inform social decision-making and can help guide the process by which we adjudicate competing and conflicting claims. In other words, I am claiming for historians of science a sort of "privileged knowledge", much as my father had, although we can and should share it. I quite realize that some will find such claims objectionable. But please note that I am not claiming for historians of science the sole right of arbitration or the ultimate or only valuable knowledge. Rather, historical knowledge can enlighten and improve our social decisions in ways open to multiple sets of values and goals. I will outline some examples before offering suggested audiences for history and philosophy of science. I will explore especially the value for promoting public understanding of science, informing science education, and advancing social goals.

Promoting the Public Understanding of Science

With few exceptions, we accept public understanding of science as a good thing. We can argue about what this really means, about how much understanding is best, about who qualifies as the public and such things, but only the most extreme science bashers would resist the abstract goal. There have been times when elitists have argued that the public did not need science and was better kept in the dark, but with the accepted public support of science in this century, this reasoning has virtually disappeared. While there are questions about the legitimacy and risks of secrecy in particular cases, the dominant view in recent decades has been that we should tear down the barriers, and teach the public about science and its power. In the post-Sputnik era, we generally assume that scientists are necessary to do good things, and that the public should have access to this work.

So, given that the public understanding of science is a good thing, who should promote it, and how? In part this should take place in the schools, through science education, though we do not always do this or do it well. We also need a wider effort addressed at adults, going beyond the schools to reinforce and extend the message outside academia. Science – and technology – serve as a central force of change for society, and increasingly many decisions will be based on scientific claims. Assuming that we would rather make wise and informed choices when we have the option, we need for at least the decision-makers to have some understanding of the science on which the issues and their decisions must be based.

For example, Members of the United States Congress (and apparently every other country's leaders) make many decisions about science and based on science. Yet only two members of the United States House of Representatives are Ph.D. scientists, although there is also a handful of physicians and people with related technical degrees. The one Ph.D. historian of science, Bob Filner (Democrat from San Diego), has focused on Veterans issues that dominate his district, rather than directly on science. Yet some have suggested that at least half the votes in Congress either require scientific knowledge or directly influence the doing of science. Since Congressmen are not scientists themselves, they must rely upon others for expertise – either their personal staffs (few of whom are trained as scientists, and many of whom openly admit that they tried to avoid science courses in college), committee staffs (and even in the House Science Committee, many do not have any scientific training), lobbyists (who have their own goals, although some, such as Research!America, lobby for all biomedical scientific research) or nonprofit organizations (well-intentioned, though often incomplete and directed at some particular, selected set of political goals).

The experience of serving as science advisor for my Congressman for the two years of the 105th Congress (1997–1998) was instructive. The Congressman representing the university district approached our College of Liberal Arts and Sciences, from which he had graduated as an English major with fluency in Mandarin Chinese. A first term conservative Republican who supported Newt

Gingrich's "Contract with America", he was appointed to the House Science Committee. He came, asking how he could learn from the many scientists working at the Research I university in his district. Significantly, he wanted to avoid making unwise decisions that would antagonize the university and its scientific core. So, Arizona State University developed a plan to place a faculty member "on assignment" to work with him, while also developing projects and information of value to the university. The university agreed to pay salary, travel costs and living expenses in Washington, D.C. Both the Congressman and the university preferred someone with broad scientific interests, and preferably a senior faculty member, with the understanding that the person should remain politically neutral. That was how I came in.

Technically, I became a Congressional Fellow and in that role served as Congressman Matt Salmon's science advisor. I was not really "staff" in the usual sense – assigned to cover a particular set of my own issues – but instead worked with all the regular government employees on issues related to science, technology, research, and education. This was a tremendous learning experience for me and also, evidently, for the Congressman. And here is the point of this story: my status as an historian and philosopher of science was important in gaining credibility because I was not seen as serving any one narrow or particular political or academic interest (this is, in part, for the rather depressing reason that nobody thinks of history and philosophy of science as being a field important enough to have interests to be served). Someone who reflects upon science more generally, and someone willing to work at communicating about what is at issue in a range of particular cases asks about ways to go about making decisions informed by past experience, using the best available present knowledge – that was what everyone wanted. Being a senior faculty member and older (and more liberal) than anybody else in the office, (including the Congressman) provided an amusing twist at times, and gave me credibility that I might not otherwise have had – as long as I was eager to listen and learn as well as to teach.

Historians of science have special opportunities to play roles as Congressional Fellows. Other historians of science who have served as Fellows (Ron Overmann, who worked with Senator Mitchell from Maine, and Rivers Singleton, who worked with the late Congressman George Brown of California) agree that we have special opportunities in this area. Indeed, when I met Congressman Brown, who had chaired the Science Committee with distinction for the Democratic majority, he immediately asked "why are you working with Matt Salmon? Why aren't you working with me?" He said that he would like an historian of science in his office to provide perspective. (Unfortunately, Congressman Brown, a great supporter of science, has died recently and will be greatly missed). Congressman Vernon Ehlers (a Republican physicist from Michigan) agrees that professionals with a wider view of the sciences, and those willing to learn, are indispensable for informed and intelligent decision making – and are all too rare. Let us look at some differences we can make.

On several occasions, a heated topic provoked lively and partisan discussion. Cloning was such a case, in which the immediate reactions were shaped by analogies with fertility cases, complete with the fears and misunderstandings that all reproductive issues bring. Hearings invoked parallels with fetal tissue and abortion, cited fears of "throw-away" humans cloned to harvest body parts, and enthused about cloning as a solution to the world's infertility problems. An historical perspective could have been helpful, by placing research into a larger context, and clarifying what is really new and what an appropriate response might be. History could have defused the initial hysteria, while also showing that there are questions deserving serious debate. Similarly, historians could have contributed depth to the President's National Bioethics Advisory Committee (NBAC) report, which had an unfortunate tendency to treat every issue as if it were completely and drastically new.[1] Discussions at the American Association for the Advancement of Science (AAAS), the National Academy of Sciences (NAS), and other sites of science in Washington brought a calming effect as they began to place cloning in perspective. Historians gave science writers a framework, and the *New York Times* called upon the history of embryology from the 1920s and 1930s in its presentation of the new results.[2] Who ever thought the long-deceased embryologist Hans Spemann would make the front page of the *Times*? This larger perspective began to influence Congressional staff who, recognizing the complexity of the case, backed off from restrictive legislation. They even backed away from Congressman Ehlers' modest bill which would have limited federal funding for human cloning if the goal were to create a human being.[3]

Another Congressional example shows the history of science very directly and actively at work. Metropolitan Life Insurance Company was concerned about lawsuits with potentially far-reaching liability consequences for scientists, academic institutions, and those who insure them. One lawsuit sought to hold researchers liable for the consequences of their research – even when it was others who had adopted their results. The lawsuit contended that researchers should have known better, obviously did bad science, and should therefore be held accountable. The underlying assumption was that scientists should know better, and should only publish that which is true for all time. Thrown out on technicalities, the suit alerted MetLife to potentially expensive and far-reaching problems. They hired Lindley Darden, historian and philosopher of science at the University of Maryland, to produce a summary paper explaining how science works. She made clear that what is perfectly good science one day, given the constraints of the time, should be *expected* to give way to other results. This does not mean that the work was bad, but rather that things simply change – as they should, indeed, as we would be surprised if they did not. Darden outlined several examples of Nobel Prize winners and other scientific leaders whose work was later supplanted.[4] Science, while fallible, is self-correcting over time. A Congressional hearing discussed Darden's presentation. Darden's academically unorthodox assignment had impact, and made a real difference. There are other such opportunities, and we should accept them when they arise. This does not mean that every individual must take on such tasks every time, but collectively we should view such opportunities as being integral to our profession. We should then carry them out with as much dedication as Lindley Darden and Everett Mendelsohn have brought to the task.

The AAAS, as the world's largest umbrella organization for general science, coordinates efforts to promote the public understanding of science in the United States, especially in Washington, D.C. Through its science policy directorate, the AAAS coordinates a Congressional Fellows program, provides training for academics to work effectively with Congress and the federal government, and helps scientists and science policymakers alike to interpret the workings of science. Interestingly, Albert Teich not only heads the policy directorate of the AAAS, but also oversees the AAAS archives, and recognizes the importance of the history of science. The AAAS reaches a wider audience through *Science* and other publications and on-line products, its annual award for public understanding of science, its mass media fellowship program, and other activities. Advancing science by promoting public understanding is what the AAAS is about, and we historians can work with the AAAS to keep a broader historical perspective on science.[5]

Educating about Science

Science education has been a significant public concern in the United States since Sputnik jolted Americans from their post-war complacency. The Third International Mathematics and Science Study (TIMSS) results released over the past few years have brought this concern to the fore once again.[6] Virtually everyone agrees that we must do something about science education, but there is considerable disagreement about what that something should be, who should do it, and who should pay. Is education a matter for federal or state control? Should we set and enforce standards for everybody, or just some? Who should set and enforce these standards? How do we reach prospective students, teachers, and others to effect improvement in curricula textbooks, and classrooms?

While every relevant agency or organization now wants to play a role, the AAAS became an early and strong leader through two complementary programs. The Education and Human Resources Directorate, headed by biologist Shirley Malcom, promotes science education for all Americans. Based on the conviction that education produces strong citizens, and that scientific and technical education produces a strong workforce, Malcom has focused on providing access to science through partnerships with schools, black churches, Girl Scouts, science centers, government agencies, and other nonprofit organizations – wherever there is sufficient willingness, energy, and enthusiasm to generate optimism and results. The Peabody Award-winning radio adventure show "Kinetic City Science Supercrew" makes clear the message: "Science is fun!" Along with "Bill Nye the Science Guy" and his television show, the AAAS provides the best inspiration available in America for students and teachers.

Another AAAS program, Project 2061, provides the framework, guidelines, and resources for teachers to develop curricula and to improve classroom instruction in science, while also outlining goals and examples for textbook writers. *Science for All Americans,* which appeared as the first volume of a series in 1990, explains that "Education has no higher purpose than preparing people to lead personally fulfilling and responsible lives". It goes on to explain that science education "should help students to develop the understandings and habits of mind they need to become compassionate human beings able to think for themselves and to face life head on".[7] James Rutherford, director of the project, worked with physicist/historian Gerald Holton, and consulted a number of historians of science from 1985 through a second (*Benchmarks*) and third volume (*Resources*).[8] All three volumes stress the goal of cultivating scientific "habits of mind" and promoting "scientific ways of knowing" for all. Science is a process, and it changes over time. Historical examples demonstrate this scientific process, while depicting scientists as normal human beings and not (necessarily) as supernerds. Historians of science have played – and I argue, ought to play – valuable roles by contributing to such projects.

Of course, historians are not always going to be invited to assume these roles, so we need leaders to serve as public intellectuals, making clear how important the historical perspective is for science and policy-making. And to continue making that point, as often as possible and even when it seems like talking to an empty room. Eventually, someone will hear. We must contribute to the formulation of National Research Council (NRC) reports on science and science education, for example, and accept invitations to comment on such reports. Several of these projects have produced stronger results because of their historical input, and a larger number (including the National Science Foundation (NSF) report on undergraduate education, "Shaping the Future") would have been more effective had they been given a larger historical perspective.[9] As biologist/educator Anthony Lawson notes, "the history of science has much to offer in terms of helping us identify 'natural' routes of inquiry, routes that past scientists have taken and routes that present students can also take – routes that should lead to scientific literacy. That is, to students who know what science is and how to do it".[10]

To frame better how we think about science, historians could point to Thomas Henry Huxley's views on science education. In an after-dinner speech on science education in 1869, Huxley noted "that no boy or girl should leave school without possessing a grasp of the general character of science, and without having been disciplined, more or less, in the methods of all sciences; so that, when turned into the world to make their own way, they shall be prepared to face scientific problems, not by knowing at once the conditions of every problem, or by being able at once to solve it; but by being familiar with the general current of scientific thought, and by being able to apply the methods of science in the proper way ...".[11] Or we might point to a century-old address by geologist Sir Archibald Geike to students of science. Observing that only a few would achieve jobs in their scientific fields, and that most would likely end up doing things other than scientific research, Geike argued: "To those who may ultimately be thus situated it will always be of advantage to have had the mental training given in [the sciences], and it will probably be your own fault if, even under unfavorable conditions, you do not find, from time to time, chances of turning your scientific acquirements to account".[12]

The point is not that there is eternal recurrence, or that the more things change, the more ... they don't. Rather, we need to be more aware of the deep underlying factors shaping how things work throughout time. That "cutting edge" that we hear so much about must cut something blunted by years of tradition and experience. And the cutting tools will be sharper if we understand the evolving context and patterns of change.

Adults also need science education. Since the *Daubert v. Merrell Pharmaceuticals, Inc.* ruling by the U.S. Supreme Court, for example, judges have been officially responsible for selecting experts to adjudicate among competing scientific claims in the courtroom.[13] Historians can help, as Jan Witkowski has shown effectively in his program for federal judges at Cold Spring Harbor Laboratory's Banbury Conference Center. Judges, who often come from a political science or history background, typically have relatively little scientific training. The historical perspective they receive from experts in science and the history of science actually makes the scientific claims they sort through more accessible and less intimidating. Since the *Daubert* decision makes judges more legally responsible for understanding the technical and scientific content of their cases, this accessibility has become increasingly important. Science-trained "experts" should be available and willing to serve judges, as well as plaintiffs and defendants – despite what is likely to be minimal pay.

Another example: Disney needs technical experts for their movies and for the "Edu-tainment" business. Why not historians of science, who can bring in truly bizarre examples from past science? Museums, science centers, and zoos all can benefit, though most of them do not yet realize it. So we have to educate them all. An opportunity for historians of science lies in educating and creating a market for historical thinking about science. How can we know what is "cutting edge" today if we have no sense of where we have been and in what direction the edge might be pointed now? Historians know that science is not simple. We know that it is dangerous to claim that science always makes progress, or that we can unfailingly tell what is cutting edge at any given time, and we must use our knowledge to advance realistic science education, based on science as science really is done. We want neither a vision of science-as-savior nor an acceptance of a science-less demon-haunted world.[14] We can help to ensure that superstition does not always – or ever easily – win out over science.[15] We should help to promote a realistic view of science as a process, undergoing change, carried out by real people, and always in a context of values, constraints, and opportunities.

One final example. Recently, an Arizona Board of Education-appointed committee sought to keep references to "evolution" out of the state's science standards. Our courageous State Superintendent of Public Instruction, Lisa Graham Keegan, led the Board in deciding that each member would appoint a representative to a special review committee. Ultimately, the presence of an informed historian of science, working closely with an evolutionary biologist, carried the day on that review committee, and successfully crafted a set of standards that brings evolution back into the science standards. This was accomplished by acknowledging that what we know now may change in response to new evidence and argument. The process, carried out in a civilized and respectful way that kept returning to the historical examples, demonstrated that political give and take can sometimes work to produce the "right" results.

The state Board and its well-intentioned public officials did not know how to include evolution in a non-controversial way, and they were happy to let an historian draft the state standards and to suggest questions for state-mandated proficiency tests. And here, both the historical perspective embraced in AAAS's Project 2061 and the concrete suggestions of the National Research Council's report on teaching evolution were very helpful.[16] Once again, the lesson is that, in cases where historians have the opportunity, we need those who will accept the challenge. Equally important, we need department chairs, administrators, and colleagues who will not devalue those who assume such roles and serve the public interest by contributing to committees, serving on school boards, running for office, or playing a range of administrative and popularizing roles. If we all insist upon sitting in our ivory towers and pontificating about the academic virtues of remaining insulated from the murkiness of the real world, then (1) there will be fewer of us, and (2) we will be making a mockery of all that the history and philosophy of science shows about how knowledge is produced.

Advancing Social Goals

While there is room for disagreement about which social goals deserve attention, some have achieved universal acceptance. Some of these are based on scientific principles or historical claims in which historians and philosophers of science can play specific roles in espousing the "right" view, or debunking unjustified claims. Two different examples are racism (and by analogy, sexism and other forms of discrimination based on unsubstantiated claims about biological differences and destiny), and freedom of speech. I raise these together.

A rich literature has emerged in the history of science surrounding the roles played by women, and politicized views of gender. Even more attention has been paid by historians of science to racism—and the use of science in promoting and legitimizing racism.[17]

Historians of science played a valuable role when Hernnstein and Murray's *The Bell Curve* appeared in 1994. Several scholars, including African-American geneticist-historian Joseph Graves, pointed out what was wrong with their warmed-over statistical manipulations and underlying assumptions.[18] Others pointed to the authors' history of controversy and showed that the new offering was not very new. At around the same time, Graves, a group of his students, and historian of science Bonnie Blustein used historical evidence and argument to show that J.P. Rushton's racist poster presentation on presumed differential racial IQ (which he had injected into a AAAS annual meeting by changing what he had originally proposed and the AAAS had accepted) was part of a continuing campaign to justify racism on pseudo-scientific grounds. Robert Proctor's close studies of Nazi eugenic practices and their underlying "scientific" claims afford another example of an historian's clarifying why scientific claims are sometimes problematic.[19] Racism flies in the face of the best available scientific evidence now, but other historians show that this has not always been true. Therefore, claims of racial differences in IQ, and in other traits and behaviors are based on appeals to other sets of values. Historians help to demonstrate this. And they help to remind us to remain skeptical and not to be seduced by the exaggerated claims that have lured people into supporting false doctrine in the past.

In each case, historians have shown how particular examples fit into a larger historical pattern. They have shown what constitutes a legitimate scientific claim and what is disputed and therefore an illegitimate basis for social action. In cases like these, they are able to say "this is wrong because it is based on an indefensible foundation". By accepting the opportunity to speak out, even if going beyond their own particular specialized expertise to draw on the collective knowledge of the professional community, they have made a difference.

I am not suggesting that historians have a special claim to insight on every issue. Of course, many claims are not illuminated by examples from the history of science. In these cases, both the interests of science (which depend upon the free exchange of scientific ideas) and the interests of a democracy in protecting free speech, call for caution. (It is not appropriate for anyone to shout down others because of disagreements over views that are reasonably held, for example.) And it is certainly not acceptable to do so in the name of science. There have been such examples – rare, but troubling examples.

The most widely-cited example was that of biologist E.O. Wilson, who met an infamous reaction at a AAAS meeting in the mid-1970s. Some tried to keep him from speaking, and one person doused him with water while he spoke. His opponents argued that his sociobiological ideas, based upon genetic determinist assumptions, were "wrong", socially unjust, hence immoral. Perhaps these accusations were valid, (although I would not condone attempting to stifle someone's speech when we ought to listen to his case and argue it on the basis of evidence and logic). But here we have disagreements over values. Some critics went on to claim that we know by virtue of studying the history of science that Wilson's work is "bad". I deny that we knew that it was bad in this moral sense in 1975, when his *Sociobiology* first appeared.[20] In fact, the history of science shows a pattern of dispute about genetic determinism and biological causation of traits and behaviors, that demonstrates its social, political, and moral motivations. We do not know enough to know how much is genetically-driven, and can only express preferences or beliefs about what the outcome will be. When we know much more, we can only know in many cases that we prefer – for extra-scientific reasons – to have one outcome over another.

In its context, Wilson's view was neither surprising, nor was it "bad" science. Neither were the alternatives. Historians of science know that. So historians might have legitimately poured water on the protestors' heads as well as on Wilson's, if the point was that his claims were not scientifically "proven", and that the social implications mattered. But, then, we would be pouring a lot of water on a lot of heads because, as we know, much of science is tentative. As the history and philosophy of science show, it takes time to adjudicate among competing claims. The history of science argues for tolerance and for embracing a range of views at any given time, then providing a rich context in which they can play out. It should not make quick judgements, nor produce social or moral claims in the name of science that the policy process seeks.

This discussion is part of the fuss about science studies and the so-called "science wars". Neither Wilson nor James Watson nor . . . whoever should make exaggerated social claims in the name of their sciences, any more than historians and philosophers of science should make exaggerated social claims. The long view, carefully considered, is desirable. We should make decisions as informed as possible.

Conclusion

While overzealous attempts to silence those who disagree with any given view are bad, utter silence is far worse. We have an excellent opportunity to speak out about how science works, to develop our scholarship, and to show what conclusions follow from recognized patterns and processes of change.

Everett once said that, in order to receive a Ph.D. degree, every student should have to explain his or her thesis to someone in the street around Harvard Square. I heard him say this in the 1970s, and at the time I thought it was cheating to choose Harvard Square. Nonetheless, the point about communication is well taken, and holds for the entire field. I would not argue that every student, in order to be a "card carrying" historian of science, must go to the head of NSF or a Congressman and explain how science works and why history matters. But I would argue that it is vitally important for the continued good of the community and for the public generally that historians of science do this. To make a difference requires that we value those who do so. We must reward rather than marginalize, emulate rather than decry and teach graduate students that social action, on issues where we act by virtue of our scholarship, is good and important. We must explain why, when, where, and how history matters in order to understand why, when, where, and how science matters.

There is also pragmatic value in this form of increased influence, jobs, and new niches for historians of science. We can – and should – do better for our students than fight among ourselves for the few academic positions in the United States. Positions in honors or interdisciplinary studies programs also exist, and we can argue that we, or our students, can fill those jobs. They will do better if we help prepare them with broader perspectives. Some find opportunities to work with government and education administrators. Some of these opportunities call for established senior scholars to step up and take the lead. Senior scholars often have the resources to help everyone in the field, and thus benefit the community and the profession as a whole.

If we do not engage in this discussion, others will. There are many who are only too willing to do so. Wilson has offered to subsume all natural science and social science into "consilience" based on genetic determinants.[21] Neither allowing him to be the only spokesman on issues of social behavior and morality, nor just dropping water on his head, is effective social action. We have an opportunity, and an argument can be made that not to exercise such an opportunity is a moral failure. If we do speak out, we can make a difference.

Rather than engaging in science wars, let us seek cooperation and understanding through effective communication. Along the way, we will create more opportunities. And as a result, we can help to promote the legitimate social goals of peace and social responsibility – as Everett inspires us to do.

Arizona State University

NOTES

[1] National Bioethics Advisory Committee, "Cloning Human Beings". June 1997 U.S. Government report. See website at <http://www.bioethics.gov/pubs.html>.

[2] Michael Specter with Gina Kolata, "After Decades and Many Missteps, Cloning Succeeds", *New York Times*, March 1997, pp. A1 and A8. Also see follow-up: Gina Kolata, *Clone: The Road to Dolly, and the Path Ahead* (New York: William Morrow and Co., 1998).

[3] See legislative bills proposed during the 105th Congress, including House Resolution (HR) 922 and 923.

[4] Lindley Darden, "The Nature of Scientific Inquiry", March 30, 1998. Discussed at a House Science Committee hearing and available on her website at: <http://www.inform.umd.edu/EdRes/Colleges/ARHU/Depts/Philosophy/homepage/faculty/LDarden/sciinq/>.

[5] Sally Gregory Kohlstedt, ed., Michael M. Sokal, Bruce V. Lewenstein, *The Establishment of Science in America: 150 Years of the American Association for the Advancement of Science* (New Brunswick: Rutgers University Press, 2000).

[6] National Science Board, Task Force on Mathematics and Science Achievement, "Preparing Our Children. Math and Science Education in the National Interest", March 1999. On TIMSS, see especially, *A Splintered Vision. An Investigation of U.S. Science and Mathematics Education*, William H. Schmidt, Curtis C. McKnight, Senta A. Raizen, eds. (Dordrecht: Kluwer, 1997); and William H. Schmidt et al., *Facing the Consequences. Using TIMSS for a Closer Look at U.S. Mathematics and Science Education* (Dordrecht: Kluwer, 1999).

[7] AAAS Project 2061, *Science for All Americans* (New York: Oxford University Press, 1990), p. xiii.

[8] AAAS Project 2061, *Benchmarks for Science Literacy* (New York: Oxford University Press, 1993); *Resources for Science Literacy*,(New York: Oxford University Press, 1997).

[9] Melvin George, chair, "Shaping the Future. New Expectations for Undergraduate Education in Science, Mathematics, Engineering, and Technology", National Science Foundation, 1996.

[10] Anthony Lawson, "What Should Students Learn About the Nature of Science and How Should We Teach It?", *Journal of College Science Teaching* (May 1999), pp. 401–411, quotation p. 411.

[11] Thomas Henry Huxley, "Scientific Education: Notes on an After-dinner Speech", in *Science and Education Essays, Collected Works*, Vol. III (London: Macmillan, 1869, 1893), pp. 111–133.

[12] Archibald Geikie, "Science in Education", an address to students at Mason University College, Birmingham, *Nature*, 59, 1898, pp. 108–112.

[13] See Daryl E. Chubin, Edward J. Hackett, and Shana Solomon, "Peer Review and the Courts, or When Scientists 'Get Real'", *Accountability in Research*, 4, 1994, pp. 1–8.

[14] Carl Sagan, *The Demon-Haunted World. Science as a Candle in the Dark* (New York: Random House, 1997).

[15] John Burnham, *How Superstition Won and Science Lost* (New Brunswick: Rutgers University Press, 1987).

[16] National Academy of Sciences, Working Group on Teaching Evolution, National Academy of Sciences, "Teaching About Evolution and the Nature of Science" (Washington, D.C.: National Academy of Sciences Press, 1998).
[17] Some of the best examples that address racism are Stephen Jay Gould, *The Mismeasure of Man* (New York: W.W. Norton, 1996); and Kenneth R. Manning, *Black Apollo of Science. The Life of Ernest Everett Just* (New York: Oxford University Press, 1984).
[18] Richard J. Hernnstein and Charles Murray, *The Bell Curve. Intelligence and Class Structure in American Life* (New York: Free Press, 1994); Joseph L. Graves and T. Place, "Race and IQ Revisited: Figures Never Lie, But Often Liars Figure", *Sage Race Relations Abstracts*, 20, 1995, pp. 4–50.
[19] Robert N. Proctor, *Race Hygiene: Medicine Under the Nazis* (Cambridge: Harvard University Press, 1988).
[20] Edward O. Wilson, *Sociobiology* (Cambridge: Harvard University Press, 1975).
[21] Edward O. Wilson, *Consilience. The Unity of Knowledge* (Cambridge: Harvard University Press, 1998).

SHIRLEY A. ROE

RADICAL NATURE IN THE *ENCYCLOPÉDIE*

Abstract

This paper offers an analysis of the role radical biological ideas played in the French *Encyclopédie* in the mid-eighteenth century. Based on new observations and experiments on the microscopic world, these biological ideas supported a radical view of active matter and threatened to destroy the traditional view of an unchanging, hierarchical social order. This radical thread is traced through several articles from the *Encyclopédie*, and is shown to have played a key role in the furor that erupted in the early 1750s after the first volumes were published and then, more decisively, when the whole enterprise was shut down as being too subversive.

Introduction

"NATURALIST, n.m., is said to be a person who studies nature and who is well versed in the knowledge of natural things, particularly those that concern metals, minerals, stones, plants, and animals". So begins the definition of a naturalist given in the eleventh volume of the *Encyclopédie* – a conventional definition for the mid-eighteenth century. But the article then continues, "One also gives the name *naturalist* to those who do not admit God, but who believe that there is only one material substance, bearing diverse qualities that are as essential to it as length, size, weight, and in consequence of which everything that we see in nature necessarily happens; *naturalist* in this sense is synonymous with *atheist, spinozist, materialist*, etc".[1] This latter part was certainly not a definition most naturalists would have been comfortable with. Penned by Denis Diderot, these last lines were added to the first definition taken from Chambers's *Cyclopaedia*, which had been the stimulus for the whole *Encyclopédie* project in the first place.[2]

Diderot's daring characterization of the naturalist might be viewed as an anomaly in an encyclopedia that was supposed to capture the entire present state of knowledge in all fields. Many of the articles on natural history, anatomy, physiology, and medicine are in fact quite conventional. But in a

G.E. Allen and R.M. MacLeod (eds.), Science, History and Social Activism: A Tribute to Everett Mendelsohn, 37–59.

number of other articles, some on topics seemingly far removed from the activities of the naturalist, one finds the more radical view expressed by Diderot. In this article, I will tease out this radical thread and show how it played a key role in the furor over the *Encyclopédie* that erupted in the early 1750s after the first volumes were published and then, more decisively, after the ill-fated seventh volume appeared.

The thread I am particularly interested in following is one that presents nature as dynamic and self-creative. A passive nature, with forces and activity added by God, was a mainstay of the mechanical philosophy, and had been so since the late seventeenth century. Challenges to this view arose primarily in the life sciences, in generation theories that attributed an active, creative role to matter in the developmental process. These theories were seen as promoting atheism and materialism, which meant at that time that the world and all of its inhabitants were the product of material causation alone. The reason for the connection between atheism and notions of creative matter has to do with the way these and more traditional theories depicted the subvisible world. Was it an ordered world of microscopic organisms arising from preexistent eggs in a predetermined fashion? Or was it a world of ceaseless activity, of attractive or vegetative forces, of particles with desire, memory, or innate sensitivity? Was nature calm, hierarchical, and prearranged – on all levels – or rather active, creative, even chaotic?

These opposed visions of nature automatically translated themselves into opposed visions of society because the common element in nature and in society is of course human beings. At once the manifestation, at the highest level, of the subvisible world underlying living phenomena, and at the same time the unit of social interaction, human beings live out the destinies of their natures. Were they to exist in a hierarchical, ordered society, where birth determined position in a "preformed", predetermined manner? Or could they possibly exist in some other relationship resulting from their interactions as active, material beings? One can easily see why the social order was perceived to be threatened, especially after 1750, by those views of nature that encompassed a more active, even equalizing, world of possibilities. These connections are made strikingly clear in the following contrasting statements; one, made by Albrecht von Haller in 1766: "Beware that it is very dangerous to admit the formation of a finger by chance. If a finger can form itself, then a hand will form itself, [and] an arm, [and] a man".[3] Juxtapose this with d'Holbach's comment made only four years later in his infamous *Systême de la nature* (System of Nature). Referring to John Turberville Needham's observations on microscopic organisms, he remarked: "would the production of a man independently from the ordinary means be more marvelous than that of an insect from flour and water?"[4]

Thus dynamic views of nature implied a model for society that was anathema to the religiously and politically powerful. A world that is dynamic is one that can change, not one in which social, political, and religious hierarchies are set from the beginning of time. Challenges to orthodox religion were dangerous because the church helped to solidify political power by promoting a moral code that served to keep the "masses" in a state of obedience and by offering an unquestionable structure for absolute monarchy. Thus the possibility of change at the level of nature threatened to lead to change at the highest levels, the church and the monarchy.

Concerns about the relationship between material activity and atheism had become a necessity with the advent of the mechanical philosophy. If matter and motion alone were all that was required for the world to operate, then what role was left for God and what defense was left against atheism? Such worries had in fact been manifested in the 1640s during the English Civil War when various "leveling" political factions based their philosophies on active nature. The eventual settlement brought in by the Glorious Revolution was intimately tied with Newtonianism and its ordered universe.[5] Natural theology and mechanism became inseparable entities. Yet the subversive views of nature and society based on materialism, atheism, or even just deism simply went underground. Several of the English deist and materialist writers became known to the French through clandestine manuscripts that circulated widely, particularly in the 1730s.[6]

After 1750 this subversive literature portraying active matter and its political and religious consequences found its way into print.[7] Often anti-clerical, anti-crown, pornographic, or all three, such literature was widely read.[8] To some extent, the more daring articles in the *Encyclopédie* simply made more public subversive ideas that had been circulating for some time. Yet the dual flowering at mid-century of the first major works of the philosophes and of "philosophical pornography", both of which criticized social mores, Christianity, and, by implication, the monarchy, indicates something new. As Robert Darnton has pointed out, with the appearance of works by Montesquieu, Diderot, La Mettrie, Toussaint, Buffon, Rousseau, Voltaire, Duclos and of the first volume of the *Encyclopédie* all between 1748 and 1751, "the intellectual topography of France was transformed".[9] It was also at this time that Joseph d'Hémery, police inspector of the book trade from 1748, began keeping records on dangerous authors, books, and pamphlets. His concern was not so much what we would call Enlightenment ideas as it was atheism and its threat to the political order.[10]

To claim that the *Encyclopédie* contained articles proposing radical ideas is, of course, not a new assertion. Much of the secondary literature on the *Encyclopédie* has focused on various aspects of controversy engendered by its publication, as has much of the literature on pre-revolutionary France. The most common topic discussed is the anti-clerical and irreligious ideas presented, and their implied challenge to the social order. Yet the radical ideas on life and matter have received scant treatment.[11] Where biological issues have been discussed, authors for the most part have relied on Jacques Roger's *Les Sciences de la vie dans la pensée française du XVIII siècle* (Life Sciences in Eighteenth-Century French Thought) for their information. Those sources that have treated life sciences or nature in the period, including Roger, have not dealt with the political or social context nor have they realized the tremendous importance biological ideas had in the political realm.[12] My purpose in this paper is to show what was presented in the *Encyclopédie* on the connection between life and matter, why this was seen as radical, and how it got caught up in – even contributed to causing – the controversy that led to the *Encyclopédie*'s suppression.

The Political Setting

The first volume of the *Encyclopédie* was published in 1751, the project having been begun by Diderot and d'Alembert in 1747.[13] The 1740s and 1750s were unstable times for France. The War of the Austrian Succession had dragged on for much of the 1740s, partly due to the indecisiveness and disinterestedness of Louis XV, who assumed direct rule of France after the death of Cardinal Fleury in 1743. This was followed in the early 1750s by increasing tensions between France and Britain in the colonies, which eventually erupted into the colonial conflict of the Seven Years War (1756–1763). This was mirrored on the European continent by the war between France and Austria, on the one side, and Britain and Prussia, on the other. Confused and factionalized foreign policy in France did not help the situation, and by the conclusion of the war France had lost its colonial empire and exhausted its treasury.

In Paris, struggles between the King and Parlement over relative power, through the revival of the Jesuit-Jansenist conflict, also dominated these decades.[14] Although ostensibly a controversy over the "refusal of sacraments", where individuals were required to renounce Jansenism before receiving the last rites, these struggles were more significantly an attempt by the Paris Parlement to become a political power. Exiled to the provinces by the King in 1753 after going on strike, the magistrates were recalled in 1754 after an agreement favorable to the Parlement was reached. Another showdown occurred in 1756, when Louis XV's declaration of power led most of the Parlement to resign in December. Within a month Paris was to witness one of the most dramatic moments of the period, the stabbing of Louis XV by Robert-François Damien in January 1757. Although the assassination attempt failed, Damien's testimony at his trial showed how deeply in society the political frustrations, religious controversy, and economic difficulties were felt. As Van Kley has noted, lack of affordable bread in Paris dovetailed with lack of confidence in the King in the "seditious talk" chronicled in the judicial archives from the period.[15]

The Natural World

These years coincided with the appearance – but not the acceptance – of some new views about the natural world and living organisms presented by Pierre-Louis de Maupertuis, Georges Louis Leclerc de Buffon, and John Turberville Needham. All three proposed theories of generation that were based on active matter and that challenged the prevailing theory of the preexistence of germs. Preexistence (also called preformation), first proposed in the late seventeenth century by Nicolas Malebranche and Jan Swammerdam, held that God had created all organisms at one time and had encased them within one another until their future appointed times of appearance.[16] Development was seen as the unfolding, as it were, of preexistent parts through automatic, physical means. The role of matter itself in this process was passive, since all had been ordained and built-in by God. As I and others have argued, this view of development was consistent with both the mechanical philosophy, which was growing in importance in the late seventeenth century, and natural theology, which became allied with mechanism for intellectual, religious, and political reasons. Challenging the preexistence of germs and the passivity of matter could easily be seen as tantamount to challenging the social order. As I argue in this paper, this was exactly what happened in mid-eighteenth-century France.

Maupertuis was the first to challenge preexistence of germs when, in 1744, he presented a theory of gradual development based on attractive forces in his anonymously published *Dissertation physique à l'occasion du nègre blanc* (Physical Dissertation Occasioned by an Albino Negro).[17] Maupertuis argued that preexistence faced too many difficulties, such as resemblance to both parents. He also felt that the two theories were equivalent with respect to religion, for, as he put it, "What has natural science lost by the idea that animals are formed successively? For God, is there any real difference between one moment in time and the next?"[18]

Maupertuis shared his views on generation with his friend, Buffon, sometime in the mid-1740s, and Buffon began to develop his own theory that also rejected preexistent germs, writing up an account of it by early 1746. He apparently read this account to Needham in 1748, and the two joined forces to carry out a series of microscopical observations on seminal fluids and on infusions. Needham published his own theory in 1748, and Buffon's was published in 1749, in the second volume of his *Histoire naturelle* (Natural History).[19]

All three theories were thus connected, through the personal contact among the authors and by their common rejection of preexistent germs. All three also presented a dynamic view of material activity in the formation of living organisms, Maupertuis postulating an attractive force of cohesion, Buffon a penetrating force, and Needham a vegetative force to guide development. Buffon also added an internal mould (*moule intérieur*) to organize the organic particles into a complex living organism. Addressing the same problem – how to organize matter into an organism – Maupertuis attributed "desire, aversion, memory" to the particles of matter out of which the organism formed.[20] The source of the offspring's organization was the Achilles heel of all epigenetic theories at this time and was at the same time one of the strongest arguments for preexistence of germs.[21] Yet even though preexistence theories dominated in the 1740s and 1750s, it was the dynamic theories of Maupertuis, Buffon, and Needham that formed the basis for the view of generation and the relationship between life and matter that were expressed in articles in the *Encyclopédie*.

Nature in the First Volumes of the Encyclopédie

It was through the conduit of Diderot that these radical biological ideas found their way into the *Encyclopédie*, initially from the impact Buffon's views had on Diderot and then as an expression of Diderot's own developing materialism. The *Encyclopédie* occupied Diderot for over twenty years, from the late 1740s to the early 1770s, the same years during which his materialist views were maturing. In 1749 Diderot spent three months in prison at Vincennes for publishing atheist ideas in his *Lettre sur les aveugles* (Letter on the Blind) and for publishing his pornographic *Les Bijoux indiscrets* (The Indiscrete Jewels). While in prison, Diderot first read the initial volumes of Buffon's *Histoire narturelle*. He apparently wrote up some notes, which he intended to communicate to the author, but they were confiscated by the prison authorities.[22] The atheist ideas Diderot had expressed in his *Lettre* were not yet rooted in biology. This shift happened during the years he was involved in the *Encyclopédie*, yet the transition began when Diderot encountered the views of Buffon.[23]

Diderot and Buffon later became friends; Diderot remarked occasionally in letters that he had seen Buffon on one of the latter's visits to Paris from his country estate in Montbard.[24] He also announced in the preface to the second volume of the *Encyclopédie* that Buffon had agreed to write the article "Nature", probably at Diderot's request. As Diderot proudly proclaimed, "We hasten to announce that M. de Buffon has given us the article 'Nature' for one of the volumes that will follow this one; a most important article, whose subject is a term that is rather vague, often used, but poorly defined, that philosophers greatly abuse and that, in order to be developed and presented in all its different aspects, needs all the wisdom, accuracy, and elevation that M. de Buffon demonstrates in the subjects that he treats".[25] This article never materialized, and when "Nature" finally appeared in volume eleven, its author was d'Alembert, not Buffon, and it dealt with the "system of the world" not with living organisms. Nor does one find the article on organic molecules cross-referenced as "Parties Organiques" in the articles "Animalcule" and "Génération"; it simply does not exist. There is no evidence to tell us why Buffon's article on nature was never included. Was it never written, and was Diderot anticipating things a bit when he claimed that he had it in hand? Or did Buffon pull it when the *Encyclopédie* came under fire? Although we will never know the answer, it could very likely have been the latter reason.

Yet although Buffon never wrote any articles for the *Encyclopédie*, his views informed many of the articles pertaining to the living world, even when no acknowledgement is given. Buffon was very definitely a "ghost writer" for the *Encyclopédie*. This is evident in articles explicitly on biological subjects as well as those in which remarks on the living world are hidden under a non-biological title. Several articles consist solely or partly of quotations from Buffon's *Histoire naturelle*, while others refer to Buffon by name or use language obviously taken from his work.[26] Halfway through the first volume of the *Encyclopédie* we encounter Diderot's article "Animal", which consists primarily of the first chapter of Buffon's treatise on generation, the "Histoire générale des animaux" (General History of Animals) (from volume two of the *Histoire naturelle*), with Diderot's comments interspersed. These comments may very well have come from those he wrote while reading Buffon's text in prison. Although only about a quarter of the text, Diderot's remarks pushed Buffon's ideas even further toward materialism. The principal point in Buffon's chapter had been to break down the distinction between animals and plants so that he could counter with his own distinction between brute matter and living matter, the latter being composed of "organic molecules" (*molécules organiques*). Buffon claimed that these organic molecules endlessly circulate in nature, are taken up in organisms' food, and are then used for growth or for reproduction (via the internal mould).

Even the fact that Diderot used Buffon's critique of traditional definitions of "animal" as the basis for his article was a challenge to accepted beliefs. The standard definition of an animal, as an organism possessing locomotion and sensation, was not a view anyone would have thought was likely to be challenged. But by using Buffon's unorthodox definition and interspersing his own comments throughout the article, Diderot carried on a dialogue with Buffon that led even further towards a materialist view of life. Buffon had argued that the gradations on the chain of being through the animal kingdom down to the plants, and the existence of borderline organisms, like the polyp, that seemed to possess both animal and plant qualities, demonstrated that there was no absolute border between the two kingdoms. Diderot pushed this further to question the existence of an absolute border between plants and minerals, thereby calling into question Buffon's separation of organic from brute matter. After quoting Buffon's statements that animals have more connections to their surroundings than plants do and that plants have more than minerals do, Diderot remarked that these connections are graded within the two kingdoms and may be so among minerals too: "One can say that there are minerals less dead than others".[27] At another point Diderot suggested that thinking and sensation exist in degrees descending down the chain of being into animals, after Buffon had stated that matter did not have feeling, sensation, or thought. Although Diderot was not directly contradicting Buffon, he certainly implied the opposite, that is, that matter may not be totally devoid of these qualities.[28] Diderot stated in another remark even more clearly that the passage between the mineral and plant kingdoms may be just as gradual as that between animals and plants.[29] Finally, Diderot let stand without any need of comment Buffon's concluding remark, "that living and animation, instead of being a metaphysical degree of beings, is [are] a physical property of matter".[30]

Diderot's "Animal" thus not only presented Buffon's very unorthodox view that animals and plants are both composed of organic material particles, but also carried Buffon's implied materialism even further. By alphabetical coincidence, the Buffonian theme is carried forward in the subsequent article, "Animalcule" (microscopic animal), written by Louis-Jean-Marie Daubenton. Daubenton had come to Paris in 1745 to serve as Buffon's assistant at the Royal Botanical Garden and had joined him and Needham in their observations on microscopic animalcules in semen and in infusions. [31]In "Animalcule" he presented as fact Buffon's claim that animalcules are not true animals, but that they are organic molecules. (Buffon had actually claimed that they are small bodies formed by chance from organic molecules, not the molecules themselves.) Daubenton also made a point of claiming Buffon's priority in these observations over Needham's. Needless to say, the common view, held by everyone who had previously observed animalcules, that they are simply microscopic animals, was mentioned only to be dismissed.

Following next in alphabetical order is the article "Animalistes", which was the French term for those who believed that the embryo was preformed in the spermatic animalcule. Written by Pierre Tarin, a little-known but major contributor to the *Encyclopédie* in anatomy and physiology, this article presented animalculist preformation only to counter it with standard objections (resemblance to the mother, prodigious number of wasted embryos, and the like).[32] To this article, Diderot added the editorial remark, "There may without a doubt be animals in these liquids; but what one takes for animals, are they always that? See *Animalcule*", thus sending the reader back to the previous article and to Buffon's explanation that animalcules in male semen were only chance combinations of organic molecules.

These first two volumes also contained several descriptive and quite traditional articles on anatomy, physiology, and natural history.[33] Yet there are no articles championing the preexistence of germs or the necessity of passive matter. This situation continues in later volumes. The radical view – that material activity is the basis of living phenomena – is dominant wherever the issue is raised.

After the second volume of the *Encyclopédie* appeared, Diderot finished a short, but very significant, work on science and knowledge, his *Pensées sur l'interprétation de la nature* (Thoughts on the Interpretation of Nature).[34] Published first at the end of 1753 and then in a somewhat expanded edition the following year, Diderot's tract expressed, this time in relation to Maupertuis's work, a similar questioning about material activity and living organisms as had his dialogue with Buffon in "Animal". Even more clearly rejecting Buffon's distinction between living and brute matter, Diderot asked, was there anything more fundamental than a relative difference between the two types of matter? Could dead matter become living matter, and vice versa?[35] By this time, Diderot had begun to develop his notion of "sensibility" (*sensibilité*), which he believed was inert in dead matter and gradually more emergent in life forms as one moved up the chain of being. These views are more clearly expressed in later articles he included in the *Encyclopédie.*

In addition to the articles that presented radical ideas on nature, there were several others that leaned rather directly toward irreligion. In 1751 some of these were singled out for criticism, although more attacks came after the second volume appeared in 1752. There is evidence that Chrétien-Guillaume de Lamoignon de Malesherbes, director of publications and therefore in charge of censorship, was instrumental in suppressing or at least toning down criticism of these first two volumes.[36] Yet some criticism did appear in print, especially from the Jesuits. Guillaume François Berthier, who edited the Jesuit *Journal de*

Trévoux from 1745 to 1762, praised the enterprise, yet expressed concern about the *Encyclopédie*'s anti-religious stance in some articles. From October 1751 until March 1752, the *Journal de Trévoux* published several articles on the first two volumes of the *Encyclopédie*. Singled out for attack were "Aius-Locutius" (Roman god of speech), in which Diderot had argued for lifting censorship for works on incredulity intended for an elite audience, and "Amour des Sciences et des Lettres" (Love of science and letters), which Berthier condemned as irreligious.[37] Some of the other articles in volumes one and two that were later to receive critical treatment, like Diderot's "Adorer" (Adoration), his "Agnus scythicus" (a plant), his long editorial addition to the abbé Yvon's "fime" (Soul), and d'Alembert's "Aveugle" (Blindness), had not yet been singled out. But the bomb that was dropped in the second volume was the article "Certitude", by a young abbé, Jean-Martin de Prades, which included laudatory introductory and concluding remarks by Diderot. The de Prades article was a flash point not only for its contents but also for its connection with a scandal that had deeply embarrassed the Jesuits of the Sorbonne and that may have actually been engineered by none other than Diderot.[38]

THE CRISIS OF 1752

On the seventh of February, 1752, the King's council issued a decree suppressing the first two volumes of the *Encyclopédie*. In part, the decree read: "His Majesty has recognized that in these two volumes several maxims have been inserted tending to destroy the royal authority, to establish a spirit of independence and revolt, and, under obscure and ambiguous terms, to build the foundations of error, of moral corruption, of irreligion, and of unbelief".[39] The second volume had appeared only a month before, but it contributed to the issuance of this decree. Several factors led to this unfortunate event (for Diderot and d'Alembert, that is), but they all seem to have revolved around the religious/political struggles between the Jesuits at the Sorbonne and at court and the Jansenist Parlement. Dale Van Kley has called attention to Diderot's article "Autorité politique" (Political Authority) in the first volume of the *Encyclopédie* as one of the first political utterances of the maturing French Enlightenment.[40] It was this article, he claims, along with the uproar caused by the Sorbonne's unwitting acceptance of de Prades' thesis, that drew attention to the *Encyclopédie* project and led to the suppression.[41]

Diderot's "Autorité politique" was indeed a bold article, justifying all political authority by the consent of the governed and by an implicit contract between the people and the ruler. Unlimited authority residing in any human being would, Diderot claimed, undermine the authority of God. Although he also supported continual obedience to the King, Diderot found himself and his article attacked, particularly by the Jesuit *Journal de Trévoux*. The Jesuits had been hostile to the *Encyclopédie* from the very beginning of the project, and there is evidence that they were behind the suppression with a plan to take over the enterprise themselves.[42] The diarist Barbier wrote at the time that "this whole storm against this fine Dictionary comes by medium of the Jesuits".

Similarly, the former secretary of state, the marquis d'Argenson, predicted in his memoirs that the Jesuits would become the new editors.[43]

In November 1751, the abbé de Prades had successfully defended his thesis in an oral examination by theologians at the Sorbonne. The thesis was full of praise for "natural religion" and doctrines derived from d'Alembert's "Discours préliminaire" (Preliminary Discourse) from the *Encyclopédie* and Diderot's *Pensées philosophiques* (Philosophical Thoughts). How the thesis got approved is something of a mystery, but it did not take long for the storm to break. De Prades' thesis was denounced on the floor of the Jansenist Paris Parlement, which led the Sorbonne to establish a committee to examine the thesis once again.[44] Their report was issued only days before the suppression of the *Encyclopédie*, and it claimed that the thesis was part of an infamous plot against both religion and the state. What made matters worse was that the second volume of the *Encyclopédie*, appearing in January 1752, contained a lengthy article "Certitude", by none other than de Prades himself. The article discussed grounds for believing in miracles, and even though it claimed to support faith it also contained numerous quotations from Diderot's *Pensées philosophiques*. Appended to the article was an editorial note by Diderot praising de Prades' love of truth in defense of religion. Meanwhile, an arrest warrant was issued for de Prades by the Paris Parlement, and he fled to Holland.

Six days after the Sorbonne condemned de Prades' thesis, the *Arrêt* suppressing the *Encyclopédie* was issued. Powerful advisors at Court, among them the Dauphin's tutor, convinced the King that the *Encyclopédie* was devoted to irreligion and undermining his authority. Coupled with the King's growing struggle with the Jansenist Parlement and alliance with the Jesuits, these factors resulted in the decree against the *Encyclopédie*.[45] There is even a story that the King's Jesuit confessor told him that God might spare his daughter, who was gravely ill, if he would suppress the *Encyclopédie*![46]

There is no evidence that, in the crisis of 1752, ideas about active matter and its role in living phenomena played any role in the dispute. Partly this was due to the fact that several of the more radical articles had yet to appear. Partly it was because substantial analyses of articles in the *Encyclopédie*, including those in the first two volumes, which began to call attention to materialist and atheist doctrines found therein, did not begin to appear until the late 1750s. The crisis of 1752 died down, due to the efforts of Malesherbes, who was responsible for overseeing the book trade. Malesherbes was devoted, as much as one could be at the time, to freedom of the press. The result was that volumes three through seven appeared at yearly intervals from 1753 through 1757. Their contents, in conjunction with other publications and growing political tensions, were to provoke the second and even more fatal *Encyclopédie* crisis in 1758. Before turning to that crisis, let me detail some of the materialist and anti-religious articles that appeared in these volumes.

MATERIALISM AND NATURE IN VOLUMES THREE THROUGH SEVEN

Let me first look at what these volumes presented on life and matter in articles on biological topics. In volume seven, we find the article "Génération" by Arnulphe d'Aumont, one of the main contributors on medicine, particularly in these middle volumes. (After the condemnation of the *Encyclopédie* in 1759, he no longer allowed his articles to be attributed to him.)[47] D'Aumont had studied medicine at Montpellier, but his contributions to the *Encyclopédie* show, according to Roger, that he leaned toward mechanism.[48] His article on generation is a cautious survey of theories on the subject from Plato to Buffon. Defining generation as "the formation of an individual similar in its nature to those from which it draws its origin by reason of preexisting principles that it received from them, that is to say from the proper matter and the disposition of a particular form that the generating beings furnish for the preparation, development and growth of the germs that they produce or that they contain", d'Aumont did not take a clear stand on whether these "germs" preexist before generation or are formed in the process.[49] His discussion of ovist preformationism is lengthy, yet he concluded that resemblance to both parents and the superfluity of germs that never develop made this theory untenable. This led, he claimed to the "mixture theories" of Maupertuis and of Buffon, which he presented in some detail. Yet the problem of how a new organism could be organized by forces or out of organic molecules made their theories unlikely as well. His lengthy article began and ended characterizing generation as a mystery that may be as impenetrable as other first causes that we know about only from their effects.[50] But his survey of theories of generation from the ancients to Buffon was clearly informed by Maupertuis' own historical survey, as d'Aumont acknowledged, and the *Vénus physique* (Physical Venus) and *Histoire naturelle* were two of the works credited at the conclusion of the article as major sources.

Among the cross-references in d'Aumont's article, to articles like "Oeuf" (Egg), "Ovaire" (Ovary) and "Semence" (Semen), only "Fetus" (by Louis de Jaucourt) can be found in these middle volumes. All of the articles referred to are of a descriptive nature, with no discussion of any theories of generation. Thus I have concluded that the only presentation of the theory of preexistence – including the strongest arguments for adopting it – was in d'Aumont's "Génération", which can hardly qualify as an endorsement, even though he claimed that "most physicians and other physicists are more strongly attached to the system of eggs furnished by the testicles of the female".[51] Although d'Aumont's article ended up not supporting any theory, his criticisms of preexistence and demonstration of why the views of Maupertuis and Buffon had necessarily developed would hardly have pleased most naturalists. This is undoubtedly why the staunch preformationist, Albrecht von Haller, contributed the article on generation to the *Supplément à l'Encyclopédie* (Supplement to the Encyclopedia), published twenty years later.

Diderot's own more radical views on the relationship between active matter and life can be found not in articles on obvious topics, but rather in an unlikely place, his article on "Ethiopiens". This article, along with others to which one was led to by a series of cross-references included in "Ethiopiens", ended up playing a major role in the second, and more serious condemnation, of the *Encyclopédie* in 1759. Diderot had alerted readers to a subversive function of cross-referencing in his lengthy article, "Encyclopédie". Here he admitted that the editors purposely subverted opinions expressed in one article by referring to another one that said exactly the opposite. This method, he claimed, "is the art of tacitly deducing the most radical conclusions. If these cross references of confirmation or refutation are foreseen far ahead of time and prepared with skill, they will give to an encyclopedia the character that a good dictionary ought to have; this character is to change the general way of thinking".[52] Although scholars agree that there were actually very few of these skillful cross-referencing schemes, nevertheless the whole idea, revealed so candidly here by Diderot, was radical in its own right. Sometimes one did not even need to go to the referred article to get the point, as when at the end of the unsigned "Providence" one is sent to "Manichéisme" or when an article on cannibalism ("Anthropophages") cross-referenced "Eucharist" and "Communion".[53]

One of the most skillfully devised chains of articles, the sum of which expressed a radical materialism, began with Diderot's "Ethiopiens". Following a straightforward article by Jaucourt on "Ethiopie", Diderot's article was rather less factual. At one point, he claimed, "The Ethiopians take themselves to be more ancient than the Egyptians, because their country has been more strongly struck by the rays of the sun, which give life to all beings. Whence one sees that these people are not far from regarding animals as the development of earth put into fermentation by the heat of the sun, and to conjecture in consequence that species have undergone an infinity of diverse transformations, before becoming the form that we see them in".[54] At the conclusion of this paragraph, Diderot added a cross-reference to the article "Dieu" (God). When we turn to "Dieu", we find several interesting things. This article was based on Jean-Henri-Samuel Formey's notes for his own encyclopedia, which were purchased early on in the *Encyclopédie* project and used by Diderot and d'Alembert for over a hundred articles.[55] After an opening discussion about how clear it is that God exists, we find a cross-reference to "Démonstration", by d'Alembert. There, under a section on a posteriori demonstration, d'Alembert claimed that it is much easier to demonstrate the existence of God through his works than through a priori arguments. "In the eyes of the people, and even of philosophers, an insect proves [the existence of] a God better than all metaphysical reasonings; and in the eyes of the philosopher, the general laws of nature prove even better the existence of God than an insect".[56] A cross-reference to d'Alembert's "Cosmologie" further underlined his deistic argument: there, using Maupertuis' argument for the existence of God based on the law of least action, a similar point was made. Supporting deism was often seen as tantamount to promoting irreligion, because it denied any direct contact with God through revelation.

Returning to the article "Dieu" we read that three parts of the article, on metaphysical proofs for the existence of God, on historical proofs, and on physical proofs, were drawn from the work of Samuel Clarke, Isaac Jaquelot, and Bernard le Bouvier de Fontenelle, respectively. Turning to the third section, drawn from Fontenelle, we find the whole discussion is about whether the first animals of each species were formed by chance interactions of matter or by the will of God.[57] Fontenelle concluded that it was the latter, but principally by arguing that if animals had been formed by chance, why was it not still happening? A paragraph claiming that generation from corruption, or spontaneous generation, had been shown to be false by modern experiments ended in a cross-reference to "Corruption".[58] This leads us to another d'Alembert article where Buffon's theory of organic molecules was once again presented. D'Alembert highlighted not only Buffon's claim that microscopic organisms can be formed by the fortuitous combination of organic molecules but also that tiny "eels" (larvae) form in flour paste in the same manner.[59] Although d'Alembert admitted that most cases that appear to be spontaneous generation are actually regular generation from eggs, he queried, "but is it demonstrated in all cases that corruption can never engender an animated body?"[60]

These articles in the middle volumes of the *Encyclopédie* thus presented a very strong argument for the material basis of life and called into question the need for God's involvement even in life's creation. In addition, Diderot's article "Chaos" criticized the notion that matter and motion alone could not have created the world and challenged interpreting the book of Genesis literally, concluding "one can say of *chaos* whatever one likes".[61] Other articles furthered the attack on orthodox religion, such as those promoting toleration, those impartially explaining the ideas of unorthodox religious sects, those expounding on the views of non-Christian religions (which attacked Catholicism by comparison), and those attacking Christianity and the clergy outright.[62] Political unorthodoxy found expression in Diderot's "Droit naturelle" (Natural Right) and in Jaucourt's "Gouvernement". Yet from 1753 through 1757, few criticisms appeared in print.[63] It was probably not that these articles went entirely unnoticed; rather, some of the vehement attacks on the *Encyclopédie* that began to appear in 1758 were quite lengthy and took some time to prepare. But the more significant reason for the explosion of criticism in 1758 and 1759 has to do with the growing political tensions of the 1750s.

The Crisis of 1758–1759

On January 23, 1759, the Attorney General Jean-Omer Joly de Fleury, rose before the Paris Parlement to condemn the *Encyclopédie*, as well as Claude-Adrien Helvétius's *De l'Esprit* (On the Mind) and six other books.[64] He opened his speech with the dire pronouncement: "Society, the State, and Religion present themselves today at the tribunal of justice. ...Their rights have been violated, their laws disregarded. Impiety walks with head held high. ... Humanity shudders, the citizenry is alarmed".[65] What was the source of this fear and disquiet? A conspiracy, claimed Joly de Fleury, one that existed for the purpose of destroying society: "can one conceal from oneself that there is a project formed, a Society organized, to propagate materialism, to destroy Religion, to inspire a spirit of independence, and to nourish the corruption of morals?'[66]' Remarking that it was sad to think what posterity will think of their century, Joly de Fleury claimed that it had fostered "a sect of so-called Philosophers who ... imagined a project ... to destroy the basic truths engraved in our hearts by the hand of the Creator, to abolish his cult and his ministers, and to establish instead Deism and Materialism".[67] Joly de Fleury then devoted his thirty-one-page harangue to an analysis of the ways in which the *Encyclopédie* and *De l'Esprit* were the very embodiment of this plot.

What was the cause of Joly de Fleury's vicious attack? The *Encyclopédie* had been appearing, one volume a year, quite peacefully since the crisis of 1752. What events had led to this renewed attention to and concern over the impact of the *Encyclopédie*? To answer this, we need to look again at the political and social events of the late 1750s, the attacks on the *Encyclopédie* that appeared in 1758, and the increased concern over "*mauvais propos*" (seditious talk) among the people of Paris.

The year 1757 opened with Damiens' attempt on the King's life. The interrogations that followed of Damiens and many others in preparation for his trial reveal a deep dissatisfaction with the monarchy. Dale van Kley has argued that an increasing "desacrilization" of the monarchy, evidenced at both the popular and elite levels, marks the decades preceding the French Revolution. Van Kley has pointed especially to the Damiens affair as both a cause and result of this process.[68] There was certainly an increase in surveillance of seditious publications and talk, as evidenced in the volumes of information collected by Joly de Fleury, attorney general, and his brother, Guillaume-François-Louis Joly de Fleury, the procurer general, during and in the wake of the trial.[69] This vigilance is clear at the intellectual level as well, as can be seen in both the Joly de Fleury and the d'Hémery archives. For our purposes, the important point is that the atmosphere of the late 1750s was so charged that it would probably have taken very little to set off a reaction against the philosophes. Following the assassination attempt, a draconian law was passed about subversive literature. "Anyone who is convicted", the new law read, "of having composed ... writings tending to attack religion, to stir up spirits, to endanger our [the King's] authority, and to disturb the order and tranquility of our state, will be punished with death".[70] Although this law was apparently never enforced, its passage indicates that the King's ministers had become extremely hostile to the philosophes.

The occasion for renewed outcry against the *Encyclopédie* was provided by volume seven, which appeared in October 1757 and contained d'Alembert's article "Genève" (Geneva). In hindsight, publishing this article was a tactical error on the part of Diderot and d'Alembert, for it led not only to a barrage of criticism but also to a falling out between the two, which resulted in d'Alembert's resigning from the project altogether. But what was the problem with "Genève"? Although d'Alembert praised democracy in Geneva and other aspects of their enlightened culture, he also "praised" the Genevan clergy for their supposed deist beliefs. This set off a reaction both in Geneva and in France, for the Genevan Calvinist pastors were incensed and the French clergy found the not-so-subtle criticism by comparison equally galling.[71] One can also detect the hand of Voltaire, with whom d'Alembert had spent three weeks at his home in Geneva prior to writing the article.[72]

A storm of criticism was unleashed after this volume came out. 1757 saw the first appearance of the fictional "Cacouacs", a newly discovered "savage tribe" living at the 48th degree of latitude (the latitude of Paris), whose main weapon was the power of their words. First revealed in an "Avis Utile" (Useful Notice) published in the *Mercure de France* in October 1757, the Cacouacs were the creation of the abbé Giry de Saint-Cyr, tutor to the Dauphin. Soon thereafter Jacob Nicolas Moreau published the *Nouveau mémoire pour servir à l'histoire des Cacouacs* (New Memoir on the History of the Cacouacs), a much more thorough treatment of the dangers of the philosophes portrayed by the fictional capture and indoctrination at the hands of the Cacouacs of an innocent young man. Although Moreau's memoir did not make too many references to the *Encyclopédie*, its significance for my purposes here lies in the role biological ideas played in the indoctrination into the ideas of the philosophes the young man received, a subject to which I shall return in a moment. Moreau's piece was followed by Giry de Saint-Cyr's *Catéchisme de décisions de cas de conscience à l'usage des Cacouacs* (Catechism of decisions in matters of conscience for the use of Cacouacs), which was a hodge-podge of quotations from works by Diderot, Rousseau, La Mettrie, and others, including articles from the *Encyclopédie*, all set out under topics of a "catechism", beginning with "What is God?" Finally, in November 1758 there appeared Abraham Joseph Chaumeix's massive *Préjugés légitimes contre l'Encyclopédie* (Legitimate Prejudices against the Encyclopedia), the first two volumes of which were devoted to a vicious attack on the *Encyclopédie*. It was clearly from the *Préjugés légitimes* that Joly de Fleury drew much of the material for his attack on the *Encyclopédie* before the Paris Parlement.

The last straw came in July 1758, when Helvétius' *De l'Esprit* burst upon the scene. Based on a sensationalism drawn from Condillac and a materialism from La Mettrie, *De l'Esprit* presented an educational and an ethical system that ignored the church and challenged the Crown.[73] Provoking a scandal in its own right, Helvétius' book only added fuel to the growing controversy surrounding the *Encyclopédie*. Apparently, it was even rumored that Diderot himself was the real author of *De l'Esprit*, which had not even been published anonymously.[74] The reality of a conspiracy devoted to overthrowing organized religion and endangering the Crown thus seemed proven beyond a doubt, and *De l'Esprit* was included, along with the *Encyclopédie*, in most of the critical pamphlet literature of 1758 and 1759, while also engendering separate attacks of its own.

A further complication for the philosophes was their connection with Frederick the Great of Prussia. Frederick had founded the Berlin Academy of Sciences in 1744, enticing Maupertuis to assume its presidency in 1746. Many of the philosophes were members, including Diderot, and some, like La Mettrie, fled to Berlin rather than face persecution in France. D'Alembert received a pension from Frederick, and Frederick even tried, unsuccessfully, to get Voltaire to move to Berlin. The problem for the philosophes was that, from 1756 on,

France was at war with Prussia. The encyclopedists were thus in a suspect position because of their close ties to the Berlin court and because several articles in the *Encyclopédie* that attacked despotism in favor of an enlightened monarchy could be seen as indirect support for Frederick the Great. The political situation, made worse by the fact that France had lost several key battles in 1757 and 1758, culminating in decisive losses in 1759, only served to heighten the sense of urgency in dealing with the philosophes. There is even some evidence that the King's chief minister, the Duc de Choiseul, in a delicate situation politically because the war was going so badly, decided to sacrifice the *Encyclopédie* in order to keep the Parlement on his side.[75]

Joly de Fleury, The Cacouacs, and Materialism

Let me return now in more detail to Joly de Fleury's attack on the *Encyclopédie* and the role materialism, particularly based on evidence from the living world, played in the critique of the philosophes. After opening with the harangue against the dangerous sect of philosophers that were out to destroy religion and the state, Joly de Fleury turned to an analysis of the condemned works. He was clearly incensed not only by the content of the *Encyclopédie* but also by the strategy of the authors and editors. He complained, for example, that articles often presented both "the pro and con; but the con, when it is a matter of religion, of morality, of authority, is always set forth clearly and with affectation".[76]

After singling out several irreligious articles, like Diderot's "Adorer, honorer, revérer" (To Adore, Honor, Revere) with its blatant deism, Joly de Fleury turned, following Chaumeix, to the editors' pernicious system of cross-referencing.[77] Rather than continuing to point out more "detestable principles" in various articles, Joly de Fleury remarked, "we believe it is interesting and necessary to stop now to show you that this Dictionary is the fruit of impious reflection". Although not everyone involved was part of the "conspiracy to attack the foundations of the state and of religion", he continued, those who were part of the plot "undoubtedly are afraid not to appear to be what they are, dangerous writers, men without decency, enemies of authority, and of Christianity, to which they have vainly sworn ruin".[78] But the "morsel that is the most singular by its audacity" and was "the key to their system", he proclaimed, was the section of Diderot's "Encyclopédie" (quoted in part earlier) where the system of cross-referencing is revealed as a way to undercut orthodox articles and to "change the general way of thinking".[79] Of all the examples he could give, Joly de Fleury claimed, the article "Ethiopiens" proved their method, for "one finds there these Authors' system on the primitive formation of animals and in particular of man". The Ethiopians are represented as "regarding animals as developments from the earth put into fermentation by the heat of the sun" (a quotation from Diderot's article).[80] Joly de Fleury then followed the path of cross references discussed earlier. Reacting to this challenge to divine creation, Joly de Fleury responded, "All the proofs of the existence of God protest against this system of atheism".[81] Yet Joly de Fleury noted that the opposite conclusion was the one the editors wanted the reader to draw, for one is sent to the article "Dieu". There the editors tried to weaken the metaphysical proofs for the existence of God by referring to "Démonstration", the physical proofs by cross-referencing "Corruption", and the moral proofs by sending the reader to "Athéisme", which was written, he claimed, "almost entirely to cast doubt on the existence of God and to openly combat religion".[82]

Since I have already analyzed the contents of these articles, I will not delve into them further here. But what is of paramount importance, in my opinion, is that it was a discussion of the creation of life out of matter that caught the eye of Chaumeix and that was seen as dangerous enough to be highlighted in Joly de Fleury's denunciation of the entire project. Diderot did not claim that the Ethiopians were atheists; yet this was Joly de Fleury's conclusion, based on their alleged views on the creation of life from matter. And in his mind, being an atheist was synonymous with being a danger to the state. The connections between biological materialism, atheism, and dangerous politics was just too obvious at the time to be ignored when evaluating the cost of allowing publications like the *Encyclopédie* to continue.

After Joly de Fleury's speech before the Paris Parlement, the four publishers of the *Encyclopédie* were instructed not to sell any more copies, and a committee was appointed to examine in detail the first seven volumes of the *Encyclopédie*. The other works, including Helvétius' *De l'Esprit*, were simply condemned outright. By issuing these orders, the Parlement was openly criticizing Malesherbe, the director of the book trade, for allowing such publications to appear. By March 1759, the decision was made: the *Encyclopédie*'s permission to publish was revoked because, as the official *Arrest du Conseil* put it, "the advantage that one could draw from a work of this type for the progress of the sciences and the arts could never balance the irreparable harm that would result for morality and religion".[83] Diderot feared arrest, hid some of his manuscripts with Malesherbes, but vowed to continue the project.[84]

I would like to offer one further example of the tie between materialist views of life and the dangers to society implied by them. As I mentioned earlier, the Cacouac episode began at the same time as the seventh volume of the *Encyclopédie* was published and continued, with the publications of Moreau and Giry de Saint-Cyr, through the turbulent year of 1758. In Moreau's *Nouveau mémoire* we again find the connection between materialist biology and the dangerous threat the philosophes represented to society. The memoir opened with the capture of the young hero of the piece, who was eventually able to tell his story after he escaped and returned to Paris. The Cacouacs, he reported, lived in tents to signify their freedom, had no government, regarded ethics as a matter of convention, and did not believe in the existence of God. Although I do not want to describe the young hero's adventures in detail, the key episode for my purposes here was when he was interrogated by a group of Cacouacs in preparation for his induction into their society. His first question, from a venerable old man, was "if dead matter could combine with living matter? How does this combination come about? What is the result?"[85] A woman continued, asking, "If moulds are the principal forms? What is a mould? Is it a real and preexisting being, or is it only the intelligible limits of a living molecule united to dead or living matter ... ?"[86] (Footnotes in the text are to Diderot's *Pensées sur l'interprétation de la nature*.) Under the influence of incense, the young man began "to understand everything marvelously", and he was told he could now regard the Cacouacs as his brothers. Eventually our young hero was rescued; he returned home only to find "dangerous and ridiculous" Cacouacs there as well. "I found", he reported, "that they had been given the name *Philosophes*, and that their works were being printed".[87]

In Giry de Saint-Cyr's follow-up work, the *Catéchisme de décisions*, which was a compilation of quotations from the most subversive of the philosophes' works, we again find biology playing a foundational role. In the second section on the creation of the world and the formation of beings, quotations from Diderot are used to claim that the universe had formed by the chance combination of atoms and that living organisms formed as well from particles that already possessed "desire, aversion, memory, and intelligence" (a reference to Maupertuis). In an even more revealing quotation, again from Diderot's *Pensées sur l'intérpretation de la nature*, the Cacouac teacher explained, "the embryo, formed out of these elements, has passed by an infinity of organizations and developments; ... it has in succession movement, sensations, ideas, thought, reflection, conscience, feelings, passions, signs, gestures, ... language, sciences and the arts".[88]

While I am not going to analyze the Cacouac episode any further here, I simply want to point out that here again we see biological ideas connecting life and matter, drawn from the work of Buffon, Maupertuis, and Diderot, being attributed as foundational to the philosophes' whole enterprise. It was only by understanding these issues that one could become a true Cacouac/philosophe, at least in the eyes of those who found them so dangerous. Of course, the use of these questions also heightened the satirical quality of the narrative, since they sound even more absurd out of context. Yet I do not believe that Moreau used these simply because they would sound hilarious. Rather they were definitional and seen by Moreau and others as providing a foundation for the philosophes' dangerous undermining of religion and society.

The Final Volumes

The remaining ten volumes, on which Diderot and Jaucourt continued to work tirelessly, finally appeared together in 1765, with the tacit permission of the government. In these volumes, we can find further evidence of Diderot's commitment to materialism. John Lough has claimed that Diderot, after suffering such criticism from the first two volumes, drew back from including articles openly expressing materialism in the next five volumes and waited until the last ten to "let himself go" in half a dozen articles or so (like "Matérialiste", "Naturaliste", and "Spinoziste").[89] While this statement is not entirely true with regard to the middle volumes, which contained, as we have seen, the statements in "Ethiopiens" and the articles cross-referenced, it is certainly the case that Diderot slipped several very direct materialist comments into the last ten volumes, even though these volumes suffered censorship from the publisher's pen. In addition to identifying a naturalist with being an atheist in "Naturaliste", quoted at the beginning of this paper, Diderot further developed his ideas on life and matter in "Naître" (Birth). There he claimed that life and death are not absolute states, but successive states of the same matter, with the emergence of sensibility as the key factor. In an editorial addition to "Matérialistes", probably by Diderot, materialists were identified with Spinozists. With regard to the latter, he claimed in "Spinoziste", "The general principle [of Spinozists] ... is that matter is sensitive [*sensible*]; they demonstrate this by the development of the egg, an inert body that by only the instrument of heat gradually passes to a sensing and living state ... .From this they conclude that there is only matter and that it suffices to explain everything".[90] This was the materialist theory that would appear again in Diderot's *Rêve de d'Alembert* (Dream of d'Alembert), written in 1768. Because the *Rêve* and other writings expressing biological materialism were not published, these articles in the later volumes of the *Encyclopédie* contained the only version in print of Diderot's mature theory.

The other biological articles in these last ten volumes are primarily descriptive. "Oeuf" (unsigned), "Ovaire" (Jaucourt), and "Semence" (unsigned) contain nothing controversial. Even the article on Trembley's polyp ("Polype", unsigned), the organism that had caused such a stir throughout Europe because it was able to regenerate entire new polyps when cut into several pieces, did not include controversial issues. Only in "Reproduction" (unsigned), which is a short article about the regeneration of the legs of crabs, do we find a comment that this is a phenomenon that "does not square at all with the modern system of generation, by which one supposes that the animal is entirely formed in the egg".[91]

Diderot had apparently been very much affected by the condemnation of the *Encyclopédie*, and he published little more for the remainder of his life.[92] Many of his masterpieces, such as the materialist *Rêve de d'Alembert*, date from this post-1765 period, but they found their way into print only after his death. Yet the thread of his radical thinking, based on a subversive view of life and matter, can be found woven throughout the *Encyclopédie*. It is beyond the scope of this paper to show more directly the tie between these views and his thoughts on society and natural morality, yet they very definitely existed. Along with d'Holbach, Naigeon, Damilaville, and other contributors who were outright atheists, Diderot wanted to reorder both knowledge and society independently from religion.[93] For this he and the project "to create a revolution in men's minds" were condemned. Yet, as Diderot rather prophetically remarked in 1771 after chancellor Maupeou's contemptuous dismissal of Parlement, "Once men have dared in some way to attack the barrier of religion, the most formidable and most respected barrier that there is, it is impossible to stop. When they have cast a hostile glance over the majesty of heaven, they will not hesitate the next moment to cast one over earthly sovereignty".[94]

University of Connecticut

NOTES

[1] Denis Diderot, "Naturaliste", in *Encyclopédie, ou dictionnaire raisonné des sciences, des arts et des métiers, par une société de gens de lettres*, 17 vols. Denis Diderot and Jean Le Rond d'Alembert, eds. (Paris: Briasson, David, Le Breton, Durand [vols. 1–7]; Neufchâtel: S. Faulche [vols. 8–17], 1751–1765), vol. 11, p. 39B. For the attribution of this and other articles to Diderot and on the whole complex question of the authorship of unsigned articles, see Denis Diderot, *Oeuvres complètes*, Herbert Dieckmann, Jean Fabre, and Jacques Proust, ed. (Paris: Hermann, 1975–), vol. 5, pp. 1–12.

[2] Ephraim Chambers, *Cyclopaedia or an Universal Dictionary of Arts and Sciences*, 2 vols. (London: James and John Knapton, 1728). The original intent had been to publish a French translation of Chambers' work. This idea was abandoned by the time Diderot wrote the prospectus for the new project in 1750, although one finds portions of articles from the Chambers encyclopedia in a number of *Encyclopédie* articles. See John Lough, *The Encyclopédie* (New York: David McKay, 1971), pp. 2–3.

[3] *The Correspondence Between Albrecht von Haller and Charles Bonnet*, Otto Sonntag, ed. (Bern: Hans Huber, 1983), p. 498; letter of 27 May 1766.

[4] [Paul Henri Thiry, baron d'Holbach], *Systême de la nature, ou des loix du monde physique et du monde moral, par M. Mirabaud* (London, 1770), p. 23 n. 5.

[5] See Margaret C. Jacob, "The Materialist World of Pornography", in *The Invention of Pornography: Obscenity and the Origins of Modernity, 1500–1800*, Lynn Hunt, ed. (New York: Zone Books, 1996), pp. 161–163 (article 157–202). See also Margaret C. Jacob, *The Newtonians and the English Revolution, 1689–1720* (Ithaca: Cornell University Press, 1976); James R. Jacob and Margaret C. Jacob, "The Anglican Origins of Modern Science: The Metaphysical Foundations of the Whig Constitution", *Isis*, 1980, 71:251–267; and Margaret C. Jacob, *The Radical Enlightenment: Pantheists, Freemasons, and Republicans* (London: George Allen & Unwin, 1981).

[6] Cyril B. O'Keefe, *Contemporary Reactions to the Enlightenment (1728–1762)* (Geneva: Libraire Slatkine, 1974), pp. 30–32.

[7] See Ira O. Wade, *The Clandestine Organization and Diffusion of Philosophic Ideas in France from 1700 to 1750* (Princeton: Princeton University Press, 1938); O'Keefe, *Contemporary Reactions to the Enlightenment* (cit. n. 6); J.S. Spink, *French Free-Thought from Gassendi to Voltaire* (London: Athlone Press, 1960); N. Jacob, "Materialist World of Pornography" (cit. n. 5); and Robert Darnton, *The Forbidden Best-Sellers of Pre-Revolutionary France* (New York: W. W. Norton, 1995).

[8] Darnton, *Forbidden Best-Sellers* (cit. n. 7), pp. 85–91.

[9] Darnton, *Forbidden Best-Sellers* (cit. n. 7), p. 90.

[10] In 1749 d'Hémery described Diderot as "very dangerous; speaks of holy mysteries with scorn". See Robert Darnton, "A Police Inspector Sorts His Files", in *The Great Cat Massacre and Other Episodes in French Cultural History* (New York: Basic Books, 1984), p. 187. See also Dale Van Kley, *The Religious Origins of the French Revolution: From Calvin to the Civil Constitution, 1560–1791* (New Haven: Yale University Press, 1996), p. 248.

[11] Lough's *Encyclopédie* is an exception, in that he follows out the cross-referencing from the article "Ethiopiens" (discussed later in this paper).

[12] Jacques Roger, *Les Sciences de la vie dans la pensée française du XVIII siècle: la génération des animaux de Descartes à l'Encyclopédie* (Paris: Armand Colin, 1963). See, for example, Emile Callot, *La Philosophie de la vie au XVIIIe siècle* (Paris: M. Rivière, 1965) and Colm Kiernan, *The Enlightenment and Science in Eighteenth-Century France*, Studies on Voltaire and the Eighteenth Century, 59A (Oxford: The Voltaire Foundation, 1973).

[13] Lough, *Encyclopédie* (cit. n. 2), Chs. 1–2, pp. 1–38.

[14] Van Kley, *Religious Origins* (cit. n. 10); Van Kley, *The Damiens Affair and the Unraveling of the Ancien Régime, 1750–1770* (Princeton: Princeton University Press, 1984); O'Keefe, *Contemporary Reactions to the Enlightenment (cit. n. 6)*; John Rossiter, *Louis XV and the* Parlement *of Paris, 1737–1755* (Cambridge and New York: Cambridge University Press, 1995).

[15] See Van Kley, *Religious Origins* (cit. n. 10), pp. 182–184, 188–189; Van Kley, *Damiens Affair*, pp. 226–265; Arlette Farge, *Subversive Words: Public Opinion in Eighteenth-Century France* (University Park: Pennsylvania State University Press, 1994), pp. 164–175.

[16] Roger, *Les Sciences de la vie* (cit. n. 12); Shirley A. Roe, *Matter, Life, and Generation: Eighteenth-Century Embryology and the Haller-Wolff Debate* (Cambridge: Cambridge University Press, 1981).

[17] This was published in an expanded edition the next year as *Vénus physique*. See Mary Terrall, "Salon, Academy, and Boudoir: Generation and Desire in Maupertuis' Science of Life", *Isis*, 87 (1996), pp. 217–229; David Beeson, *Maupertuis: An Intellectual Biography*, Studies on Voltaire and the Eighteenth Century, 299 (Oxford: The Voltaire Foundation, 1992), pp 171–182, 206–215.

[18] [Pierre-Louis Moreau de Maupertuis], *Vénus physique* (n.p., 1745). Quotation from *The Earthly Venus*, trans. Simone Brangier Boas, with notes and an introduction by George Boas (New York: Johnson Reprint Corporation, 1966), p. 42.

[19] For Buffon's conversations with Maupertuis, see John Turberville Needham, "A Summary of Some Late Observations upon the Generation, Composition, and Decomposition of Animal and Vegetable Substances", *Philosophical Transactions of the Royal Society of London*, 45 (1748), p. 633. Buffon's account formed the first five chapters of his *Histoire naturelle, générale et particulière*, 31 vols. (Paris: L'Imprimerie Royale, 1749–1789); see vol. 2, p. 168. On Needham and Buffon, see Shirley A. Roe, "Buffon and Needham: Diverging Views on Life and Matter", in *Buffon 88: Actes du colloque international*, Jean Gayon, ed. (Paris: Librairie Philosophique J. Vrin, 1992), pp. 439-450, and Roe, "John Turberville Needham and the Generation of Living Organisms", *Isis*, 74 (1983), pp. 159–184.

[20] Pierre-Louis Moreau de Maupertuis, *Systême de la nature*, in *Oeuvres* (Lyon: Bruyset, 1756), vol. 2, p. 147.

[21] This was also the problem that preformationists called attention to at every opportunity. See especially Albrecht von Haller, *Réflexions sur le systême de la génération, de M. de Buffon* (Geneva: Barrillot, 1751), and Charles Bonnet, *Considérations sur les corps organisés* (Amsterdam: Marc-Michel Rey, 1762). See also Roe, *Matter, Life, and Generation* (cit. n. 16), and *Science Against the Unbelievers: The Correspondence of Bonnet and Needham, 1760–1780*, Studies on Voltaire and the Eighteenth Century, 243, Renato G. Mazzolini and Shirley A. Roe, eds. (Oxford: The Voltaire Foundation, 1986), pp. 7–52.

[22] Although these notes no longer exist, they probably formed the basis for Diderot's *Encyclopédie* article "Animal". See Roger, *Les Sciences de la vie* (cit. n. 12), p. 599; see also Roger, *Buffon: A Life in Natural History*, Sarah Lucille Bonnefoi (trans.) (Ithaca: Cornell University Press, 1997), p. 199; Denis Diderot, *Correspondance*, Georges Roth and Jean Verloot, ed. (Paris: Editions de Minuit, 1955–1970), vol. 1, p. 96; Diderot, *Oeuvres complètes* (cit. n. 1), vol. 5, p. 382 n. 2.

[23] On Diderot's developing materialism, see also Wilda Anderson, *Diderot's Dream* (Baltimore: Johns Hopkins University Press, 1990), pp. 11–76. Roger, in *Les Sciences de la vie* (cit. n. 12) (p. 601), did not see as much of a development in Diderot's thinking between 1748 and 1753 as I do. See also Jacques Roger, "Diderot et Buffon en 1749", *Diderot Studies*, 4 (1963), pp. 221–236.

[24] Roger, *Buffon* (cit. n. 22), pp. 199–200, 214.

[25] *Encyclopédie*, vol. 2, p. i.

[26] Natural history articles that consist solely or mainly of quotations from Buffon's *Histoire naturelle* include, in addition to "Animal", "Espèce", "Humaine espèce", and "Homme (Hist. Nat.)". Searching the on-line version of the *Encyclopédie* for "Buffon" (<www.lib.uchicago.edu/efts/ARTFL/projects/encyc>) yields 233 times his name appears. Several of these mentions occur in geological or forestry articles. See also James Llana, "Natural History and the *Encyclopédie*", *Journal of the History of Biology*, 33 (2000), pp. 1–25.

[27] *Encyclopédie*, vol. 1, p. 469B.

[28] This implication was not lost on one of the *Encyclopédie*'s most assiduous critics, Abraham Joseph de Chaumeix, who in his eight-volume *Préjugés légitimes contre l'Encyclopédie, et Essai de réfutation de ce dictionnaire* (Brussels, Paris: Herissant 1758–59), made this same comment (vol. 1, p. 214).

[29] *Encyclopédie*, vol. 1, p. 471B.

[30] *Encyclopédie*, p. 474A. Buffon, *Histoire naturelle*, vol. 2, p. 17. Chaumeix took off from this comment to devote a whole section to "The faculty of thought is, according to the encyclopedists, a property of matter", *Préjugés légitimes* (cit. n. 28), vol. 1, p. 224.

[31] On Daubenton's participation, see Buffon, *Histoire naturelle*, vol. 2, p. 171. On Daubenton, see Frank A. Kafker and Serena L. Kafker, *The Encyclopedists as Individuals: A Biographical Dictionary of the Authors of the Encyclopédie*, Studies on Voltaire and the Eighteenth Century, 257 (Oxford: The Voltaire Foundation, 1988), pp. 88–91.

[32] On Tarin, see Kafker and Kafker, *Encyclopedists* (cit. n. 31), pp. 360–362.

[33] For a discussion of the epistemological radicalism that ran through some of the articles on natural history see Llana, "Natural History" (cit. n. 26).

[34] In his introduction in Diderot's *Oeuvres complètes* (cit. n. 1), Jean Varloot described the *Pensées sur l'interprétation de la nature* as a second *Discours préliminaire* to the *Encyclopédie* (the first one having been written by d'Alembert); see vol. 9, p. 5. In a similar vein, P.N. Furbank, in *Diderot: A Critical Biography* (New York: Alfred A. Knopf, 1992), remarked that Diderot's *Pensées sur l'interprétation de la nature* was "meant as some sort of theoretical complement to the *Encyclopédie*" (p. 109).

[35] Diderot, *Oeuvres complètes* (cit. n. 1), vol. 9, LVIII, pp. 95–98.

[36] Lough, *Encyclopédie* (cit. n. 2), p. 109.
[37] John N. Pappas, *Berthier's Journal de Trévoux and the Philosophes*, Studies on Voltaire and the Eighteenth Century, 3 (Geneva: Institut et Musée Voltaire, 1957), pp. 181–182, 184–185. See also O'Keefe, *Contemporary Reactions to the Enlightenment* (cit. n. 6), pp. 88–89, 94–95; Lough, *Encyclopédie* (cit. n. 2), pp. 107–108; Arthur M. Wilson, *Diderot* (New York: Oxford University Press, 1972), pp. 152–154.
[38] Wilson, *Diderot* (cit. n. 37), pp. 154–158; see also Furbank, *Diderot* (cit. n. 34), pp. 89–96.
[39] *Arrêt du Conseil d'État du Roi ... du 7 févier 1752*, in Diderot, *Oeuvres complètes* (cit. n. 1), vol. 5, p. 37.
[40] Van Kley, *Religious Origins* (cit. n. 10), p. 191.
[41] See also Lough, *Encyclopédie* (cit. n. 2), pp. 113–115, Wilson, *Diderot* (cit. n. 37), pp. 154–158; Furbank, *Diderot* (cit. n. 34), pp. 90–96.
[42] Pappas, *Berthier's Journal de Trévoux* (cit. n. 37), pp. 185–186.
[43] See Wilson, *Diderot* (cit. n. 37), pp. 158–159; Edmond-Jean-François Barbier, *Journal historique et anecdotique du règne de Louis XV*, 4 vols. (Paris, 1847–1856), vol. 3, pp. 344; René-Louis de Paulmy, Marquis d'Argenson, *Journal et mémoires*, 9 vols. (Paris, 1859–1867), vol. 7, pp. 56, 71–72.
[44] Furbank, *Diderot* (cit. n. 34), p. 89.
[45] Wilson, *Diderot* (cit. n. 37), p. 159.
[46] Furbank, *Diderot* (cit. n. 34), p. 92.
[47] Kafker and Kafker, *Encyclopedists as Individuals* (cit. n. 31), pp. 16–18.
[48] Roger, *Les Sciences de la vie* (cit. n. 12), p. 631n.
[49] *Encyclopédie*, vol. 7, p. 559B.
[50] D'Aumont's language is nearly identical to a similar passage written by Buffon. See *Encyclopédie*, vol. 7, pp. 560A, 573B; Buffon, *Histoire naturelle*, vol. 2, p. 3.
[51] *Encyclopédie*, vol. 7, p. 567B.
[52] *Encyclopédie*, vol. 5, p. 642B.
[53] See Van Kley, *Religious Origins* (cit. n. 10), p. 3.
[54] *Encyclopédie*, vol. 16, p. 55A.
[55] Kafker has raised the question whether revisions were made in Formey's articles before they were printed. He has also pointed out that Formey never made such a claim. See Kafker and Kafker, *Encyclopedists as Individuals* (cit. n. 31), pp. 141.
[56] *Encyclopédie*, vol. 4, p. 823B.
[57] This section is almost a verbatim reprinting of Fontenelle's "De l'existence de Dieu", which according to Roger dates from 1724. See Roger, *Les Sciences de la vie* (cit. n. 12), p. 365 n. 224.
[58] Fontenelle probably had in mind Francesco Redi's and Jan Swammerdam's observations in the 1680s disproving spontaneous generation of flies from rotting meat and from plant galls.
[59] This latter work was actually Needham's, but it was reported in Buffon's *Histoire naturelle* as well as directly by Needham. See Roe, "Buffon and Needham" (cit. n. 19); Roe, "John Turberville Needham. (cit. n. 19)".
[60] *Encyclopédie*, vol. 4, p. 278B.
[61] *Encyclopédie*. vol. 4, p. 159B.
[62] See Lough, *Encyclopédie* (cit. n. 2), pp. 196–270.
[63] The one notable exception was the *Réflexions d'un Franciscain sur les trois premiers volumes de l'Encyclopédie* (Berlin, 1754); see Lough, *Encyclopédie* (cit. n. 2), pp. 115–116.
[64] The other books condemned were [Louis de Beausorbre], *Le Pirrhonisme du sage* (Berlin, 1754); [Jean Baptiste d'Argens], *La Philosophie du bon sens* (La Haye, 1755); [Voltaire], *La Religion naturelle. Poème ... par M.V.* (Geneva, 1756); [J.B. Pascal], *Lettres semi-philosopiques du chevalier de "au comte de"* (Amsterdam & Paris, 1757); [Diderot], *Étrennes aux esprits forts* (London, 1757) (an edition of his *Pensées philosophiques*); and [abbé G.F. Coyer], *Lettre au R. P. Berthier, sur le matérialisme* (Geneva, 1759). Joly de Fleury had apparently originally intended to include Diderot's *Lettre sur les aveugles* (1749), *Lettre sur les sourds et les muets* (1751), and *Pensées sur l'interprétation de la nature* (1753–1754), along with Condillac's *Traité des sensations* (1754), Rousseau's *Discours sur l'inégalité*, J.F. de Bastide's *Les Choses comme on doit les voir* (1757), and Voltaire's *La Pucelle d'Orléans* (1755), but he apparently changed his mind, as these titles are crossed out in the first draft of his indictment. See Bibliothèque Nationale, Collection Joly de Fleury, vol. 352, dossier 3807; and D.W. Smith, *Helvétius: A Study in Persecution* (Oxford: Clarendon Press, 1965), pp. 40–41.
[65] *Arrests de la Cour de Parlement, portant condamnation de plusieurs Livres & autres Ouvrages imprimés. Extrait des Registres de Parlement. Du 23 Janvier 1759*, pp. 1–2.
[66] *Arrêts*, p. 2.
[67] *Arrêts*, p. 3.

68 See Van Kley, *Religious Origins* (cit. n. 10), pp. 172–175, 179–183; Van Kley, *Damiens Affair* (cit. n. 14), pp. 246–265.

69 Van Kley, *Religious Origins* (cit. n. 10), pp. 180–190. See also Farge, *Subversive Words* (cit. n. 15), pp. 161–175.

70 Quoted in Lough, *Encyclopédie* (cit. n. 2), p. 23.

71 Lough, *Encyclopédie* (cit. n. 2), pp. 221–223; Furbank, *Diderot*, pp. 168–169.

72 Furbank, *Diderot* (cit. n. 34), pp. 167.

73 Smith, *Helvétius* (cit. n. 64), pp. 13–15.

74 Furbank, *Diderot* (cit. n. 34), p. 180.

75 Furbank, *Diderot* (cit. n. 34), pp. 183–184.

76 *Arrests* (cit. n. 65), p. 13.

77 Joly de Fleury also discussed the unsigned "Christianisme", "fime" by the abbé Yvon with Diderot's addition, Jaucourt's "Conscience (Liberté de)", Diderot's "Aius-Locutius", Diderot's "Autorité politique", and the abbé Yvon's "Athées".

78 *Arrêts* (cit. n. 65), pp. 16–17.

79 *Arrêts*, ibid., p. 18.

80 *Arrêts*, ibid.

81 *Arrêts*, ibid.

82 *Arrêts*, ibid., p. 19.

83 *Arrêts du Conseil d'Etat du Roi, qui révoque les Lettres de privilège obtennues pour le Livre intitulé: Encyclopédie ou Dictionnaire raisonné des Sciences, Arts & Métiers, par une Société de gens de Lettres. Du 8 Mars 1759*, p. 2. Bibliothèque Nationale, Collection Joly de Fleury, vol. 572, fol. 288; reprinted in Diderot, *Oeuvres complètes*, vol. 5, pp. 43–44.

84 Wilson, *Diderot* (cit. n. 37), p. 339.

85 [Jacob Nicolas Moreau], *Nouveau mémoire pour servir à l'histoire des Cacouacs* (Amsterdam, 1757), p. 24. This and the next question are direct quotations from Diderot, with one misquote. Where Diderot had written "La matière vivante se combine-t-elle avec de la matière vivante?" the quotation in Moreau's memoir reads "La matière morte se combine avec la matière vivante?" Diderot asked a similar question about dead and living matter at another point in the same section, so the misquotation does not distort Diderot's ideas. See Diderot, *Pensées sur l'Interprétation de la nature*, in Diderot, *Oeuvre complètes* (cit. n. 1), vol. 9, pp. 97–98, LVIII, 11, 14. I have used the Slatkine Reprint edition of Moreau's memoir, which includes as well Giry de Saint-Cyr's *Catéchisme de décisions de cas de conscience à l'usage des Cacouacs* (Geneva: Slatkine Reprints, 1968).

86 Moreau, *Nouveau mémoire* (cit. n. 85), p. 24.

87 Moreau, *Nouveau mémoire* (cit. n. 85), p. 52.

88 [Giry de Saint-Cyr], *Catéchisme de décisions de cas de conscience à l'usage des Cacouacs* (Cacopolis, 1758), p. 14.

89 Lough, *Encyclopédie* (cit. n. 2), pp. 170–179.

90 *Encyclopédie*, vol. 15, p. 474A.

91 *Encyclopédie*, vol. 14, p. 149B.

92 Diderot's publications from the post-1765 period include his *Salon* of 1767, 1769, 1771, 1775, and 1781; his *Supplément au voyage de Bougainville* (1773–1774); and several essays. See Furbank, *Diderot* (cit. n. 34), pp. 479–483.

93 Franz A. Kafker, in *The Encyclopedists as a Group: A Collective Biography of the Authors of the* Encyclopédie, Studies on Voltaire and the Eighteenth Century, 345 (Oxford: The Voltaire Foundation, 1996), pp. 74, 75 n. 5, identifies nineteen (fourteen percent) of the contributors to the *Encyclopédie* as atheists, skeptics, or deists. In addition to Diderot and d'Alembert, this list includes Nicolas-Antoine Boulanger, Étienne-Noël Damilaville, Alexandre Deleyre, César Chesneau Du Marsais, Étienne-Maurice Falconet, Louis-Jacques Goussier, d'Holbach, Jean-Baptiste de La Chapelle, Nicolas Lenglet Du Fresnoy, Charles-Louis de Secondat, baron de Montesquieu, Jean-Denis de Montlovier, Jacques-Andrés Naigeon, Augustin Roux, Jean-François de Saint-Lambert, François-Vincent Toussaint, Anne-Rober- Jacques Turgot, and Voltaire.

94 Diderot to Princess.

FREDERICK B. CHURCHILL

AUGUST WEISMANN AND FERDINAND DICKEL: TESTING THE DZIERZON SYSTEM

Abstract

I examine and place into context a thirteen year correspondence (1897–1910) between the zoologist August Weismann and the apiarist Ferdinand Dickel. Their exchange concerns the validity of Johannes Dzierzon's mid-century claim that drone bees are the outcome of parthenogenesis. What emerges is a story that reflects upon the relationship between professional and amateur, and between hard and soft inheritance. The story also compels us to examine the authority of microscopy, Weismann's collaboration with his students, and the details of his germ-plasm theory at the turn-of-the-century.

Introduction

The story I wish to present is about a modest episode in the history of science: an episode that flashed for four brief years at the turn of the last century and the embers of which flickered in the literature until 1913.[1] By then the individual events were being eclipsed by the bright glare of Mendelian genetics and the new science of chromosomal genetics with its powerful tools for mapping chromosomes, describing the effects of genes and manufacturing hybrid organisms upon demand. Modest episodes in our profession, however, do not have to end in triumph to be instructive. In fact, this story is just the opposite, for it is about collaboration and disenchantment. It is about the reconfirmation of views accepted in biology for the preceding forty years. Its protagonists moreover were not to become the emerging heroes of the new century: one, Ferdinand Dickel, was a strong willed and persistent amateur in science who had the ability to stir things up, but who was unable to see his own shortcomings. The other, August Weismann, was a major zoologist of the nineteenth century whose contributions and career were drawing to a close and who, in a different way, was unable to see the shortcomings of his own life's work. Nevertheless, their combined investigation of a well defined, but in retrospect complex phenomenon, offers a window with a new perspective onto a broad range of events that were changing biology exactly a hundred years ago.

G.E. Allen and R.M. MacLeod (eds.), Science, History and Social Activism: A Tribute to Everett Mendelsohn, 61–76.

The Dzierzon System

The backdrop for this four year episode consists of a trail of dispute that began in 1845 with a publication by the beekeeper Johannes Dzierzon and extended beyond the turn of the century to its conclusion in 1913 with the doctoral dissertation of the nuclear cytologist and classical geneticist, Hans Nachtsheim.[2] Central to the dispute was the issue of sex determination in the honey bee, *Apis mellifera.*[3] Dzierzon had claimed the following three principles in his important text on beekeeping, *Theorie und Praxis:*[4]

1) Queen and worker bees, both female, were the products of normal fertilization.

2) Drones were the product of an unfertilized egg, that is, they developed by parthenogenesis.

3) Each caste was normally raised in specially sized cells. The queens developed in larger cells and differed morphologically and physiologically from the worker bee because they received pepped-up nutrition known as "royal jelly". The worker bees, also known as the neuter caste, hatched and matured in the smallest cells of the hive and were the developmental outcome of a less sophisticated diet. The drones, raised in cells differently capped from the worker cells, may have received different food, as well.

Dzierzon based his system on a lifetime of beekeeping experiences. He rejected a range of classical explanations about the status and origins of the three castes of bees and above all, challenged the widely held assumption that sexual reproduction was the law-given way for animals to reproduce.[5] As an experienced beekeeper Dzierzon was familiar with a range of well-known abnormal situations that he described in support of his system. He knew that unmated queen bees laid nothing but drone eggs, no matter in what sized cells they were placed. He knew that old and nearly exhausted queens laid nothing but drone eggs. He also knew that if the queen died or failed to produce workers or queens, some workers would lay eggs from which only drones came. Under any of these circumstances the hive would decline if the queen was not fertilized or replaced by a younger, healthier queen. Furthermore, Dzierzon recognized that hybrid queens produced drones that had the coloring and behavior of the female lineage. This suggested to him that there could be no paternal input to the drones heritage.[6]

Dzierzon's system was soon endorsed by Baron August von Berlepsch, a trained lawyer, the master of an estate in Thuringia, and an enthusiastic beekeeper. In the early 1850's "the Bienenbaron" sought to demonstrate the parthenogenetic origins of drones by immobilizing the spermatozoa in the seminal receptacle by icing the queen's abdomen and so presumably forcing her to lay only drone eggs.[7] More important for the confirmation of the Dzierzon system was Berlepsch's logistical support of the zoologists Rudolf von Leuckart and Carl Theodor von Siebold. Both recognized that the recently discovered micropyle of the insect egg provided an opportunity for demonstrating with the microscope the presence or absence of spermatozoa within the chorion of the egg. In 1855 both independently visited von Berlepsch with the intent to examine the freshly laid eggs he put at their disposal. Von Siebold was the more successful in his endeavors, for he found spermatozoa within the chorion of 58% of the fifty-two worker eggs he examined, while he failed to find spermatozoa in any of the twenty-seven drone eggs.[8] The parthenogenetic origins of drones appeared to be confirmed.

The microscope, however, was not the only arbiter of the Dzierzon theory. Experiments performed in the 1860's by Hermann Landois, professor of zoology at the university of Münster, provide examples of attempts to transfer eggs from one kind of cell to another. Landois recognized that worker bees often undid such efforts; so he developed the counter measure of cutting out a newly laid egg from the bottom of the cell and transferring the cell's bottom with its egg into a cell sized for the opposite sex, i.e. drone eggs into worker bee cells and worker eggs into drone cells. The consequence, he claimed, was that drone destined eggs remained unharmed by workers and so produced larvae that were inadvertently fed worker bee food and became worker bees. In a similar fashion the worker destined eggs were inadvertently fed drone food and became drones. Clearly, Landois reasoned, "The determination of which sex the larva developed into, depended upon the physical conditions of life, namely upon nutrition".[9] He added that the comparative study of many other parthenogenetic insects, including some *Lepidoptera* and *Diptera* and above all, aphids, also revealed the sex determining effects of nutrition. Landois's experiments and conclusions were immediately challenged by Dzierzon supporters, such as Siebold and Emil Bessels, but one finds quickly that the facts of such experiments depended not just on the authority of scientific leaders but equally on the craft of beekeeping.[10]

Another test of the Dzierzon system came in the form of hybridization experiments. These were exemplified by the 1878 work of the French comparative anatomist and zoologist Jean Pérez, who described attempts to cross an Italian queen (now identified as *A. mellifera ligustica* Spin.) with drones of the northern European or German race (*A. mellifera mellifera* L.). The former race possesses a light colored abdomen in contrast to the latter, which possess a darker abdomen. The Dzierzon system predicted that in this case of hybridization all the abdomens of the F1 drones should be of the light Italian color. Pérez, however, countered that of the 300 drones of the F1 generation he examined, 151 were of pure Italian coloration, 83 showed the pattern of the northern race and 66 exhibited a transitional pattern between the two. Pérez's experiments were controversial even in his day. As Nachtsheim later reflected, the purity of the original queen had not been established according to rigorous standards and, even if the experiments had been carried out in the most isolated of locations, it was not possible for the zoologist or beekeeper of the day to guarantee that the queen mated only with German drones. Nevertheless,

Pérez received a sympathetic ear from beekeepers and in addition found effective ways of arguing against the microscopic demonstrations of Siebold and Leuckart.[11] A brief biographical account of Pérez may be found in the *Archives biographiques françaises.*

Thus three different approaches, a) microscopy, b) experiments designed to test the effects of nutrition and c) hybridization of contrasting races, were marshalled as evidence for and against Dzierzon's claim that drones, unlike the queens and workers, were produced parthenogenetically. The results, however, were inconclusive. Zoologists and beekeepers alike found holes in the evidence for and against the Dzierzon system. Each of the approaches required special expertise. The microscopy of the period represented advances in zoology and botany that were associated with what both Siebold and Leuckart called "*wissenschaftliche zoologie*", and it is no accident that a number of the papers already cited appeared in a scientific journal with that expression in its title. The other two approaches – i.e. swapping of eggs within the hive and hybridization between different races of *Apis mellifera* – were more in the province of experienced beekeepers. Dzierzon, von Berlepsch and many others understood and could handle far better the difficulties and complications of these latter two areas of investigation.[12]

By the 1880s the balance between the three approaches and the authority of the zoologist vis-à-vis the beekeeper shifted. New techniques and equipments began to change rapidly the scene of microscopy. The oil immersion lens, microtomes that would permit fine and serial sectioning, aniline stains and new methods of conservation and preparation all helped focus attention upon the nucleus, chromosomes and other micro-organelles of the cell. Such refinements brought extraordinary advances to biology from Otto Bütschli's, Oscar Hertwig's and Hermann Fol's demonstrations of the events and figures of gamete union; to Walther Flemming's, Eduard Strassburger's and others' demarcation of the stages of indirect cell division (i.e. mitosis), and to Éduard van Beneden's, August Weismann's and Theodor Boveri's persistent examination of events of gametogenesis, including polar body formation (i.e. meiosis). These were the phenomena that would become the ultimate arbiters of sexual and parthenogenetic reproduction but at the cost of creating an epistemological divide between the professional zoologist, on the one hand, and other biologists and many professional beekeepers, on the other.

Ferdinand Dickel and A New System

It was an experienced beekeeper who in the 1890's again galvanized attention to the possible inadequacies of the Dzierzon system. Ferdinand Dickel (1854–1917) taught school in Darmstadt, was an avid beekeeper and editor of the local *Nördlinger Bienenzeitung.*[13] As reported in a biographical note Dickel formally presented his challenge at the 1898 meetings of the annual Itinerant Gathering of German and Austro-Hungarian Beekeepers [*Wanderversammlung deutscher und österreichisch-ungarischer Beienenwirte*]. His argument was founded not only upon his own experiments but upon the experience of many of the outstanding apiarists of the day. It caused a sensation!

> The presentation made a deep impression and the applause didn't want to come to an end. 'It showed esteem not only for the contents, but for the bold courage of the young revolutionary, who had broken the proscription that has weighed on the gathering' – wrote a participant of this meeting.[14]

In the same year Dickel also published his ideas in a seventy page pamphlet entitled *Das Prinzip der Geschlectsbildung.* It is to this document that we owe a detailed account of Dickel's early work and challenge to the Dzierzon system.[15] His attack was two-fold. First, he criticized the earlier investigations of von Berlepsch, Leuckart and Siebold. He was particularly critical of Siebold's microscopical efforts that were carried out at the end of August on a hive that was forced to maintain an aging queen. As every beekeeper knows, queens under such conditions would only lay drone eggs. Siebold's evidence, moreover, was limited to the examination of only twenty-seven drone eggs, and given the circumstances, this hardly seemed sufficient.[16] Both criticisms reflected a deeper concern. Nature, Dickel argued, was subject to uniform laws, and when Dzierzon maintained that bisexual organisms could reproduce both sexually and parthenogenetically, he appeared to violate this uniformity. According to Dickel, the Dzierzon system:

> declares ... drones are derived from unfertilized eggs and moreover stands an unquestionably well established law of science on its head when at the same time it asserts: the contribution of the father during the act of reproduction prevents the rise of sons, since among bees the successful mating by the father determines that only female beings can come into existence![17]

His is perhaps a clumsy sentence, but Dzierzon's apparent violation of common sense and Dickel's outrage against the insult to his manly pride is all too clear.

Second, Dickel presented a range of observations and a few experiments that militated toward the conclusion that worker bees rather than the *Anlagen* or germinal constitution of the egg determined the sex and caste of bees. He observed that worker bees spent considerable time tending to the cells of developing larvae. This would not be necessary, he declared if the sex of the larvae were already determined. He noted that worker bees have four pairs of secretion glands; so one might suppose that they served different feeding functions. Moreover, the worker bee was sensitive to the different sizes of cells, which were known to give the cue for the normal feeding routine of the larvae inside. Dickel carried out exchange experiments of the kind Landois had done, both with the transfer of eggs and of larvae, and came to the conclusion that given appropriate steps, one could force any of the three castes from the worker cells.

What emerged from Dickel's lengthy and discursive broadside was a system, whose principles contrasted with Dzierzon's:

1) Dickel urged, with a nod to Weismann's germ plasm theory, that the normal male or female zygote must be hermaphroditic, for it must contain both male and female *Anlagen.* He elaborated that the drone gametes carried the female *Anlagen* and the queen gametes carried the male *Anlagen.*

2) He argued that in their construction of different kinds of cells and because of their nutritional care of the larvae, the workers were the agents that through their secretions determined which of the two *Anlagen* became activated and which of three castes appeared.

3) It followed that the eggs from unmated and aged queens and worker bees could not be hermaphroditic. When such eggs developed they necessarily led to abnormal drones regardless in what cells they were placed and what nutrition the larvae received. There was thus, according to this principle, a contrast between the normal, functioning hermaphroditic drone and the abnormal, or so-called "false" drone, which was unable to contribute in any way to the perpetuation of the hive. Dickel's distinction between normal and "false" drones was based on the logic of the situation and not on morphological distinctions.[18]

There existed a balance of a beekeeper's experience and a complex argument undergirding Dickel's replacement system. There can be no doubt that Dickel was a keen observer and was constantly in touch with the writings of other professional apiarists.

Critical of the academic zoologist encroaching upon his field of expertise Dickel wrote sarcastically about Siebold's naivete in investigating bees. We also know, however, that Dickel had studied for "a few years" at the Darmstadt Technische Hochschule, where he had been exposed to zoology and chemistry under professors Gottlieb von Koch and Wilhelm Städel, respectively. This experience may have encouraged him to comment in a general manner at the end of his text about the importance of finding a unified theory of reproduction that would connect his ideas about sex determination in bees with that of higher organisms.[19] For this reason he may have recognized that observations and arguments alone would not suffice and that he needed microscopical verification of his ideas, which he himself could not provide. In the summer of 1897 he contacted Rudolf Leuckart, who responded with encouragement but declined to become involved.[20] Dickel had apparently also contacted Weismann in early July of the same year, and this overture initiated a period of four years of collaborative research and thirteen years of correspondence. Unfortunately, we have only one side of the correspondence; nevertheless, the twenty-four letters by Weismann to Dickel, copied in Weismann's copybooks, reveal how very important this issue became for the Freiburg zoologist.[21]

The Dickel–Weismann Collaboration

At first, Weismann begged off involvement with Dickel's project to demonstrate the erroneous nature of Dzierzon's system, perhaps because of his ailing eyes, which no longer could tolerate extended microscopical research. Nevertheless, he encouraged Dickel to keep him informed and to send him copies of his relevant publications.[22] A year and a half later Dickel obliged him with copies of his *Bienenzeitung* and the recently published *Prinzip der Geschlechtsbildung*. Weismann replied that he found some of its arguments quite convincing. "I consider the main point as proven", he responded enthusiastically, "... It seems to me almost certain that the queen fertilizes the drone eggs. You have thus done science a great service, whether or not many of your further conclusions prove to be correct".[23] It also becomes clear that Weismann had not forgotten the issue during the intervening eighteen months, for he had put a student, probably Wilhelm Paulcke, on the project in the summer of 1898. Because it was late in the season and they could only obtain older eggs, this first study was unsuccessful. The second letter also gave Weismann the opportunity to arrange a collaboration with Dickel to examine freshly laid eggs (less than an hour old) under the microscope.

During the season of 1899 Weismann's student, now clearly identifiable as Paulcke,[24] supplied Dickel with vials of conserving fluid. Dickel placed freshly laid worker and drone eggs into separate and labeled vials, and sent them back to Freiburg. Paulcke then used the most up-to-date techniques of serial sectioning and staining in order to examine the eggs for evidence of sperm penetration through the vitelline membrane, for the male pro-nucleus within the cytoplasm of the egg, and for aster formation at the time of its union with the egg pro-nucleus. This kind of evidence should be able to confirm definitively Dickel's conviction that the drone eggs were also fertilized. "... The conclusion", Weismann pointed out in his next letter, "can only be made with the micro.[scopical] investigation of fresh eggs".[25] At the same time Weismann had been in correspondence with Hans Butel-Reepen, an experienced apiarist and a student at the zoological institute in Jena, who must have read about Weismann's encouragement to Dickel in Dickel's *Bienenzeitung*. It seems that he wrote Weismann a letter of caution. For one thing, he informed Weismann that queens are known to err frequently by laying inappropriate eggs in worker and drone cells. Weismann would have immediately recognized that this common error might distort Dickel's findings.[26] Weismann also claimed to have recognized from the outset that "in fact, the theoretical labyrinths of Herr. Dickel are scarcely worthy of attention". Nevertheless, he intimated to Reepen that "... since Siebold, Leuckart and Dzierzon we have proceeded in error. Probably not certainly"![27] Thus Weismann was clearly playing a double role. On the one hand, he encouraged Dickel and wished to collaborate with him because he recognized a diligent observer and experienced bee keeper who might well have put his finger on a serious problem with the accepted Dzierzon doctrine. On the other hand, he was not enthralled by Dickel's own system, which in a cumbersome manner merged multiple conjectures into a general system.

The collaboration indeed took place. By May 1899 Weismann reported that Paulcke's microscopical investigations were going well. After further exchanges, encouragements and urging for ever fresher eggs, Weismann in July reported to Dickel some exciting news: "We are in the clear with the matter. The eggs from drone cells are in fact, as you have already long since concluded from your experiments, fertilized". He explained that Paulcke had found sperm pro-nuclei on five different occasions and at the same time he had located the polar-bodies of the egg. Both Weismann and his first assistant, Valentin Haecker, had verified the preparations; all that was needed to be done was to see the astral and chromosomal figures of sperm and egg pro-nuclei fusion. Weismann ended his glad tidings with cheer and enthusiasm: "It pleases me especially that this theoretically very important question is now finally decided and I wish you luck moreover in removing this fatal error about the sex-determining nature of the spermatozoa from science".[28]

Any celebration that Dickel might have had, however, was short-lived. The next day Weismann wrote again: "Unfortunately this morning I had to retract by telegram my glad tidings of yesterday's letter".[29] A new series of slides had indicated that the Freiburg team had confused the egg pro-nuclei with sperm pro-nuclei. No matter which arrived first – the letter of the 10th or the telegram of the 11th – one can well imagine the disappointment Dickel must have felt. He evidently tried to offer some explanation as to why Paulcke could not get consistent results, for by the end of the month Weismann had to explain that it was impossible for the two pro-nuclei to have formed a zygote before the egg was deposited. Further letters brought more discouraging news. Over fifty different series of observations failed to reveal any male pro-nuclei in drone eggs; while the control series on worker eggs demonstrated the effectiveness of Paulcke's technique of searching for the male pro-nucleus. On further occasions Dickel must have objected, but each time Weismann explained how they could not avoid the microscopical implications of Paulcke's work. "With regret", Weismann reported on 20 August on the end results of the season's investigations. "This is entirely different from what we had expected from your numerous and careful experiments.[30]

The outcome of Paulcke's microscopical investigations was important not only to Dickel but to Weismann as well. For a number of reasons, Weismann felt strongly enough about the uncertainties of the Dzierzon system to commit yet another season and another talented student, the Russian emigre Alexander Petrunkevitch, to the project. He also took pains to lay out in a detailed fashion to Dickel what the issues were. Petrunkevitch would determine whether the drone eggs were fertilized or not, in fact, "... that is possible only through the microsc[opical] investigation". No matter what the answer to that question might be, Petrunkevitch would be unable to respond to the second question through the microscope, i.e. the determination of sex. "What determines the sex here", Weismann penned, "is then *your* affair, and your investigations to date speak strongly for a determination by the worker".[31] Weismann went on to urge Dickel to serve science by continuing to provide the institute with fresh eggs. As Petrunkevitch's work got successfully underway, Weismann even held out the possibility that a microscopical difference in the development of worker laid drone eggs and of queen laid drone eggs, discovered by Petrunkevitch, might support Dickel's belief in a distinction between "true" and "false" drones.[32]

Despite the encouragement, Dickel must have been on his guard about a project he could not control nor fully understand. At one point he tested Petrunkevitch's procedure by reversing the labels on the returned vials. Petrunkevitch quickly detected a problem and executed a hurried trip to Darmstadt to sort things out. Weismann's reaction to the episode is of interest; instead of scolding Dickel for the deception, he reasoned that "I find it very natural that you wanted to convince yourself whether we were truly in the position to unearth sound results and whether in addition this was really the right approach".[33] As events unfolded, Petrunkevitch's microscopical studies built up a decisive refutation of Dickel's claims that drone eggs were fertilized, and in the process he developed a new conserving fluid and found the subjects for both his doctorate and *Habilitationsschrift*.[34]

By mid-July Weismann promised to publish a brief report on the results of three years of research. He was good to his word, wrote the report over his summer vacation, and sent it to the *Anatomischer Anzeiger* in early October.[35] It was a straight-forward report addressing the core of the Dzierzon theory, i.e. whether drone eggs were fertilized. Appreciative lines scattered through the text made clear his debt. He commended Dickel for recognizing the weakness in Siebold's and Leuckart's forty-five year-old demonstrations and for his energetic pursuit of the problem through egg-swapping experiments, and he thanked Dickel for supplying both Paulcke and Petrunkevitch with fresh eggs laid at various stages during the maturation process. He pointed out, however, among other demonstrations that Petrunkevitch had found the sperm aster in 100% of sixty-two worker bee eggs at the stage of second polar body formation, but by way of contrast, he had found no evidence of the sperm nucleus or its aster in 272 drone eggs at the same stage.[36] Weismann turned his attention to Dickel's nutrition theory and argued that too little was known about the worker secretion(s) to make any claims about their effects. It was clear, however, that worker bees with their secretions could not counteract the deleterious effects on the hive of aging queens that produced only drone eggs. Finally, Weismann did not dismiss out of hand Dickel's concept of true and false drones. After all, he pointed out, wasps often produced different castes of males as well as females. In brief, Weismann did not debunk Dickel's complex theory. He treated it with respect but insisted that microscopical inspection had once again vindicated Dzierzon's original claim about the parthenogenesis of drones.

The ever-resilient Dickel responded in print with an attack on the trustworthiness of Petrunkevitch's method and new conserving fluid. He cited once again the well-known results of hybridization experiments and elaborated upon his secretion theory. The editor of the *Anatomischer Anzeiger*, Karl Bardeleben, an anatomist himself, permitted Weismann a short rebuttal and Dickel a final rejoinder.[37] The ground had now been tilled and exhausted on both sides and could no longer bear fruit. Weismann explained to Bardeleben, who had apparently asked him about Dickel, that "He [Dickel] had also written *me* earlier often long letters about sex and reproduction and other problems, which all showed that he has an original and restless, active mind, which is, however, deficient in training in reasoning and above all in a knowledge of biological fundamentals".[38] The general tenor of this comment was to be repeated to other professional correspondents over the next few months.

Dickel, however, would not be satisfied. He responded to the publication of Petrunkevitch's dissertation with another defense, this time in the *Zoologischer Anzeiger*.[39] In 1903 he wrote a systematic account, complete with a short history of the controversy over Dzierzon and a synopsis of the experiments that supported his position. What was really at stake, besides the recital of his experiments, came out in a general comment at the outset: "In my opinion the determination in this theoretically so important a question cannot be achieved through the microscope, but only through experimental investigation". It was appropriate that Dickel by this time had turned to the *Archiv für der gesammten Physiologie* as a venue for his defense.[40] Its editor, the Bonn physiologist Eduard Pflüger, added a supportive note later in the year.[41] The physiologist Albrecht Bethe and the physicist Porphiry Ivanovich Bachmetjew also weighed in on Dickel's side.[42]

Weismann's reaction to this flood of support for Dickel was based on the fundamental principle that had been established in zoology in the 1870s when it was shown that *material* from the spermatozoa joined *material* from the egg in the process of fertilization. This principle underwrote the importance of nuclear microscopy. Some insight into Weismann's response to what he considered outside kibitzers can be gleaned from a letter he wrote to Buttel-Reepen, who by 1903 had received his doctorate in zoology in Freiburg and had returned to Oldenburg to continue his interests in the biology of bees. After thanking his former student for a reprint, Weismann continued:

> I would, however, prefer to dispense mixing in this apiarian confusion; at least for now. I could only repeat what I have said at the beginning. How do *Bethe* and *Pflüger* come to their non-zoological points of view? *Bachm. [Bachmetjew]*, is a physicist, obviously doesn't know about the process of fertilization, otherwise he would be unable to operate with "half" fertilization.[43]

Weismann's side of the correspondence with Dickel dropped away for a while, but three letters toward the end of Weismann's copy books finished the exchange. Dickel apparently continued to send Weismann reprints, manuscripts and many letters: Weismann wrote to Petrunkevitch in 1907 that he had received 12 long letters from Dickel that year alone.[44] Finally, in 1910 Weismann insisted to Dickel that his personal conclusions had not changed and complained that he could no longer understand the beekeeper's arguments. "I want to suggest to you", he counseled in reply to two further reprints, "that you temporarily let these essays rest. If they contain good ideas and promote science, they will without doubt come into their own".[45] By the last letter in the series, however, Weismann had lost his patience. "I cannot ascribe any scientific value to the scientific fantasies you write me ... Please do not write me anymore".[46]

Lessons From The Narrative

This story, charged with hope and excitement at the beginning, ended in a pathetic and lamentable fashion. Nevertheless, it reveals a good deal about competing authorities in zoology and about Weismann and his germ plasm theory.

First, the story as a whole indicates that Weismann was both methodical and cautious in arriving at scientific conclusions, and that he depended upon his students for the microscopical work which, toward the end of his life, he could no longer perform. In fact, he continued to interact with both Petrunkevitch and Buttel-Reepen long after they had left Freiburg and established their own professional careers.

Second, his letters to Dickel revealed Weismann's recognition of the important role that amateurs could play in zoology. We know from other correspondence that Weismann was in touch with many others.[47] He was not hesitant to learn from them as well as to suggest strategies for their further research. In this episode it happened that the authority of the *Ordinarius* and *Wissenschaft* eventually collided with the novel ideas of the amateur, but it was neither an unsympathetic nor reactionary establishment that met the revolutionary upstart. It was, rather, a man who saw an opportunity to advance science in a way that might have benefitted both of them. He insisted first and foremost, however, that nature alone must hold the final verdict even though, as we shall see, she did not benefit the cause of either.

Third, from the beginning Weismann was never a neutral participant. Consider that he spent the time and energy to write twenty-four letters, many of them lengthy, to Dickel at a time in his life when his eyesight was poor and deteriorating; consider, too, that he committed the time and resources of his institute and steered two of his best students to assist in the four year project (i.e. 1897–1900)[48] in attempting to document Dickel's challenge to Dzierzon, and one has to conclude that this project was high on Weismann's agenda. But why?

This may be answered, in part, by a fourth lesson. This story touches upon the nature of German zoology at the end of the century, which had over the previous thirty years separated itself from the medical faculty and found in the combined subjects of heredity, development, natural history and evolution a domain to justify that independence.[49] Its authority, and to a lesser extent the authority of comparative and microscopical anatomy, could prevail in this domain because of the rapid advances that had been made since 1870 in professional microscopy. Weismann's insistence that the nuclear cytology of fertilization should be the final arbiter of drone parthenogenesis rather than hive-side experimentation, reflected this conviction, and it explains both his snide remarks about Dickel's later appeal to and the intervention of physiologists and a physicist into the controversy.

Finally, the story forces us to confront the nature of the germ plasm theory, which Weismann had developed in the 1880s when his research was at the forefront of nuclear cytology and which he molded into a formidable though controversial theory of evolution in the 1890s. Two aspects of his achievements, however, were not readily compatible with the Dzierzon claim of the parthenogenesis of drones. First, Weismann shared with most of his generation an assumption that sex was determined embryologically.[50] The conviction was derived from the details of descriptive embryology, which had shown that vertebrates, particularly the early embryonic stages, had the structures of both sexes and were, in fact, hermaphroditic.[51] As he developed his germ plasm theory, Weismann was known to justify this embryological perspective with well-known facts about the transmission of sexual traits through parents of the opposite sex and well-studied cases found throughout the animal kingdom of teratological hermaphrodites. To account for these phenomena in his *Keimplasma* of 1892 Weismann had spoken of "double determinants". A nutritional competition between them helped account for the normal exclusion of half of the sexually dimorphic traits. He furthermore speculated that there must be some kind of germinal linkage between determinants to explain the combinations of sexual traits that showed themselves in one or the other sex. Weismann's explanation for the determination of sexual characters was patterned after the developmental nature of his explanation for polymorphism in general.[52] The second aspect of the germ plasm theory that conflicted with the implications of the Dzierzon system was Weismann's overall conception of the chromosome, or "idant" in his terminology. Weismann believed that the chromosomes were a lineal assemblage of packages of ancestral determinants, i.e. "ids" in his terminology. By theory, the germ plasm contained the Anlagen for many potential variations, and each id contained all the determinants for a complete individual. Weismann's understanding of reduction division therefore was a mechanism for eliminating half the packages of ancestral determinants before fertilization brought their quantity back to the normal number – a move from diploidy to haploidy back to diploidy – except that Weismann was more focused on the packages of ancestral plasms (i.e. the ids) than the number of chromosomes (i.e. the number of idants).

At the same time Weismann attributed a continuity and an "individuality" to the chromosomes through cell divisions, but not in the sense of classical genetics. Here lies the crux of the difference between him and the genetics that soon followed. For Weismann the individuality of the chromosomes was to be found in the unique ancestral lineage of each chromosome. With classical genetics, individuality was to be found in the unique combination of factors determining certain traits. Weismann's conception was dependant upon a general acceptance of recapitulation as articulated by Ernst Haeckel in the 1860s.[53] The conception of classical genetics was derived from the brilliant polyspermic experiments of Theodor Boveri, who demonstrated that at least one member of each chromosomal pair was necessary for normal and complete development.

So on two counts the Dzierzon theory of the parthenogenesis of drones presented Weismann with unavoidable complications. He had to explain through hypothesized developmental processes how the lack of fertilization determined the male. His overall theory, however, prevented him from invoking as an explanation the presence or absence of sex determining chromosomes or factors. Given Weismann's 1892 account of caste determination in bees, it is no wonder that Dickel felt comfortable writing to Freiburg in 1897. Also, given the details of his germ plasm theory, it is not surprising that Weismann believed it to be in his interest to re-examine drone parthenogenesis with the new tools of microscopy. Thus a bargain was struck between the professional zoologist and the beekeeper to collaborate in testing the Dzierzon system.

In the short run Weismann, with the help of his students, convinced himself and many other zoologists that drone eggs were not fertilized and that Dzierzon had been right all along. Dickel would not accept the results of the new microscopy and continued to urge in various ways his secretion theory of drone determination. In the long run, however, I believe Weismann was to have the harder time of it. Long before 1913, when Hans Nachtsheim presented the definitive demonstration of the accuracy of the Dzierzon system by correcting significant details in Petrunkevitch's work, the advent of Mendelian genetics and a new chromosomal genetics associated with the names of Theodor Boveri, Walter Sutton and Thomas H. Montgomery had revealed the inadequacies of Weismann's germ plasm theory in ways that the Dzierzon system had only suggested. The professional zoologist, more keenly aware of the limitations to his scientific arguments, had to face a more unsettled future than did the amateur beekeeper. But that is another story.

Indiana University

Notes

1 The skeleton of this story is one I related to Everett Mendelsohn nearly thirty-years ago at the International Congress of the History of Science in Moscow. That was a time when Everett wisely encouraged a group of young western historians of biology to associate with our peers in the Soviet Union, but it was also a time when many older Soviet biologists were still unfriendly to any history that might remotely reflect on their Lysenko ordeal and legacy. Little did I realize at the time how I might touch on those raw nerves, and little did I understand why my innocently crafted historical report on Weismann/Dickel was crudely disrupted by a front-row mafia. The reworking of this story has allowed me to understand better all kinds of hidden parallels between two periods of biology and how the writing of history, when done without a full and honest social and intellectual contextualization, may be misconstrued and misused.

2 Hans Nachtsheim, "Cytologische Studien über die Geschlechtsbestimmung bei der Honigbiene (Apis mellifica L.), *Archiv für Zellforschung*, 11 (1913a), pp. 169–240.

3 During the period covered by this paper *Apis mellifica* L. was the scientific name used for the common honey bee.

4 Johannes Dzierzon, *Theorie und Praxis des neuen Bienenfreundes* (1845). Actually Dzierzon had originally presented his ideas about the nature of the drone egg in the *Eichstädter Bienenzeitung*. Both the journal and the privately printed first-edition of *Theorie und Praxis* are difficult to find. A

second edition (Nördlingen: Deutsche Bienenverein, 1849) is a more readily accessible work. Subsequent authors nevertheless use 1845 as the effective beginning point of Dzierzon's ideas.

5 The notion of reproduction without sex was just beginning to be understood by a few biologists. Richard Owen coined the term "parthenogenesis" in 1849 to designate what he called "Lucina sine concubitu". Although Owen embellished his term with a rather idiosyncratic theory about a "spermatic force" influencing multiple generations of asexual reproduction, his term survived and soon designated the process of development of the egg without fertilization. For the history of parthenogenesis, see Otto Tashenberg, "Historische Entwickelung der Lehre von der Parthenogenesis", *Abhandlungen der Naturforscher Gesellschaft. zu Halle*, 17 (1892), pp. 365–454. More recent historical accounts may be found in John Farley, *Gametes & Spores. Ideas about Sexual Reproduction 1750–1914* (Baltimore and London: Johns Hopkins University Press, 1982) and Frederick B. Churchill, "Sex and the single organism: biological theories of sexuality in mid-nineteenth century", *Studies in the History of Biology*, 3 (1979), pp. 139–177.

6 Modern zoologists recognize that there are many complexities in bee reproduction; for example, the queen undergoes multiple matings, which take place in a "drone congregating area" several kilometers from the hive and at heights of 10 to 40 meters. A good modern source on the reproduction and behavior of bees is Mark L. Winston, *The Biology of the Honey Bee* (Cambridge, Mass. and London: Harvard University Press, 1987).

7 An English translation of his work, "The Dzierzon Theory" appeared in the *American Bee Journal*, 1 (1861), pp. 5–6, 25–27,49–51,73–76, 97–99, 121–25, 145–48, 169–72, 199–202, 223–26. A biographical sketch of von Berlepsch may be found in Erich Schwärzel, *Durch Sie Wurden Wir. Biographie der Grossmeister und Förderer der Bienenzucht im deutschsprachigen Raum* (Giessen: Die Bienen, 1985), pp. 17–19; see also *Neue Deutsche Biographie*, 1953, 2, pp. 94–95.

8 Carl Theodor von Siebold, *Wahre Parthenogenesis bei Schmetterlingen und Bienen. Ein Beitrag zur Fortpflanzungs-geschichte der Thiere* (Leipzig: Wilhelm Engelmann, 1857). Siebold reproduced lengthy passages from Dzierzon's text and examined Leuckart's work before describing his own. His technique was from a later perspective quite crude. He gently pushed down with a coverglass on the end of the egg with the micropyle and then scanned the extruded fluid with his microscope for spermatozoa. The account of his investigations may be found on pp. 109-120. Further details of Leuckart's and Siebold's studies may be found in Nachtsheim (n. 2) and Churchill (n. 5).

9 Hermann Landois, "Ueber das Gesetz der Entwickelung der Geschlechter bei den Insecten", *Zeitschrift für wissenschaftliche Zoologie*, 17 (1867), pp. 375–379. Quotation is on p. 376.

10 This conclusion is drawn from the description of the experiments not only by Landois, but by Emil Bessels in "Die Landois'sche Theorie widerlegt durch das Experiment", *Zeitschrift für wissenschaftliche Zoologie*, 18 (1868), pp. 124–141.

11 Hans Nachtsheim, "Die parthenogenesis bei der Honigbiene. Ein historischer Ueberblick über den Kampf um die Dzierzonsche Theorie", *Bienenwirtschaftliches Centralblatt*, 49 (1913), pp. 298–328.

12 A survey of local beekeeper journals would be important in order to evaluate the full reception of these three approaches by the beekeeping community at large. The author unfortunately did not have access to satisfactory runs of such journals.

13 For a brief overview of Dickel's life, see E. Schwärzel (n. 7), pp. 45–46. I do not intend to enter into the extensive and important literature on amateurs in science. In the German context of this episode I refer to Dickel as an amateur because he was a well-read and enthusiastic apiarist but did not have a university degree in science nor, for that matter, in any other subject. In other words, he was not a "Wissenschaftler" or experienced scholar practiced in "scientific" methodology.

14 Schwärzel mentions six approving apiarists by name. The quotation comes on p. 45. Dickel also repeatedly named fellow beekeepers who supported his position.

15 Ferdinand Dickel, *Das Prinzip der Geschlectsbildung bei Tieren geschlechtlicher Fortpflanzung entwickelt auf Grundlage meiner Bienenforschung* (Nördlingen: C.H. Beck, 1898) Later on Dickel was to defend and slightly alter his ideas in print.

16 Dickel failed to mention that Siebold also examined fifty-two worker bee eggs and found spermatozoa in 58%.

17 Dickel (n. 15), pp. 13–20. Quotation is on pp. 14–15.

18 A summary of Dickel's theory may be found on pp. 20–22. It is interesting to note that Dickel must have relied on reduction division as a mechanism for segregating the % of *Anlagen* – and so had Weismann.

19 Schwärzel (n. 7), pp. 45–46; Dickel (n. 13), pp. 66–67.

20 Dickel (n. 15), p. 20. Leuckart died the following February.

21 All twenty-four letters are printed in full in Frederick B. Churchill and Helmut Risler, *August Weismann. Ausgewählte Briefe und Dokumente*, 2 vols. (Freiburg im Breisgau, Universitätsbi-

bliothek, 1999). These letters present an unusual picture of Weismann's interaction with a fellow investigator. Despite his reluctance to write lengthy letters, Weismann discussed with Dickel empirical details and procedures, matters of disagreement, modes of scientific documentation and broader evolutionary issues, such as hive selection and Lamarckian inheritance, in a way that was uncharacteristic in his other letters written in the last two decades of the century. We are here concerned simply with the testing of the Dzierzon system. All letters cited are reproduced in this collection. See also, Churchill, "The Weismann-Dickel Correspondence on the Parthenogenesis of Drones", *Proc. XIII International Congress of the History of Science, Moscow, Aug. 18–24, 1971*, Section IX, pp. 84–89 for a preliminary but limited report on these letters.

[22] Weismann to Dickel, 9 July 1897.

[23] Weismann to Dickel, 3 Dec. 1898.

[24] Wilhelm Paulcke was an experienced student, having worked in Weismann's institute since 1895, having spent a year in Arnold Lang's institute for zoology and comparative anatomy in Zürich and having worked at the Naples Zoological Station for two months. The work complemented his dissertation, "Ueber die Differenzirung der Zellelemente im Ovarium der Bienenkönigin (Apis mellifica &)", *Zoologischer Jahresbericht II. Abt.*, p. 14 (1901), pp. 77–202.

[25] Weismann to Dickel, 18 February, 1899.

[26] One can infer the cautionary contents from Weismann's letter to Reepen, 15 January, 1899. Reepen soon expanded his name to von Buttel-Reepen and became a student at Weismann's institute. He was an experienced apiarist but for professional reasons did his dissertation on parasitic trematodes. After his promotion he could speak with authority as both a zoologist and beekeeper. He became one of the noteworthy interpreters of bee biology of his generation. See Karl von Frisch account in *Neue Deutsche Biographie*, 3 (1957), p. 80 and Schärzel (n. 7), pp. 35–36.

[27] The full sentence reads: "Wenn ich Hrn. Dickel den betreffenden Passus der Anerkennung geschrieben u. ihm durch das Recht, denselben abzudrucken zugestanden habe, geschah es, weil mir seine Versuche, die in der Bienenzeitung veröffentlicht wurden, in der That recht warscheinlich machten, daß wir uns seit Siebold, Leuckart u. Dzierzon in einem Irrtum bewegt haben. Wahrscheinlich nicht gewiß!" Weismann to Dickel, 15 January 1899.

[28] Weismann to Dickel, 10 July 1999.

[29] Weismann to Dickel, 11 Juli 1899. The mistake may indicate how badly Weismann, Haecker and Paulcke wanted to refute the Dzierzon system.

[30] Weismann to Dickel, 20 August 1899.

[31] Weismann to Dickel, 24 April 1900.

[32] Weismann to Dickel, 20 June 1900.

[33] Weismann to Dickel, 20 June 1900.

[34] Alexander Petrunkewitsch [Petrunkevitch], "Die Richtungskörper und ihr Schicksal im befruchteten und unbefruchteten Bienenei", *Zoologische Jahresbericht. Abt. II*, 14 (1901), pp. 573–608 and "Das Schicksal der Richtungskörper im Drohnenei. Ein Beitrag zur Kenntniss der natürlichen Parthenogenese", *Ibid.*, 17 (1903), pp. 481–516. When he married an American and moved to the United States, Petrunkevitch anglicized the spelling of his name. After some short-term positions, including that of an independent researcher, he became a internationally recognized expert on the morphology and taxonomy of arachnids and a distinguished professor of zoology at Yale University. For biographical information see G. Evelyn Hutchinson, "Alexander Petrunkevitch. An appreciation of his scientific works and a list of his published writings", *Transactions of the Connecticut Academy of Arts and Sciences*, 30 (1945), pp. 9–24.

[35] August Weismann, "Ueber die Parthenogese der Bienen", 18 (1900), pp. 492–499. A slightly altered version with a different title appeared at the same time in *Die Biene* (Giessen), the official journal of the *Wanderversammlung*.

[36] Petrunkevitch (1901; n. 34), p. 581; Weismann (1900), p. 495. Petrunkevitch found that it was easier to locate the much larger sperm aster than the sperm pro-nucleus in his serial sections. This dictated that Dickel supply him with eggs at the stage of second polar body formation.

[37] Ferdinand Dickel, "Meine Ansicht über die Freiburger Untersuchungsergebnisse von Bieneneiern", *Anatomischer Anzeiger*, 19 (1901), pp. 104–108; August Weismann, [Erwiderung auf Ferdinand Dickel], *Ibid.*, pp. 108–110; Dickel, "Thatsachen entscheiden, nicht Ansichten", *Ibid.*, pp. 110–111.

[38] August Weismann to Karl Bardeleben, 13 January 1901.

[39] Ferdinand Dickel, "Über Petrunkewitsch's Untersuchungsergebnisse von Bieneneiern", *Zool. Anzeiger*, 25 (1902), pp. 20–27. Dickel emphasized that Petrunkevitch was not justified in assuming that the absence of the astral rays in drone eggs was equivalent to the absence of the much smaller sperm pro-nucleus. He further questioned the efficacy of his staining methods and doubted that the queen could open and close the sphincter of her spermatheca in a systematic manner.

[40] Ferdinand Dickel, "Die Ursachen der geschlechtlichen Differenzierung im Bienenstaat. (Ein Beitrag zur Vererbungsfrage)", Pflüger's *Archiv für die gesamte Physologie des Menschen und der Tiere,* 55 (1903), pp. 66–106. Quotation appears on p. 70.
[41] Eduard Pflüger, "Ueber die jungfräuliche Zeugung der Bienen", *Archiv für die gesamte Physologie des Menschen und der Tiere*, 99 (1903), pp. 243–244.
[42] Albrecht Bethe, "Dürfen wir den Ameisen und Bienen psychische Qualitäten zu schreiben?" *Archiv für die gesamte Physologie des Menschen und der Tiere*, 70 (1898), pp. 15–100; Porphiry I. Bachmetjew [Bakhmetev], "Ein Versuch, die Frage über die Parthenogese der Drohnen mittels der analytisch-statistichen Methode zu lösen", *Allgemeine Zeitschrift für Entomologie*, 8 (1903), pp. 37–44.
[43] Weismann to Buttel-Reepen, 24 Nov. 1903. Pflüger, Bethe and Bachmetjew all carried out extensive research in the physiology of insects. Weismann was not opposed to physiology per se; many of his students were encouraged to minor in the subject at Freiburg. Two years before his comment he had also written a complimentary "Preface" for the first volume of Bachmetjew's, *Experimentelle entomologische Studien*, 2 vols. (Leipzig: Wilhelm Engelmann).
[44] Weismann to Petrunkevitch, 29 Sept. 1907. There is internal evidence in later correspondence to Dickel that Weismann wrote Dickel at least once in 1904. This letter has not been identified.
[45] Weismann to Dickel, 27 February 1910.
[46] Weismann to Dickel, 18 May 1910.
[47] For example, Weismann carried out extended correspondence with Max Standfuss, Emil Fischer and Frederic Merrifield, all lepidopterists by avocation, and with Otto von Staudinger of Dresden, a commercial collector by trade.
[48] As previously indicated the actual microscopical examinations began in 1898. Petrunkevitch devoted another three years to completing his dissertation and Habilitationsschrift on the subject.
[49] Lynn K. Nyhart, *Biology Takes Form. Animal Morphology and the German Universities, 1800-1900* (Chicago and London: University of Chicago Press, 1995). The transfer of Weismann's position from the medical to philosophical faculty and the administration of the zoological institute occurred at separate times, and both were accomplished in 1874. E.Th. Nauck, *Zur Vorgeschichte der Naturwissenschaftlich-Mathematischen Fakultät der Albert-Ludwigs-Univeristät Freiburg im Breisgau. Der Vertretung der Naturwissenschaften durch Freiburger Medizinprofessoren* (Freiburg im Breisgau, 1954), pp. 37–8.
[50] Jane Maienschein, "What determines Sex? A study of converging approaches, 1880–1916", *Isis*, 75 (1984), pp. 457–480; and Scott F. Gilbert, "The embryological origins of the gene theory", *Journal of the History of Biology,* 11 (1978), pp. 307–351.
[51] The recognition of the hermaphroditic nature of vertebrates may be found in the embryological work of Johannes Müller and others in the first half of the century, but it is conveniently associated in the last third of the century with Wilhelm Waldeyer's *Eierstock und Ei. Ein Beitrag zur Anatomie und Entwicklungsgeschichte der Sexualorgane* (Leipzig: Wilhelm Engelmann, 1870).
[52] August Weismann, *Das Keimplasma. Eine Theorie der Vererbung* (Jena: Gustav Fischer, 1992), pp. 494–497. Weismann did not discuss in a detailed fashion the parthenogenetic nature of drones, but he assumed the Dzierzon theory and explained the rise of all three castes in terms of sets of double female and single male determinants. He spoke of the male gamete containing the double female determinants and the female gamete containing the male determinants. According to his understanding, the male determinants prevailed embryologically only in the case when the egg was not fertilized, and so they did not have to compete physiologically with the double female determinants. At the time, he confessed that this was more a mental exercise than a strong assertion, but his explanation was similar to the strategy he used in explaining cases of seasonal and geographic polymorphism. With respects to the determinants there was, of course, a similarity been Weismann's and Dickel's accounts.
[53] Weismann was not a dogmatic follower of Haeckel's "biogenetic law", which was an exaggerated extension of recapitulation, pushed by Haeckel in the early 1870s. When Weismann conceived of a reduction division in the 1880s, he was in essence postulating the elimination from the germ line of half the phylogenetic past for each offspring – this is hardly what Haeckel had in mind.

RAPHAEL FALK[1]

MENDEL'S HYPOTHESIS

Abstract

Mendel's 1866 treatise "Experiments on Plant Hybrids" became the driving force of the modern science of genetics. It is suggested that Mendel's motivation as well as his methodology were guided by his conviction that Nature's design is "writ in the language of mathematics", and that these were sustained by his training in physics. Although there is no way of proving this directly, evidence of Mendel's predilection for algebraic maneuvers is drawn from Mendel's paper and from his correspondence with Nägeli.

Introduction

Gregor Mendel's report to the Brünn (Brno) Natural History Society on his hybridization experiments with varieties of garden peas and other plant species was published in the *Proceedings* of the Society in 1866.[2] Hybridization, besides being of interest to breeders, was at the time an important experimental tool in the discussions on the constancy and variation of species. Still, the significance of Mendel's paper was realized only in 1900, after Hugo de Vries noted that Mendel's interpretation tallied with his own current theory of intracellular pangenesis.[3] It was, however, William Bateson who was largely responsible for the exertion of the full impact of Mendel's paper and its promotion to that of the founding article of a new science of heredity.[4] In the century that followed, Mendel's paper became an inspiration to researchers and a model of experimental research to historians and philosophers of the life sciences. Curt Stern, a leading experimental geneticist, also interested in the history of science, introduced his 1966 "Mendel Source Book" as following:

> Gregor Mendel's short treatise "Experiments on Plant Hybrids" is one of the triumphs of the human mind. It does not simply announce the discovery of important facts by new methods of observation and experiment. Rather, in an act of highest creativity, it presents these facts in a conceptual scheme which gives them general meaning, Mendel's paper is not solely a historical

G.E. Allen and R.M. MacLeod (eds.), Science, History and Social Activism: A Tribute to Everett Mendelsohn, 77–86.

> document. It remains alive as a supreme example of scientific experimentation and profound penetration of data. It can give pleasure and provide insight to each new reader and strengthen the exhilaration of being in the company of a great mind at every subsequent study.[5]

What was it that made Mendel's research as presented in his 1866 paper so mission-oriented? This question has become the more significant not only because Mendelism provided the foundation for the modern science of heredity, but also because it has been the most analytical reductionist branch of the life sciences in the twentieth century.[6] Over the years a voluminous literature has accumulated on Mendel's possible motivation. The answers that were advanced range from those who suggested that Mendel had broad and practical agricultural, rather than theoretical interests; that he was interested in disproving Darwin's recently publicized theory of evolution by natural selection, or alternatively, to complement it by providing a theory of heredity; or that he was simply interested in a theory of heredity.[7] These propositions obtained a new turn in 1936, when Ronald A. Fisher demonstrated in his paper "Has Mendel's work been rediscovered?" the improbability of Mendel's experimental results.[8] Not only Mendel's motivation, but also his methodology, were now scrutinized in detail, bringing up suggestions that the report was intentionally or unintentionally faked,[9] or that it indicated a judicious and ingenious pedagogical interpretation of data. Although it will necessarily remain a speculation, the question, what was "the real object of Mendel's paper?"[10] is as relevant at the centenary of Mendel's "rediscovery" as it ever was.

Mendel's Motivation

In the centennial commemoration talk of Mendel's pioneering experiments, Leslie C. Dunn suggested that Mendel had a well formulated theory before he reported his data, and that "the excessive goodness of fit to a theory which runs through his data certainly indicates that he had the theory in mind when the data as reported were tallied".[11] Dunn also suggested that the roots of this theorizing were in Mendel's algebraic reasoning:

> His recognition of the binary behavior of such elements, *A* and *a* always splitting in the hybrid *Aa* to enter different gametes of which equal numbers were therefore produced, was clearly evident in his application to them of the binomial principle and the laws of combination based upon the assumption of integral character of elements. This kind of character or behavior had not heretofore been imputed to biological units.[12]

Indeed, Ronald A. Fisher stressed that his 1936 paper was not meant to discredit Mendel, but rather "to celebrate his power of abstract reasoning".[13] Fisher claimed that "any abstract thinker in the middle of the nineteenth century" could have come up with the modern genetical system "if he were led to postulate that inheritance was particulate, that the germinal material was structural, and that the contributions of the two parents were equivalent. ... [Mendel's] experimental programme becomes intelligible as a carefully planned demonstration of his conclusions".[14]

How, then, was Mendel led to postulate that inheritance was particulate? How did he know what he was looking for? I suggest that Mendel, the student of philosophy, mathematics and physics, as well as of theology, was looking for the plan of the Creator for eternal transcendent realities. Like Johannes Kepler, nearly two hundred and fifty years earlier, he had a notion of the mathematical analysis of reality into its ultimate units; unlike Kepler, he included the algebra of probability in this notion.[15] Furthermore, as indicated by Vitezslav Orel, young Mendel benefited from the advice of his fellow monk Matous Klácel, whose philosophy was one of Hegelian idealism.[16] Mendel was thus acquainted with Hegel's notion that all that exists must be mental, and consequently that truth must be mind imposed. To the Hegelian, truth meant a unique whole system that develops through the dialectic of contradiction. This complete system is the one and only reality, a reality which must be true.

It should be kept in mind that only at the beginning of the 19th century did research of living creatures succeed in adopting the reductionist notion and methodology that were so successful in the physical sciences. However, even then, reduction was believed to apply primarily to physiological studies, and later to (experimental) embryological research.[17] To the extent that empirical hybridization was concerned with notions of the constancy and change of species, these were mainly concerned with whether it was the role of the structural concept of the archetype or that of teleological functional adaptation that acted as the driving idea of nature.[18]

As noted by Orel, the Brno Society for Natural History was not merely an association of farmers interested in immediate practical goals, but rather had the pretensions of a scientific society, many of whose members genuinely strove to understand the Laws of Nature. Mendel, a member of the Society, educated primarily in mathematics and physics, researched plant breeding as only one of his scientific interests. His interests included other meteorological observations and bee breeding. He endeavored to provide his local agricultural community with sound scientific foundations as tools for their breeding efforts.[19] He was acquainted with the biological literature on experimental hybridization that provided all the guideposts needed for the design of his experiments. Mendel knew of results obtained by Andrew Knight, Augustin Sageret and Karl Friedrich von Gärtner, and had the work of Alexander Seton and John Goss called to his attention.[20] Mendel, who raised and bred bees, undoubtedly knew the papers of Johann Dzierzon, a fellow cleric and bee-breeder from nearby Upper-Silesia, published eleven years before his own paper.[21] Dzierzon discovered parthenogenesis in the honey-bee in which females are produced from fertilized eggs but the males develop from unfertilized eggs (see Churchill paper, this volume). He crossed German and Italian bees and noted that the unmated hybrid queens produced German and Italian drones in equal number, thus laying the foundations for the notion of character segregation.[22] In short, the rationale of Mendel's work was that of providing answers to practical problems of hybridization by applying the experience of the empirical sciences to his notion of Nature that follows the orderly laws of mathematical regularities. As noted by botanist and historian Conway Zirkle:

> Mendelism was the creation of an investigator who both hybridized plants and bred bees ... But even if he had read all of the original contributions he still had found no traces of definite numerical ratios in the different types because the earlier plant hybridizers seemed to have paid no attention whatever to relative numbers of segregating forms.[23]

Consequently, I agree with historian Jan Sapp that most of the controversy over the sources of the improbability of Mendel's results as presented in his paper has been misdirected:

> The reason why data are considered to be less true the closer they reach theoretical expectations is based on the idea that geneticists *should be* studying a random sample. It assumes that experiments should be carried out independently of the law or theory the observer is using for explanation. In other words, it appeals to naive empiricism and ignores the theory-ladenness of observations. The theory itself informs the experimenter about what kind of experiment to perform, what kind of phenomena to examine and how the results are to be understood; it also tells the experimenter when the experiment is over.[24]

Put differently, Mendel's experiments were as theory-laden as any experimental work in science. However, his university and ecclesiastical educational background provided him a vantage point at variance with that common among biological researchers of the day. What evidence beyond simply reasonable speculation exists for such a reconstruction of Mendel's real intentions?

As noted repeatedly, Mendel left very few clues to resolve this question. he kept no diary and wrote little about himself. The whole story of the development of the new theory is usually claimed to be given in the 44 printed pages of Mendel's report presented at the Brünn (Brno) Natural History Society in 1865.[25] However, as noted by Sapp "Scientific papers are not diaries .... Scientific papers are often designed so as to give a veneer of objectivity and 'matter of factness' to published claims". Consequently "establishing the motives and intentions of scientists is a precarious business at the best of times but in few cases do we have to rely so much on logical constructions of a scientific paper as we do in the case of Mendel".[26] As a matter of fact, there exists another source, which reflects Mendel's state of mind nearly as well as a diary. These are the ten letters sent by Mendel to botanist Carl von Nägeli between 1866 and 1873.[27]

Mendel's Algebraic Proclivities

Corcos and Monaghan detected Mendel's bias for algebraic maneuvers even in his major paper: after Mendel summed up the results of his experiments with two- and three-trait hybrids he then turned to what they term "one of those little mathematical disgressions [sic] he seemed to have enjoyed".[28] This predilection is evident in Mendel's correspondence with the botanist Nägeli. Here, interspersed in the depressing and frustrating descriptions of his experiments with *Hieracium* (the hawkweed), dictated by Nägeli's interests, he repeatedly developed a detailed algebraic explanation for the results he obtained and those he anticipated in possible further experiments. For example, in his second letter of 18 April 1867, he digressed into a quite detailed analysis of the seeds from his pea experiments in the packets that he sent to Nägeli. The origin of the pollen of the prospective outcome of fertilization of pea-plants grown from green and angular seeds may be predicted with precision from the examination of the numerical ratios of plants with given ratios of seeds with the characters obtained. He ended the analysis noting:

> I have not performed this experiment myself, but I believe, on the basis of similar experiments, that one can depend on the result indicated.[29]

Here Mendel clearly anticipated results of experiments before he carried them out, if he ever did. It is difficult to avoid the conclusion that Mendel's fascination with algebraic maneuvers was more than "little mathematical disgressions he seemed to have enjoyed" and toward which he drifted inadvertently while reporting on specific experiments. Mendel seems to have had a quite clear *Weltanschauung* in which mathematical order played a pivotal role.

An especially interesting digression from the description of the experiments with *Hieracium* occurs in his third letter, of 6 November 1867. Mendel describes the elaborate technique he used to obtain hybrids between *Hiereacium* species. In one case he succeeded and got a hybrid plant that produced 1044 flowers on 14 heads (74.57 flowers/head). "Some remarks on those characters of the hybrid which are rather difficult to determine in the dried plants", which he obviously included in his letter, are added. After describing bristles, hairs and bracts on the stem, the sheath of the heads and the color of marginal flowers, he says: "On the average, the number of flowers per head is (according to 14 different counts) 39 in *praealtum*, 145 in *stoloniflorum*, and 75 in the hybrid. The latter number thus does not represent the mean, which should be 92; it is, however, almost the exact geometric mean; since 75^2 is approximately 39×145".[30] 156 plants developed from the seeds of this hybrid. Naturally, Mendel looked forward with great hope to showing that the laws of inheritance he observed for the pea hybrids could be extended to the *Hieracium* hybrid: "Whether they will retain the characteristics of the hybrid type, or whether they will show variations, will be determined by next year's observation".

Indeed, in his fifth letter, of May 4, 1868 he refers again to the progeny of that hybrid. About 100 seedlings survived the winter. Unfortunately, it turned out that these F_2 plants were "uniform in the structure and the hairy covering of the leaves, and resemble the hybrid seed". However, there was still hope since the plants were still very small, and Mendel awaited "their further development with some suspense". Shortly thereafter, in his sixth letter, of June 12, 1868, after complaining of the damages inflicted on his experimental plants by the incompetence of his gardener during an inspection tour he had to make, Mendel mentions the 112 flowering plants, the first generation of last year's hybrid *H. praealtum + flagellare*. This is the *H. praealtum* × *H. stolloniflorum* of last year.[31] As far as he can judge, "all plants are alike in the essential traits" and do not meaningfully deviate from those of the original hybrid plant. Mendel refers once more to the "third generation" progeny of this hybrid in his eighth letter of July 3, 1870:

> Again the hybrids do not vary in these generations. On this occasion I can not resist remarking how striking it is that the hybrids of *Hieracium* show a behavior exactly opposite to those of *Psium*. Evidently we are here dealing only with individual phenomena, that are the manifestation of a higher, more fundamental law.[32]

Obviously, Mendel followed with great hope the progeny of this one hard-wrung successful hybrid of *Hieracium*. His expectation that these would allow him to extend his lawful *interpretation* of inheritance of characters, failed. This, however, did not diminish his belief in the *existence* of such "fundamental laws". On this background, I wish to suggest that Mendel's mention that the observed number of flowers per head of the hybrid plant is the geometric mean rather than the arithmetic mean of the numbers of flower per head of the parental types – a fact which was of little interest to his correspondent – is highly revealing rather than being "merely a coincidental agreement", as Correns put it in a footnote to his publication of Mendel's correspondence with Nägeli.[33]

The geometric mean, or "the central proportion of the two parental values", as Mendel put it, is the number of flowers per head expected when the trait of the progeny is the result of the trait of both parents being expressed independently, i.e. exactly what is expected by the two parental factors maintaining their particulate identity in the hybrid. This would mean that each parent contributes its intrinsic value acting (interchangeably) on the value obtained by the effect of the intrinsic value of the other parent. Note that Mendel did not discriminate between the trait and the factor-for-the-trait.[34] An arithmetic mean value of a trait in the hybrid progeny corresponds also to that expected by the old notion of intermediate inheritance, where the parental contributions fuse into one middle-value, eventually non-segregating trait, in the hybrid progeny. In contrast, the geometric mean suggests the idea that factors maintain their independent work on each other and may segregate again in the next generation.

Consider the example of population growth over time. Suppose the initial population P_0 increases one year by a factor x and next year the population $P_1=P_0 \cdot x$ increases by a factor y, so that the population size after two years is $P_2=P_1 \cdot y = P_0 \cdot x \cdot y$. The magnitude that would best represent the average rate of increase for these two years is $G=\surd(x \cdot y)$. This is so because if we look for a constant rate of increase for these two years that will yield the same population size at the end of this period[35] we obtain $P_2=P_0 \cdot x \cdot y=P_0 \cdot G \cdot G$. Note that G is the constant value that when replacing x and y will give the same product ($x \cdot y=G \cdot G$). In contrast, the arithmetic mean A is the constant value that when replacing x and y will yield the same sum ($x+y=A+A$). When it comes to average growth the appropriate measure is the multiplicative mean (G) rather than the additive one (A). This average growth rate will be the same even if the rate alternates between x and y for shorter and equal time units. Even for infinitesimally short alternating time-slots, the average growth rate will stay that of the geometric mean of x and y. One may speculate that when Mendel looked for an appropriate mean as a substitute for the two parental contributions he selected the mean representing growth rate. I suggest that Mendel alluded here to an analogous multiplicative process (that may occur over some time or even instantaneously) by extending his notion of segregating factors in *Pisum*: Each parent contributes to each offspring one of its own two "growth factors" (expressed as number of flowers/head) that maintains its independence. The two factors that the progeny inherit interact so that each works on the outcome of the other, as if they were alternating instantaneously and continuously: Such a mechanism would result in the geometric mean of the two separate effects, rather than in the arithmetic mean expected of an additive non-interactive model. The *praealtum* plants might be considered as "homozygotes" for the factor p that interact to induce 39 flowers per head, $\sqrt{(p \cdot p)}$; the *stoloniflorum* plants might be considered as "homozygotes" for the factors that interact to induce 145 flowers per head, $\sqrt{(s \cdot s)}$; and the hybrid might be considered as "heterozygous" for the parental factor s and the parental factor p that interact instantaneously and continuously to induce 75 flowers per head, $\sqrt{(p \cdot s)}$.

Mendel was neither the first to discern between heredity and embryonic development, nor the first to report on inheritance in quantitative terms (notably Charles Naudin preceded him).[36] He was, however, the pioneer who put algebraic ratios at the basis of his hypothesis, with which the conformation of the empirical data should be tested, rather than to try and find a numerical model that would fit the data. Significantly, three decades later, when de Vries tried to establish his theory of intracellular pangenesis, he found important support in Quetelet's statistical analysis (published in 1870), as well as in Galton's, in establishing his experiments as a search for the *biological* laws of nature. De Vries believed Mendel's work to be a special case of his interpretation,[37] overlooking that for Mendel, contrary to him, an *algebraic* notion of heredity was the order in Nature.

Conclusion

If I am right in my interpretation, this is another indication that contrary to the laws of segregation of parental factors that maintain their independence in the transmission of characters, dominance is *not* a law of inheritance.[38] Characters may develop by constant interaction between factors, as much as by the effect of one factor overriding that of the other. In his original *Pisum* experiments Mendel reported on other characters for which observations in the hybrids were not yet finished, notably, flowering time. "However, it can already be reported that [flowering time] is almost exactly intermediate between that of the seed plant and that of the pollen plant, and that development of the hybrids probably proceeds in the same manner with respect to this trait as it does for the remaining traits".[39] Mendel knew that not all characters showed the dominating-recessive relationships in the hybrids as did the seven characters he studiously chose for his major experiments. Disregarding for a moment segregation, obtaining progeny with traits intermediate between those of both parents was not diagnostic for a particulate theory of inheritance; obtaining progeny with a trait whose value was that of the geometric mean between parents, even more than progeny where the trait of one of the parents was dominant over that of the other, was. The disappointing non-segregation of the character of flowers per plant, contrary to what he looked forward in the progeny of the rare *Hieracium* hybrid, must have been a great blow to what I suggest was Mendel's conception of truth as a unique and universal system of mathematical order in Nature.

The Hebrew University, Jerusalem

Notes

1 I wish to thank Ruma Falk for guiding me through an understanding of the geometrical mean, and to Hans-Jörg Rheinberger for his helpful critical comments.

2 A facsimile of Gregor Mendel's paper *Versuche über Pflanzen-Hybriden* (Brünn: Georg Gastl, 1866) is included in the catalogue of the *Ausstellung 1984: Johann Gregor Mendel (1822–1884)* in Salzburg, G. Czihak, Akademiestrasse, ed., 15 – A 5020 Salzburg, Austria. The English translation used here is that of Curt Stern, and Eva R. Sherwood, *The Origin of Genetics: A Mendel Source Book* (San Francisco: Freeman, 1966), pp. 1–48.

3 For recent analyses of de Vries' adoption of Mendel's paper see: Erik Zevenhuizen, "The hereditary statistics of Hugo de Vries", *Acta Botanica Neerlandica*, 47(4) (1998), pp. 427–463; Ida H. Stamhuis, Onno G. Meijer and Erik J.A. Zevenhuizen, "Hugo de Vries on heredity, 1889–1903: Statistics, Mendelian laws, pangenes, mutations", *Isis*, 90(4) (1999), pp. 238–267.

4 Raphael Falk, "The struggle of genetics for independence", *Journal of the History of Biology*, 28(2) (1995), pp. 219–246.

5 Stern and Sherwood, *The Origin of Genetics: A Mendel Source Book* (cit. n. 2),p. v.

6 Kenneth Schaffner, "Reduction in biology: Prospects and problems", in *Proceedings of the 1974 Biennial Meeting of the Philosophy of Science Association*, R.S. Cohen, et al., ed., Boston Studies in the Philosophy of Science (Dordrecht: Reidel, 1976), pp. 613–632. For a detailed discussion see Sahotra Sarkar, *Genetics and Reductionism* (Cambridge: Cambridge University Press, 1998).

7 For an analytic review of the different interpretations of Mendel's motivation see Jan Sapp, "The nine lives of Gregor Mendel", in *Experimental Inquiries*, H. E. Le Grand, ed. (Amsterdam: Kluwer, 1990), pp. 137–166.

[8] Ronald A. Fisher, "Has Mendel's work been rediscovered?" *Annals of Science*, 1 (1936), pp. 115–137.
[9] Jan Sapp, *Where the Truth Lies: Franz Moewus and the Origins of Molecular Biology* (Cambridge: Cambridge University Press, 1990), pp. 104–119.
[10] See Floyd V. Monaghan and Alain F. Corcos, "The real objective of Mendel's paper", *Biology and Philosophy*, 5(3) (1990), pp. 267–292, and the response of Raphael Falk and Sahotra Sarkar, "The real objective of Mendel's paper", *Biology and Philosophy*, 6(4) (1991), pp. 447–451.
[11] Leslie C. Dunn, "Mendel, his work and his place in history", *Proceedings of the American Philosophical Society*, 109(4) (1965, August 18), pp. 189–198, p. 194. See also Robert Root-Bernstein, "Mendel and methodology", *History of Science* 21 (1983), pp. 275–295, and Robin Dunbar, "Mendel's peas and fuzzy logic", *New Scientist*, 38 (1984, August 30).
[12] Dunn, "Mendel, his work and his place in history" (cit. n. 11), p. 195.
[13] Jan Sapp, "The nine lives of Gregor Mendel" (cit. n. 7), p. 157.
[14] Fisher, "Has Mendel's work been rediscovered?" (cit. n. 8), p. 123.
[15] See Gerald Holton, *Thematic Origins of Scientific Thought: Kepler to Einstein* (Cambridge, MA: Harvard University Press) pp. 31–74. See also p. 453. A central thematic proposition of Kepler is that the physically real world is the world of mathematically expressed harmonies which man can discover in the chaos of events, (p. 62). I am grateful to Ute Deichmann who called my attention to Holton's important work.
[16] Vítezslav Orel, *Gregor Mendel: The First Geneticist* (Oxford: Oxford University Press, 1996), pp. 49–51.
[17] See, e.g. Timothy Lenoir, *The Strategy of Life* (Chicago: The University of Chicago Press, 1989).
[18] Ron Amundson, "Typology reconsidered: Two doctrines on the history of evolutionary biology", *Biology and Philosophy*, 13(2) (1998), pp. 153–177.
[19] See especially Vítezslav Orel, *Mendel* (Oxford: Oxford University Press, 1984), but also Floyd V. Monaghan and Alain F. Corcos, "The real objective of Mendel's paper" (cit. n. 10).
[20] Conway Zirkle, "Gregor Mendel & his precursors", *Isis*, 42 (128, part 2) (1951), pp. 97–104.
[21] Dzierzon was accommodated in the monastery in Brno in 1865 and most probably came in contact with Mendel (V. Orel, personal communication).
[22] Orel, *Gregor Mendel: The First Geneticist*. (cit. n. 16), pp. 227–230. See also Zirkle, "Gregor Mendel & his precursors" (cit. n. 20). See also Frederick B. Churchill, "August Weismann and Ferdinand Dickel: Testing the Dzierson System", *this volume*.
[23] Zirkle, "Gregor Mendel & his precursors" (cit. n. 20), pp. 103 & 100.
[24] Sapp, "The nine lives of Gregor Mendel" (cit. n. 7), p. 159.
[25] Mendel, *Versuche über Pflanzen-Hybriden* (cit. n. 2).
[26] Sapp, "The nine lives of Gregor Mendel" (cit. n. 7), pp. 144–145 & 149.
[27] This is, of course, in addition to the very important source of information on Mendel's views and state of mind derived from Mendel's comments and remarks in his handwriting on different works that he read – see especially Orel, *Gregor Mendel: The First Geneticist* (cit. n. 16).
[28] Alain F. Corcos, and Floyd V. Monaghan, *Gregor Mendel's Experiments on Plant Hybrids: A Guided Study* (New Brunswick, NJ: Rutgers University Press, 1993), p. 118.
[29] Carl Correns, "Gregor Mendels Briefe an Carl Nägeli, 1866–1873", *Abhandlungen der Königlich Sächsichen Gesellschaft der Wissenschaften, Math.-phys. Kl.*, 29(3) (1906), pp. 189–265, p. 206. For the English translation see: *The Birth of Genetics*, Supplement to *Genetics*, 35 (5, part 2) (1950), pp. 1–47, p. 8; Stern, and Sherwood, *The Origin of Genetics: A Mendel Source Book* (cit. n. 2), p. 67.
[30] Correns, "Gregor Mendels Briefe an Carl Nägeli, 1866–1873", p. 214; *The Birth of Genetics*, p. 11–12; Stern, and Sherwood, *The Origin of Genetics: A Mendel Source Book*, p. 73 (all cit. n. 29). Mendel's term *Mittelzahl* was translated as 'mean', while *mittlere Proportinale zu den beiden anderen*, i.e. 'the central/middle proportion of the other two', was translated as 'geometric mean'.
[31] See Correns, "Gregor Mendels Briefe an Carl Nägeli, 1866–1873" (cit. n. 29) footnote 2, p. 211.
[32] Correns, "Gregor Mendels Briefe an Carl Nägeli, 1866–1873", p. 233; *The Birth of Genetics*, p. 22; Stern, and Sherwood, *The Origin of Genetics: A Mendel Source Book*, p. 90 (all cited n. 29).
[33] Correns, "Gregor Mendels Briefe an Carl Nägeli, 1866–1873" (cit. n. 29), footnote 1, p. 214: "*Es scheint mir hier nur ein zufälliges übereinstimmen vorzuliegen.*"
[34] This distinction between genotype and phenotype was only introduced in 1909 by W. Johannsen, *Elemente der exakten Erblichkeitslehre* (Jena: Gustav Fischer, 1909).
[35] See Ruma Falk, *Understanding Probability and Statistics: A Book of Problems,* (Wellesley, MA: A.K. Peters, 1993), "Which Mean Do You Mean?" pp. 137–138; see also G. Udny Yule and M.G. Kendall, *An Introduction to the Theory of Statistics*, Fourteenth edition (London: Charles Griffin & Company, 1950), pp. 118–120.

[36] See Robert Olby, *Origins of Mendelism* (Chicago and London: Chicago University Press, 2nd edition, 1985).
[37] Zevenhuizen, "The hereditary statistics of Hugo de Vries"; Stamhuis, Meijer and Zevenhuizen, "Hugo de Vries on heredity, 1889–1903" (all cit, n. 3).
[38] Dominance was "raised" to the status of an essential law, rather than an empirical observation, in 1900 by Hugo de Vries in his "rediscovery" papers. See Falk, "The struggle of genetics for independence" (cit. n. 4).
[39] Stern and Sherwood, *The Origin of Genetics: A Mendel Source Book* (cit. n. 2), pp. 22–23.

LILY E. KAY†

BIOPOWER: REFLECTIONS ON THE RISE OF MOLECULAR BIOLOGY

ABSTRACT

Since the 1930s the rise of molecular biology has been intimately connected with the rise of Rockefeller-inspired "biopower", the visionary program developed by Warren Weaver that aimed to promote research and application of the life sciences, including psychology, to the aims of social control. For Weaver, a physicist, control of nature meant understanding processes at the most fundamental level of organization – atoms and molecules – leading to the greater power of prediction that such understanding would give. Inspired also by the Comtean view of the unity of the sciences, Weaver sought to make biology molecular by funding the disparate areas of genetics and biochemistry. This paper focuses on the "trading zone" between these two fields, and cryptanalysis (decoding practices) developed during World War II, to trace the evolving models and metaphors of gene structure (protein or nucleic acid) and function (information transfer) between 1930 and 1965. Such models led eventually to cracking the genetic code by Marshal Nirenberg and Heinrich Mattei in 1961.

A VIEW OF MT. EVERETT

Nearly twenty years ago, as a beginning graduate student, I presented my first paper at the Joint Atlantic Seminar in the History of Biology at Johns Hopkins University. The exuberant persona of Everett Mendelsohn, one of the seminar's founders – and editor of the *Journal of the History of Biology (JHB)* – loomed large over the gathering and on the horizon of the history of science. In fact, at the height of the reign of physics, Everett had put the history of biology on the map. That seminar paper, "Life in a Black Box: The Biology of Physicist Max Delbrück", was my first attempt to integrate the cognitive and social dimensions of the rise of molecular biology, an approach that Everett has helped pioneer since the mid-1960s. He was one of a handful of historians of science who spearheaded the sociology of science at a time when "external" considera-

G.E. Allen and R.M. MacLeod (eds.), Science, History and Social Activism: A Tribute to Everett Mendelsohn, 87–102.

tions were viewed as heretical to the faith in scientific thought and progress. Everett's enthusiasm and invitation to publish my work in *JHB* (my first article) exemplified his supportive interest in new scholars; it launched my work and my twenty-year career in the history of life science.

The disciplinary terrain has diversified considerably since those days. Beyond hermeneutics and contextualism, scholars have incorporated approaches from anthropology, cultural studies, literature, feminist critiques, and poststructuralism into the history of science. Against the objections of traditionalists, and throughout the "culture wars" and "science wars", Everett has remained *au courant* of the changing scholarly landscape. Often participating in these intellectual expeditions himself, he has supported controversial modes of knowing and doing, both in his national and international commitments. He towers as an ambassador of knowledge and good will, an energetic teacher and scholar at the summit of studies of science, technology, and society. It is as a tribute to Everett's foresight and adventurous spirit that I offer these reflections on the rise of molecular biology and its latter-day rechristening as "genomics".

A View From The Top

Biology, gone molecular, has come to occupy a privileged place in knowledge and in social and economic affairs. Beyond its leaps from inanimate matter to the exquisite diversity of life – from thermophiles (archaic form of bacteria living in high-temperature environments) to *Archeopteryx* (a flying reptile with true feathers) – biology has also erected bridges to neuroscience, and thus to the bases of our perception of the physical world. As Ernst Mach, inspired by the "psychophysics" of Gustav Fechner, pronounced in the 1870s: All knowledge is a matter of sensation. This position, which formed a basis for logical empiricism, was all the more radical, coming at a time when faith in Newtonianism had reached its peak and the Enlightenment's legacy reigned supreme in the form of Comtian positivism. August Comte, aspiring, after Condorcet, to found a positive science of society – *sociologie,* as he called it in the 1830s – built his classification of the sciences upon the bedrock of physics; the 'more complicated and highly individualised' sciences of life above, and the social sciences on top. A unity of knowledge was thus attained through physics: there were only physics, organic physics, and social physics.[1]

But in recent times, notably in the post-Cold War era, the kingdom of physics has been challenged by the republic of biology, charting other genealogies of knowledge and power. The new science of genomics (simultaneously a more targeted and expansive molecular biology) has not only been altering – or geneticizing – concepts of the body, mind, identity, society, health, disease and behavior; but also changing material practices and the global relations sustaining them. Although gene-based therapeutics remain largely promissory, investments are soaring. In the United States patents on nucleic acids climbed from 4000 in 1991 to 500,000 in 1996, and keep rising. Genomics has redefined academic biology and has forced the world's largest companies to reinvent themselves, as borders between pharmaceuticals, biotechnology, food agricultural, chemical, cosmetics, environmental, energy, and computer industries erode, reconfiguring national interests and international stock markets. The merger of Ciba-Geigy and Sandoz – creating the conglomerate, Novartis, at over $100 billion – epitomizes this trend; Monsanto and Searle, Dupont and Merck, Hoechst and Schering are other billion dollar joint ventures and global trend setters. These mammoth mergers and the new industrial-academic complexes they spawn give new meaning to Peter Galison's idea of scientific "trading zones". Beyond the metaphor of contingent and informal scientific exchanges these new zones are sites for literal trade of goods, of genomic information as global capital.[2]

These trends also convey the enormous potency of Michel Foucault's concept of biopower (or biopolitics), positing that the disciplines of the body (e.g. anatomy, physiology, biochemistry, and genetics) and the regulation of populations (e.g. statistics, actuarial and insurance practices, and demographics) have constituted the two poles around which the organization of power over life has been deployed since the dawn of the modern, industrial era. Thus the disciplining of reproduction in its various manifestations – for example, genetics, eugenics, endocrinology, molecular biology, genomics, behavioral genetics, or gene-based medicine – has been at the very pivot of the biopower axis, connecting the control of bodies on the one hand with the control of population on the other.[3] Indeed, next to genomics, the fastest growing field is neuroscience. The enormous resources fueling the "Decade of the Brain" (since 1990) have already had a major impact: cognitive, medical, pharmaceutical, and social. The anticipated mergers of neuroscience and genomics are igniting the scientific and social imagination and sparking futures' portfolios.

These convergences have also rekindled euphoric visions of a Comtian unity of knowledge. Repeatedly invoked by scientists and philosophers in the past, they have been most recently revived in Edward O. Wilson's book, *Consilience: The Unity of Knowledge,* a scientific platform far more extreme and pervasive than ever perceived by former Enlightenment-enchanted champions.[4] The seeming coherence, closure, and promise that such unifying epistemic reductions offer – from the birth of stars to molecular structures, and via the genetic code, to all life forms and human behavior – seem to be particularly appealing as an atavistic escape from a fragmented and fleeting postindustrial world, which is better grasped through fractals and Prigoginian dynamics than by Newtonian laws and Laplacian models. But, ultimately, such sweeping molecular visions of life and society are flawed. Their anticipated potency is based primarily upon hindsight construction and *post hoc ergo propter hoc* – reasoning; its hegemony enabled and maintained by having bracketed out those areas and forms of inquiry that do not fit the paradigm and patronages of genomics.

This essay traces the formation of that biopower from the 1930s to the 1970s, and its transformation from a purely molecular/material to the informational/textual. I first sketch how a neo-Comtian quest – a "new Science of Man" – became epistemically, technologically, and institutionally instantiated between the 1930s and the early 1950s, and how the molecular vision of life was central to this project. Next, I draw on the history of the genetic code (1953–1970) as a key episode in the consolidation of molecular biology, in fostering an informational view of life and in sparking new biotechnologies industries. The conclusion revisits genomics as a problem of a selective unity of knowledge and power and (returning to Mach) ponders the implications of this hegemony for neuroscience, mind and consciousness.

The Molecular Vision of Life and the New Science of Man

Comte's epistemic legacy was given powerful programmatic embodiment as biopower in the 1930s, when the Rockefeller Foundation invested massively in a "new science of man", a prescriptive agenda for confronting the problem of "cultural lag". Society's ability to adapt to technological change lagged dangerously behind the imperatives of industrial capitalism, so the analysis went, thus new tools were needed for adjusting to the modern age. The new "science of man" was a technocratic vision of social control – a rational scientific-social intervention – aimed at bringing order into troubled domestic and world affairs. By rebuilding the human sciences upon the bedrock of the physicochemical sciences, the Foundation hoped to engineer a society of excellence, the kind that (in their own words) would beget an Aristotle, Newton, or Darwin. According to a statement of the Trustees in 1933:

> Science has made significant progress in the analysis and control of inanimate forces, but science had not made equal advances in the more delicate, more difficult and more important problem of the analysis and control of animate forces. This indicates the desirability of greatly increasing emphasis on biology and psychology, and upon those special developments in mathematics, physics, and chemistry which are themselves fundamental to biology and psychology.

The past hundred years marked the supremacy of physics and chemistry, the trustees observed, even from a pre-Hiroshima and pre-Holocaust vantage point. Sobered by the lessons of World War I, they concluded that the destructive potential of the physical sciences could be offset by investing in the human sciences. The hope for the future of mankind rested in the development of a new biology and psychology.[5]

In consolidating the diverse programs of the Foundation, its president, Wisconsin physicist Max Mason, explained that this concentration of resources was directed to the general problem of *human behavior*, with the aim of control through understanding. It was a giant cooperative venture of the natural, medical, and social science divisions. Biology would be rationalized – quantified and molecularized – through the tools of the physical sciences and, in turn, will be used to rehabilitate the scientifically-backward medical and social sciences. Mason's Wisconsin protégé, physicist Warren Weaver, became director of the natural sciences' division and its biology program. Originally christened as psychobiology, the program was renamed by Weaver in 1938 "molecular biology".[6]

This Comtean project was turned on full steam. In the quarter century 1932-1959 the Foundation poured about $50 million into the molecular biology program in the U.S. and Europe (equivalent to about $2 billion today). But its power to shape the life sciences transcended the dollar investment. Its effectiveness lay in facilitating transactions at the "trading zones" of life: promoting international networks, mechanisms of interdisciplinary cooperation, a system of fellowships and grants; facilitating the migrations of physicists and chemists into biology, and fostering physicochemical practices and modes of reasoning. The problem of gene action – tied to the protein view of life and to earlier eugenic visions – was a primary priority.[7]

A cautionary note. I am not spinning here a totalizing narrative. The molecularization of biology did not begin or end with the patronage of the Rockefeller Foundation. I am neither arguing for institutional determinism nor for top-down co-option of scientists by the Foundation; that would be too facile. Local scientific cultures, geographical differences, and the contingencies of history certainly played important roles in the molecularization of biology. I am merely highlighting how a molecular hegemony was enabled during a particular historical period through a cooperation between scientists and their patrons, favoring certain areas and styles of research over others.[8]

For example,Thomas Hunt Morgan, who reoriented his classical Mendelian program in *Drosophila* toward physiological genetics in the 1930s, under the aegis of the Rockefeller Foundation, clearly captured this programmatic mission when he moved to Caltech.

> It is with the desire to lay emphasis on the fundamental principles underlying the life processes in animal and plants that an effort will be made to bring together, in a single group, men whose common interests are the discovery of the *unity*of the phenomena of living organisms rather than the investigation of their manifold *diversities.*[9]

Unity not diversity. Based on this rationale, it was far more convenient to study fundamental life phenomena (e.g. reproduction, but not development) on the minimalist levels. Thus the new biology mainly employed simple experimental systems – bacteria, viruses, and fungi – as phenomenological probes. This was a line of thinking that led to Jacques Monod's notorious dictum in 1961 that what is true for bacteria is true for elephants.[10] The organism was eclipsed, becoming merely a probe into the gene, a product of the experimental system and of the formalisms of genetic logic. As we shall see, with the current return of the organism ("postgenomics?") and with the rise of neuroscience, these reductionistic premises are being challenged from several directions.

The principal experimental systems of the new biology were bacteria and viruses, especially bacteriophages. The ingenious manipulations of the phage system, matings, genetic analyses, and intricate deductive reasoning – initiated in the 1930s by physicist Max Delbruck – yielded powerful insights into the mechanisms of heredity in the form of fine-resolution maps, down to single nucleic acid bases in the 1950s.[11] These were the images of genetic logic, but shorn of organismic contingencies and complexities (which recently have come back to haunt biologists). In that period, older biological traditions, such as natural history, botany, zoology, embryology, ecology, and evolutionary biology, were gradually displaced by the molecular vision of life and marginalized by its selective disciplinary consensus.

Until around 1953, biopower and the material unity of life inhered in proteins. The conceptualization of the material basis of life in terms of proteins, which reached back to the second part of the nineteenth century, bestowed upon the protoplasm all the physical and mental attributes of life, enshrining the gelatinous substance as the source of biological diversity and the locus of material, cognitive, and social control. Various scientists and popularizers during the first three decades of the twentieth century strengthened the grip of the protoplasmic view of life on scientific thought and culture. Eugenicists in particular infused potency into the linkage of hereditary fitness, protoplasmic endowment, and social control by deploying images of stewardship over the national protoplasm.[12] Most life scientists, especially in America, believed that genetic specificity and fitness sprang out of the combinatorial arrangements of the twenty amino acids, the building blocks of proteins; DNA was viewed as a "stupid" molecule, just a repeating pattern of the four bases – adenine, guanine, cytosine, and thymine – devoid of any biological specificity (a view formalized through Phoebus A.T. Levene's notorious "tetranucleotide hypothesis"). Thus, in promoting the molecular vision of life, enormous resources went into protein chemistry; hardly any into DNA research. "All that association of phenomena which we term life is manifested only by matter made up to a very large extent of proteins", argued Warren Weaver in justifying the centrality of protein research in the Foundation's new molecular biology program:

> [Proteins] enter into nearly every vital process. They are the principal component of the chromosomes which govern our heredity; they are the basic building stuff for the protoplasm of each cell of every living thing ... Indeed many diverse scientists, each with his own special enthusiasm, would be willing to agree that these proteins deserve their names of "first substances".[13]

The renowned Caltech chemist, Linus Pauling, epitomized these scientific and institutional trends, having moved into molecular biology in the 1930s largely in response to the Rockefeller Foundation's new priorities. And as their chief advisor, Pauling assured the Foundation's officers in 1945 that "the answers to the most basic problems of biology – the nature of growth; gene and cell replication; enzyme action; physiological activity of hemoglobin, drugs, hormones, and vitamins; and neurological functions – were all hidden in the folds of the giant protein molecules, in the remaining unknown region of the dimensional forest, in the strip between $10^{-7}$and 10^{-6} cm". From a physicochemical standpoint that submicroscopic region signified the future of biology. And here the tools of physics and chemistry were essential to these goals, forming the linchpin within the dialectic of what Ian Hacking has termed, "representing and intervening".[14]

Pauling's vision of a nearing golden age of biomedicine was a move from mere palliative action: biology turning molecular, medicine maturing into an exact science, and social planning becoming rational. "It will not be enough just to develop ways of treating the hereditary defects ...", he forecasted a few years later, "We shall have to find some way to purify the pool of human germ plasm so that there will not be so many seriously defective children born ... We are going to have to institute birth control, population control ...".[15] Pauling's interventionist concepts of social control resonated with those of the Rockefeller Foundation, reflecting an emerging consensus about the biological and social body.

Thus with the rise of molecular biology one observes the stabilization of a biopower grounded in a contemporary economy of discourse, in the way that certain biosocial phenomena between the early 1930s and early 1950s came to be systematically configured together. For example, social control was aimed at rationalizing and controlling individual and group behavior; behavior – as a representation of personality and intelligence – was seen as shaped by biological processes; biological processes were increasingly conflated with notions of genetic determinism; genes were taken to be physical entities whose materiality was constituted by protoplasmic endowment; and protein action was thought best studied on the molecular level, through molecular technologies that were at once representations and interventions for bodies and behaviors.[16]

But this form of biopower was soon to be transformed. With the elucidation of the double-helix by Watson and Crick in 1953 the protein paradigm of the gene soon shifted toward DNA (though research on protein chemistry continued). The discursive economy of molecular biology altered as well. In their claim that it was the DNA code that carried genetic information, Watson and Crick exemplified and reinforced the power of the emergent discourse of information. A product of the dawning information age – the advent of computers, information theory, cybernetics, and systems theories – the information discourse came to shape the representations of nature and society across many disciplines. The informational gene became a central element within DNA-based molecular biology, redefining biological research and biopower in the second half of the twentieth century. The breaking and completion of the genetic code was a key episode in this transformation.[17]

The Genetic Code and The Book of Life

The breaking and completion of the genetic code, between 1953 and 1967, was one of the most important events in 20th-century science, a manifestation of the expansive reaches of molecular biology and its potential as genetic engineering. The so-called genetic code – actually, a table of correlations –

outlined a logic of gene-based protein synthesis, then a central problem in biology: some would call it the "secret of life". It showed how the four bases of DNA – A, T, C, G – combined three at a time, yielded 64 triplet codons, specifying the assembly of 20 amino acids into myriads of exquisitely specific proteins. It came to signify, as it does for E.O. Wilson, the unity underlying the diversity of life.

Two consecutive phases, the formalistic and material, spanned the history of the genetic code. The mathematical-genetic phase, 1953-1961, was marked primarily by the efforts of about two-dozen eminent researchers, mostly in the physical sciences, who approached the problem inferentially, as a black box, with DNA input and protein output. This theoretical approach failed, because from a technical standpoint – linguistic and cryptanalytic – DNA is not a language and the genetic code is not really a code. The second phase, 1961–1967, was initiated experimentally by the correlation of a nucleotide triplet (poly-U) with an amino acid (phenylalanine) and was mainly biochemical (but genetically informed). I will now look back at these two phases as a period in which the discursive economy and power relations of molecular biology were transformed, once again, restructuring the workings of biopower.

One of the first to respond to Watson and Crick's double-helix article in *Nature* in 1953 was the Russian emigre-astrophysicist George Gamow: cartoonist, science popularizer, and military strategist. At that time – the Cold War – the Rockefeller Foundation had already largely curtailed its support for the biomedical sciences in the U.S., due to the restructuring of American science in the post-World War II period (but was still funding European biology under the Marshall Plan); more than half the support for life science in the U.S. came from the military (though European science also benefitted from some American military support). It was a time of physicists' growing interest in biology, now more attractive after two decades of molecularization.[18]

Gamow quickly perceived the relationship between DNA structure and protein synthesis as a mathematical cryptanalytic problem. That is, how can a long sequence of four nucleotide symbols determine the assignment of long protein sequences comprised of twenty amino acid symbols. His solution: The genetic code would have to consist of a triplet of nucleotides since a doublet of two nucleotides would yield only 4^2 (=16) not enough to specify all twenty amino acids. A triplet, however, would yield 4^3 or 64 combinations, more than enough to specify all the known amino acids. In fact, the code turned out to be "degenerate", meaning that two, three, and even four codons specified a single amino acid. And it was an overlapping code, each triplet sharing two nucleotide bases with the neighboring sequence on each side, so that the sequence AGCTGAACT would consist of the overlapping combinations, AGC, GCT, CTG, TGA, etc. With this two-base overlap, the 64 codons corresponded exactly to the twenty amino acids. It seemed like a perfect fit. Dispatching a letter to Watson and Crick soon after he wrote.

> I am a physicist, not a biologist ... But I am very much excited by your article ... And think that this brings Biology over into the group of the "exact" sciences ... If your point of view is correct ... Each organism will be characterized by a long number written in quadrucal (?) system... It seems then more logical to assume that different properties (single genes?) of any particular organism are determined by different mathematical characteristics of the entire number (something like the coefficients of a Fourier series). For example, the animal will be a cat if Adenine is always followed by cytosine in the DNA chain, and the characteristics of a herring is that Guanine always appear in pairs along the chain, etc. This would open a very exciting possibility of theoretical research based on the theory of combinatorix [*sic*] and the theory of numbers! That simple.

Here was the logic, the algorithm, of life – or so it seemed. The idea of a code as an escape from entropy and universal law of life – articulated by Erwin Schrodinger in 1944 – captured the physicists' imagination.[19]

By the time Gamow began to mobilize support for his "coding scheme", it was becoming clear that DNA directed protein synthesis via an RNA intermediate (a few years later christened as the "transcript" messenger RNA). To decipher life's "RNA code" Gamow founded the RNA Tie Club, a loose group of twenty scientists, mostly physical scientists, with only a handful from the life sciences. Max Delbrück, James Watson, Francis Crick, Sydney Brenner, Richard Feynman, Edward Teller, Nicholas Metropolis; even John von Neumann and Stanislaw Ulam were lured by the challenges of the coding problem. And along with their defense projects at Los Alamos, the MANIAC computer and Monte Carlo simulations were mobilized to generate artificial protein sequences based on various proposed RNA codes – all fully or partially overlapping – and compared them to distributions of amino acids in natural protein sequences. These informal exchanges formed a trading zone, where transactions of genetic, chemical, coding and mathematical logic (and their respective languages) intersected for a period of about six years.

Yet, with all the effort of this eminent gray matter the physicists failed to "break the code". As they found out, there was a fundamental contradiction in the coding scheme, which problematized the very notion that the RNA sequence is, in fact, a code. Amino acid distributions in known proteins (e.g. Insulin) always yielded random (Poisson) distributions, implying that there were no restrictions on the order of the nucleotide triplets in the RNA. But this finding clearly contradicted the precepts of their overlapping codes: Since each triplet shared two nucleotide bases with the preceding one, there was, in fact, a restriction on the order of bases. Furthermore, by definition, codes operate on language (letters, words, sentences), yet the finding of randomness contradicted the rules of language, since in all languages there are always restrictions on the order of letters in words (e.g. you cannot have a sequence of four vowels, or five consonants). Indeed, all decoding protocols are predicated on linguistic intersymbol restrictions (thus on non-random distributions); without such strictures there can be no decoding. Either the genetic code was overlapping and contradicted the empirical data of amino acids distribution, or it was not overlapping and contradicted the rules of language and cryptanalysis, in which case, it was not, technically, a code. At this point, based on this empirical evidence, the code metaphor should have been abandoned. But the seduction of the logic of biology and the unity of life survived the contradictions and lack of empirical evidence.[20]

As it turned out, within a couple of years most schemes of overlapping codes were abandoned. The so-called genetic code was shown to be non-overlapping, which is a contradiction in terms, since with no intersymbol restrictions there is no language and no decoding. Yet the appeal of an informational and textual imagery of heredity was enormous, since, of course, it was not purely internal to the workings of molecular biology. It also derived from the conjunction of several cultural trends: the general allure of coding in a cold war culture intrigued with espionage, from the code as the eternal "word" and a symbol of natural law, and the broader circulation of the information discourse in the academy and society. These representations of protein synthesis as the "code of life" soon became a preoccupation not only of physical scientists but also of biochemists and molecular geneticists. Around 1960, as theoreticians were despairing of cracking the code strictly by correlating input with output – without opening the black box of protein synthesis – several biochemists were attempting to open that black box. Yet these scientists were not simply following in the footsteps of the theorists. Rather, they were working out the pragmatics of the older problem of protein synthesis, now increasingly informed by DNA genetics, which itself had been transformed through its various intersections with microbiology and biochemistry.

In the spring of 1961, two young biochemists at the National Institutes of Health, Marshall Nirenberg and Heinrich Matthaei (very) quietly succeeded in solving the problem of protein synthesis, or in "breaking the code". Having fine-tuned the protein synthesizing machinery of cell extracts of *E. coli* bacteria – the principal experimental system of molecular biology – they had been placing in it known synthetic RNA messenger molecules. One of these messengers was the monomer poly-uracil, and they recovered a simple monomeric protein, poly-phenylalanine. Metaphorically speaking, they deciphered the first word of the code: UUU coded for amino acid phenylalanine.[21]

It was one of the most stunning events in the history of modern science, a David v. Goliath kind of victory. The *New York Times* predicted a "revolution far greater in its potential significance than the atomic and hydrogen bomb". The coding problem was thrown open to genetically informed biochemical experimentation. State-of the art laboratory technologies, teams of principal investigators, international flows of postdoctoral fellows, technicians – hundreds of participants – were propelled by the lavish funding allocated for science in the post-Sputnik era. Within six-years the genetic code was essentially completed (1967); by then (in fact, since 1963) – well before the advent of recombinant-DNA technologies – promises of a brave new world had generated concerns about future biological engineering and fueled public debate.[22]

But there were difficulties with the "code of life", known even back then. It did not conform to any definitions of codes; it is highly degenerate – several codons specify an amino acid – and mildly ambiguous – a few amino acids can be specified by the same codon. The initiation codon, GUG, had a double meaning, signifying the amino acid methionine at the beginning of the chain and valine in the middle; thus, the so-called reading-frame of the nucleotide triplets was not specified by the code. While triplet codons worked, a doublet often sufficed, the third base having considerable latitude (this became known as Crick's "wobble hypothesis"). And these contingencies made protein translation context-dependent (positional, cellular, organismic, and environmental). Thus the code's "universality" was also context-dependent, the coding rules having some flexibility across organisms and environmental conditions. Most important, though proteins' primary structure – the order of amino acids – is predicted by the RNA sequence, its tertiary structure, protein folding (the key to most protein functioning), cannot be deduced from the sequence. Consequently, at present, protein chemistry and protein engineering, aimed at understanding the formation and dynamics of tertiary structure is a burgeoning research field, quite independent of DNA genetics. Certainly, one cannot predict the organism from a genomic sequence. But these contingencies were masked (and continue to be) by the euphoria around the perceived decoding of the "book of life" written in "DNA language". A new form of biopower had emerged: beyond control of living matter in all its organic viscosity there was now the promise of controlling the form, a mastery of life's logos on the pristine level of genetic information, the DNA sequence, the word.

The stunning breaking of the genetic code came to be viewed as a victory of the laboratory over theory, a triumph of material ingenuity over Pythagorean ideals. The first decoding phase was soon regarded as naive and misguided. But as one participant (Carl Woese) wisely reflected,

> What has not been generally appreciated is that the subsequent spectacular advances in the field, occurring in the second period (1961-1967) were interpreted and assimilated with ease, their values appreciated, and new experiments readily designed, precisely because of the conceptual framework that had already been laid.[23]

Indeed the two "decoding" phases together stand as a perfect example of trading zones in life science. In 1953 biochemists and geneticists occupied different epistemic and material spheres and spoke different languages. A few physicists and mathematicians had contact with molecular geneticists but rarely with biochemists or microbiologists. By the late 1960s the language of mathematics and physics – information, coding, degeneracy, reading frames, and texts – suffused the discourse and the ongoing fusion between biochemistry and molecular genetics. A not-so-innocent, historically contingent molecular hegemony had been forged in biology (part of the Rockefeller vision), an epistemically powerful and technically productive consensus – but at a price. As biology departments "went molecular" the organism, in all its poetic contingencies, was bracketed out. It was only through the *lingua franca* (the common communication mode) of molecular biology, and its material and institutional practices, that the diversities of life now appear to have been tamed into a unity. It is through its explanation of upward causation, namely, that all higher levels of biological organization can be extrapolated purely from the lowest level, that such unity can be perceived within the harness of molecular genetics.

EPILOGUE AND CONCLUSION: MOLECULES, GENOMICS, AND MIND

Since the 1970s biology has gathered enormous momentum and power. On the one hand, the introduction of recombinant DNA technologies and the patenting of these products and processes have fueled the rise of an academic-industrial genomics and a new form of biopower. Increasingly, molecular biologists have become invested in commercial ventures that have radically altered the academic landscape of biology, including the choice and formulation of research problems, education, and career trajectories. On the other hand, (in the footsteps of the Rockefeller dream?) several "founding fathers" of molecular biology (e.g. Pauling, Delbrück, Stent, Crick, Benzer, Nirenberg) have moved to the new frontier of neuroscience. "We may now have reached the time when a successful attack on psychobiology ... consciousness, memory, narcosis ... can be initiated", explained Linus Pauling, inaugurating his program of orthomolecular psychiatry in the 1960s.[24] At MIT, the genetic code inspired a quest for the RNA engram, for the mythical chemical code of memory and mind. Behavioral genetics was partially rehabilitated from its tainted links with an earlier eugenics and Nazism. And human genome projects – the vision of the grail, according to Walter Gilbert – promise not only gene therapy for somatic ills but also for mental disorders and social management.[25]

I will not venture here into the critiques of these simplistic genomic visions that hark back at least to the Rockefeller era – that literature is vast and was recently updated at the international conference, *Postgenomics?* at the Max-Planck Institute for History of Science in Berlin.[26] Many biologists acknowledge that these large-scale sequencing initiatives, though useful, are based on faith in the predictive powers of genomic sequences, a view presupposing a straightforward correspondence between genes, protein structures, and biochemical/physiological functions; a "genetic program" in short-hand terminology. But with the knowledge about transposons, exons, introns and splicing; and the effects of post-translation modifications of protein structure, the relation is now seen as plastic, context-dependent, and contingent. Even by experts' accounts, gene-therapy is unlikely, at least not in the foreseeable future; gene-based therapeutics are slow in coming. Moreover, only 2% of human disorders are monogenic, the rest are polygenic, complex traits, their expression often modified by environment. Genomics is now moving beyond monogenic and polygenic determinism, even beyond functional genomics, towards a phase concerned with non-linear, adaptive, properties of complex dynamic systems, where visions of linear causality will be replaced by analyses of networks interacting with the environment and operating across levels of regulation: genetic, epigenetic, morphogenetic, and organismal. The search for magic epigenetic rules that guide these complexities are likely to prove a frustrating quest.

And what of the vision of a genetically-driven neuroscience and knowledge of mind, brain, and behavior? Some have staked out great hopes in the genetic studies of behavior in *Drosophila*. It is also true that the tools of molecular biology have been fruitful in certain areas of neuroscience, primarily in biochemical studies of the brain (e.g. receptors and neurotransmitters). But the emergent properties of mind and consciousness, which arise from neural topologies, are far removed from the genetic level. The relationship must be epigenetic, after all, humans have about 30,000–35,000 genes and 10 billion neurons; yet many scientists maintain a faith in biological algorithms to catapult us across the orders of magnitude separating DNA from mind. With Hilary Putnam (*Representation and Reality, 1988)* and Gerald Edelman (*The Remembered Present, 1989)* I believe that no computational specification can provide sufficient conditions for thought, or for the intentional states behind higher-order consciousness which produce language or Mozart's Requiem.[27]

And here we come full circle from Comte's ladder of knowledge to its inversion by Mach (where all knowledge is a matter of psychology); and so to the dilemma of consciousness, self-consciousness, and mind. While most biologists concur that primary consciousness (learning, memory, dreams, the ability to form mental images) emerged through millions of years of evolution, and therefore is shared by higher animals, humans occupy a privileged position. They may not be the only conscious animals, but evidence suggests they are (with the possible exception of chimpanzees) the only *self-conscious* animals: "conscious of being conscious", or possess reflexivity, as historian Roger Smith aptly put it. To become "conscious of being conscious" a being must have a distinction of self from non-self that is to some degree time-independent. It must also possess structures for modeling the effects of past events on the self. At this point "self" becomes a term referring to such a conceptual model and not just to a biological individual (primary consciousness). Thus historically and creatively the perceiver is prior to the science of physics and to any theory of consciousness. The sequence is: matter → evolution → vertebrates → primary consciousness → humans → social communication and language → natural science. And this sequence dissociates the actual history and nature of human consciousness from any dependence upon theories based on physics.[28]

I will close with a reflection by the biochemist and Nobel laureate Albert Szent-Gyorgyi:

> I moved from anatomy to the study of tissues, then to electron microscopy and chemistry, and finally to quantum mechanics. This downward journey through the scale of dimensions has its irony, for in my search for the secret of life, I ended up with atoms and electrons, which have no life at all. Somewhere along the line, life has run through my fingers.[29]

And if this consequence of reductionism is true for cellular and organismic phenomena it is all the more so for brain, mind, and behavior. The challenge now is to retrace the steps from molecules to consciousness. Genomics is already over-determining as a way of knowing . We need multiple perspectives, different units of analysis, and other categories for organizing and interpreting human experience. And this can be attained not only through the life sciences but also the social sciences, the arts, and other venues of self-consciousness that define our humanity and creativity.

Harvard University

Editors' Note

While this volume was in preparation, we were saddened to learn that Lily Kay succumbed to a resurgence of the cancer that had plagued her for almost a decade. This article represents her final published contribution to the field to which she devoted a large part of her multi-faceted career. She will be sorely missed by her colleagues and many friends in the history of science.

Notes

[1] For example, Ernst Mach, *Popular Scientific Lectures* (La Salle, IL: Open Court Publishing Co., 1986), especially, "On Sensation and Orientation", pp. 282–308; August Comte, *The Essential Comte: Selected from the Course de philosophie positive*, S. Andreski, ed. (London: Croom Helm, and New York: Barnes & Noble, 1974). For an excellent treatment of the subject, see, Roger Smith, *The Human Sciences* (New York: W.W. Norton & Co., 1997), Chapter 12.

[2] Juan Enriquez, "Genomics and the World's Economy", *Science*, 281 (1998), pp. 925–926; Peter Galison, *Image and Logic: A Material Culture of Micro physics* (Chicago: University of Chicago Press, 1997), Chapter 9.

[3] Michel Foucault, *The History of Sexuality, Vol. I: An Introduction* (New York: Pantheon, 1978). See also, Ian Hacking, "Biopower and the Avalanche of Numbers", *Humanities in Society*, V (1982), and *The Taming of Chance* (Cambridge: Cambridge University Press, 1989); and Lily E. Kay, "Rethinking Institutions: Philanthropy as an Historiographic Problem of Knowledge and Power", *Minerva*, 35 (1997), pp. 283–293.

[4] Edward O. Wilson, *Consilience: The Unity of Knowledge* (New York: Alfred A. Knopf, 1998). The title is taken from William Whewell, *The Philosophy of the Inductive Sciences* (1840). But Wilson's thesis is not merely an epistemic manifesto. It is also meant to be a weapon in the "culture wars" blaming the relativism of the social sciences, humanities, and the postmodern turn on their lack of incorporation into the sciences.

[5] Raymond D. Fosdick, *The Old Savage and the New Civilization* (New York: Doubleday, Doran, 1928), pp. 80–81; Rockefeller Archive Center (hereafter RAC), RG3, 915, Box1.1, from N-S Section, Annual Report, 18 April, 1933; and Lily E. Kay, *The Molecular Vision of Life: Caltech, the Rockefeller Foundation, and the Rise of the New Biology* (New York: Oxford University Press, 1993), Chapter 1.

[6] RAC, RG3, 900, Box 24.184, Report of the Committee on Appraisal and Plan, 11 December, 1934, p. 25; Kay, *The Molecular Vision of Life* (cit. n. 5), p. 46; Warren Weaver, "Molecular Biology: Origins of the Term", *Science*, 170 (1970), pp. 591–592.

[7] Warren Weaver, "A Quarter Century in the Natural Sciences", *The Rockefeller Foundation Annual Report* (1958), pp. 28–34; Kay, *The Molecular Vision of Life*, Introduction.

[8] Soraya de Chadarevian and Harmke Kamminga, *Molecularizing Biology and Medicine: New Practices and Alliances, 1910s–1970s* (Amsterdam: Harwood Academic Publishers, 1998), especially, Steve Sturdy, "Reflections: Molecularization, Standardization, and the History of Science", pp. 273–292; Kay, "Rethinking Institutions" (cit. n. 3).

[9] Thomas H. Morgan, "Study and Research in Biology", *Bulletin of the California Institute of Technology*, 36 (1928), p. 87.

[10] Jacques Monod and Francois Jacob, "Teleonomic Mechanisms in Cellular Metabolism, Growth, and Differentiation", *Cold Spring Harbor Symposia on Quantitative Biology*, 26 (1961) pp. 389–401

[11] Lily E. Kay, "Conceptual Models and Analytical Tools: The Biology of Physicist Max Delbruck", *Journal of the History of Biology* 18 (1985), pp. 207–246; SeymourBenzer, "The Elementary Units of Heredity", in *The Chemical Basis of Heredity*, William D. McElroy and Bentley Glass, eds. (Baltimore: Johns Hopkins University Press, 1957), pp. 70–93; Joshua Lederberg, "Genetic Studies of Bacteria", in *Genetics in the Twentieth Century*, Leslie C. Dunn, ed. (New York: McMillan, 1951), pp. 281–292

[12] See Garland E. Allen, "The Eugenics Record Office at Cold Spring Harbor, 1910–1940: An Essay in Institutional History, *Osiris* (Seccion Series) 2 (1986) pp. 225–264.

[13] On the epistemic and social dimensions of the protein paradigm, see, Kay, *The Molecular Vision of Life*, Interlude 1 (cit. n. 5). RAC, RG2, 100, Box 170.1235, General Correspondence, Warren Weaver, 28 August, 1939 (Weaver citd T.R. Parsons, *Fundamentals of Biochemistry*, 5th ed. (Baltimore: W. Wood & Co., 1935). On the tetranucleotide hypothesis see, Robert C. Olby, *The Path to the Double Helix* (London: Macmillan, 1974) Chapter 6.

[14] RAC, RG1.1 205D, Box 4.23, grant application, Pauling to Weaver, 4 December, 1945, p. 2; Ian Hacking, *Representing and Intervening* (Cambridge: Cambridge University Press, 1983); Kay, "Life as Technology: Representing, Intervening, and Molecularizing", *Rivista di Storia della Scienza* (sec. II), I (1993), pp. 85–103.

[15] California Institute of Technology Archives, Historical File, Box 88, Pauling File; "The Next Hundred Years", KRCA-Channel 4, 13 December, 1958, p. 12

[16] On the economy of discourse, see, Michel Foucault, *Power/Knowledge: Selected Interviews and Other Writings, 1972–1977*, Colin Gordon, ed.,(New York: Pantheon, 1980), pp. 92–133; on discourse and institutions see, Hubert Dreyfus and Paul Rabinow, *Michel Foucault: Beyond Structuralism and Hermeneutics* (Chicago: University of Chicago Press, 1982), pp. 222–224; and Kay, "Rethinking Institutions" (cit. n. 3).

[17] Lily E. Kay, *Who Wrote the Book of Life? A History of the Genetic Code* (Stanford: Stanford University Press, 2000), Chapter 1.

[18] James D. Watson and Francis H.C. Crick, "Molecular Structure of Nucleic Acids. A Structure for Deoxyribose Nucleic Acid", *Nature,* 171 (1953), pp. 737–738. On Gamow, see, Richard M. Doyle, *On Beyond Living: Rhetoric of Vitality and Post-Vitality in Molecular Biology* (Stanford: Stanford University Press, 1998); Kay, *Who Wrote the Book of Life?* (cit. n. 17), Chapter 4.

[19] Sydney Brenner Files, Archive Box 2; Gamow to Watson and Crick, 8 July, 1953; see also, Lily E. Kay, "A Book of Life? How the Genome Became an Information System and DNA a Language", *Perspectives in Biology and Medicine,* 41 (1998), pp. 504–528

[20] Kay, *Who Wrote the Book of Life?* (cit. n. 17), Chapter 4 .

[21] Marshall W. Nirenberg and Heinrich J. Matthaei, "The dependence of cell-free protein synthesis in E. Coli upon naturally occurring or synthetic polyribonucleotides, *Proceedings of the National Academy of Sciences*, 47 (1961), pp. 1588–1602; Kay, *Who Wrote the Book of Life?* (cit. n. 17), Chapter 6.

[22] "New Gains Cited on Genetic Code", *New York Times, 24* January, 1962, p.35; On early reactions (since 1963) to genetic engineering based on research on the genetic code, see, Charles Weiner, "Anticipating the Consequences of Genetic Engineering: Past, Present, and Future". In *The Social Consequences of the New Genetics,* Carl F. Cranor, ed., (New Brunswick: Rutgers University Press, 1994), pp. 2–31.

[23] Carl E. Woese, *The Genetic Code: The Molecular Basis for Genetic Expression* (New York: Harper, 1967), p. 17.

[24] On the rise of biotechnology industries, see, Martin Kenney, *Bio-Technology: The University-Industrial Complex* (New Haven: Yale University Press, 1986); Susan Wright, *Molecular Politics: Developing American and British Regulatory Policy for Genetic Engineering, 1972–1982* (Chicago: University of Chicago Press, 1994); Herbert Gottweis, *Governing Molecules: The Discursive Politics of Genetic Engineering in Europe and United States* (Cambridge: MIT Press, 1999). Linus Pauling, "Molecular Theory of General Anesthesia", *Science,* 134 (1961), pp. 15.

[25] Walter Gilbert, "Vision of the Grail", in Daniel J. Kevles and Leroy Hood, eds., *The Code of Codes: Scientific and Social Issues in the Human Genome Project* (Cambridge, MA: Harvard University Press, 1992), pp. 83–97.

[26] "Postgenomics? Historical, Techno-Epistemic, and Cultural Aspects of Genome Projects (International Conference), *Max-Planck Institute for the History of Science, Berlin,* Reprint 110 (1998).

[27] On the behavioral genetics of Seymour Benzer, see, Jonathan Weiner, *Time, Love, Memory: A Great Biologist and his quest for the Origins of Behavior* (New York: Knopf, 1999), and for a thoughtful review, see, Helen Epstein, "The Fly in the DNA", *New York Review of Books,* (June 24, 1999), pp. 14–18. On neuroscience and consciousness, see Hilary Putnam, *Representation and Reality* (Cambridge, MA: MIT Press, 1988); Gerald M. Edelman, *The Remembered Present: A Biological Theory of Consciousness* (New York: Basic Books, 1989).
[28] Roger Smith, "The Human Sciences as a Human Science", Paper presented at the Boston Colloquia, December 7, 1998; Edelman, *The Remembered Present* (cit. n. 27), Chapter 15.
[29] Albert Szent-Gyorgyi, "What is Life?" in *Biology Today*, John H. Painter Jr., ed. (Del Mar: CRM Books, 1972), pp. xxiv.

II

PERSPECTIVES ON THE SOCIAL STUDIES OF SCIENCE

ANNE HARRINGTON

CHINA, "QI," AND THE CHALLENGES OF ENGAGED SCHOLARSHIP

Abstract

This essay is a meditation on the challenges of living a life of engaged scholarship. Everett Mendelsohn's commitment to this kind of life in his own career is the occasion for some reflections on the variety of types of engagement in which an historian and cultural critic of science may become involved, and the different kinds of dialectics between activism and analysis they may entail. The essay then offers a series of vignettes and choice moments associated with a particular project that posed new challenges to the author's sense of how to maintain a knife-edge balance between engagement and detachment in her own work. The events in question here begin with some research associated with an historical-ethnographic book project about the making of Western mind-body medicine; and then evolve into new project concerned with bringing Chinese meditative and medical *qigong* practices into cross-cultural dialogue with Western laboratory science.

Varieties of Engaged Scholarship

Over the course of some 20 years – beginning when he was my undergraduate teacher at Harvard – Everett Mendelsohn has embodied for me an inspiring commitment to the possibility of living a dialectic between analysis of the past and engagement in the present, between scholarly rigor and activist passion. Throughout his career, he has taught and analyzed fraught issues at the interface of science and society; and he has also been a player in the effort to make a difference to the outcome of that interface: whether through his anti-nuclear activism, or his engagement with Soviet scientists on questions of disarmament, or through public forums in which he has explored problematic aspects of emerging genetic technologies. In short, it has never been enough for Everett to function simply as part of the "Greek chorus" in the crucial plays of our times. He has always also been an engaged actor in those plays.

My own ethical and intellectual passions have evolved in ways quite different

G.E. Allen and R.M. MacLeod (eds.), Science, History and Social Activism: A Tribute to Everett Mendelsohn, 105–122.

from those of Everett, but the dialectic he has lived has emerged as one of the challenges of my own scholarly life as well. As an undergraduate, I was drawn to the sciences of the human mind and brain for the insights I thought they would give me into this complex, fraught experience we call "being human" – and very quickly came up short against the apparent incoherences and gaps I found there. With a strong previous background in literature and the humanities, it was clear to me that narratives of existential passion and ethical choice offered one kind of truth about my being; the causal necessities of neurons and biochemistry offered me another kind of truth. So far as I could tell, both kinds of truths made real claims on me, both had to be taken seriously. Nevertheless, the whole for me added up to an unintegrated picture, a recipe for a kind of epistemological vertigo. I began to suspect that there were profound issues to be clarified here about the relationship between our experienced humanness, and the logic of our ways of knowing, particularly the ways of knowing given to us by science.

The historian's trade attracted me at this point, not initially because I had ambitions to master the past on its own terms, but because I concluded that mastering the past of science would help me to understand the logic of its choices and authority in the present. The challenges and the tensions here seemed to promise both intellectual adventure and an opportunity to exercise moral and existential passion. If the job of the historian of science was to clarify how science and its truths functioned within the human world, then there might be a way also to parlay that understanding into an insistence that both science and its truths be held accountable to values we consider to be humane.[1] If the work of historical analysis began by insisting that we feel the "strangeness" of our current ways of thinking and acting, then there might be a way also to ask whether other ways of thinking and acting are conceivable within our present world, and could even be persuasive. In short, I went back to study the past because deep down I was passionate to contribute to the effort to "get things right" in the present.

So I began my own career, and I began it with the human brain. I was interested in its status as that part of ourselves that defines us most clearly and triumphantly as human, while demanding no less insistently that we reckon with all the ways in which we are lowly material organisms (however cunningly constructed), products of "mere" physiological process. In this sense, our understandings of the brain seemed to me to be an arena where our fractured understandings of ourselves could not help but meet and jostle together uneasily.

When I looked at the progress of the brain sciences from the vantage point of the present, certainly it seemed impressive on many fronts. And yet, as early as my first book, *Medicine, Mind and the Double Brain*,[2] I had already begun to wonder whether in fact things were so simple; whether there was not more slippage and inconsistency in the results of brain science than a lot of the celebratory rhetoric might lead one to think.

The problematic of *Medicine, Mind and the Double Brain*, for example, had its origins in two apparently rather simple anatomical and physiological observations that had become clear by the middle of the nineteenth century: (1) the observation that the human brain is bilaterally organized into two hemispheres, and (2) the observation (derived mostly from clinical evidence) that the two hemispheres work differently; most notably, the left hemisphere seems to be particularly involved in language production. A simple enough set of observations, but the book then tracked a process whereby some of the most deeply engrained binary oppositions of Western culture were discovered to reside in the dual structure of the cerebrum: reason versus madness, intelligence versus emotion, civilization versus primitiveness, maleness versus femaleness. The book then explored the way in which – in an era marked by the growing pains of secularization, the anxieties of colonial rule and class unrest – this literature also betrayed a telling preoccupation with the instability of the relationship between the "brute half" and the "rational half" of our brain. In intellectual arenas ranging from criminology to psychiatry to spiritualism to anthropology, we see version after version of the idea that the brain's hemispheres may sometimes achieve independence from each other; that "two minds" may struggle for control within a single skull. In seeing through this project, I began to appreciate something that still occupies me to this day: the extent to which objects of science can function as material and metaphoric resources that at the same time can work to advance questions of the laboratory and the clinic, while simultaneously participating in the cultural and moral work of a society.[3]

I worked on *Medicine, Mind and the Double Brain* during a time when much of contemporary behavioral and brain science was caught up in the provocative agendas of so-called "split-brain" and brain laterality research. In 1981, neuroscientist Roger Sperry had been awarded the Nobel Prize in medicine "for his discoveries concerning the functional specialization of the cerebral hemispheres".[4] In the years since, in fields ranging from philosophy to medicine to education to New Age philosophy, questions were buzzing about the implications of these discovered difference between our right hemisphere (seen now as "holistic", "imagistic" and "emotional") and our left hemispheres (seen now as "sequential", "verbal", and "calculating"), and the extent to which they did or did not integrate what people supposed to be their distinct "ways of knowing".

I had an idea that the nineteenth century story of the double brain, that was basically unknown at the time, should be part of the consciousness of the contemporary scene. Here, a bit of luck intervened, when a mentor facilitated an introduction to the editor of a behavioral sciences journal with an unusual format: it not only published articles, but it also published extensive "peer commentary" on each of the articles it published, with an "author's response". So it came that, in 1985, some 17 brain and behavioral scientists had an opportunity to comment in print on my argument that the past was not a foreign country: that similar metaphorical/material elisions to the ones I had seen in the nineteenth century were at work in contemporary laterality and split-brain research.[5]

This was a certain kind of engaged historical scholarship. In the late 1980s, however, I felt the call of a different kind – more ethical, more self-reflective – when I became involved in a new project concerned with early twentieth century German "holistic" mind and life science: a project that ultimately saw closure in a 1996 book entitled *Reenchanted Science*.[6] Holistic life and mind science was a movement that claimed to be all about reforming the declared inadequacies of the mechanistic sciences of the time. It was committed to developing new, more sophisticated understandings of life and mind that appreciated both in their "wholeness". Gestalt psychology, systems theory, new directions in ethology, clinical neurology, and developmental biology all found inspiration in this movement. However, like the sciences of the double brain in the nineteenth century, I found that holism turned out to be a movement that swelled with multiple meanings. *More than a machine!* was now the new clarion call in the clinic and the lab, but also no less in the culture at large. The impulse to "overcome" mechanism, and to instate a new vision of "wholeness" contained within itself not only concerns about how best to do the work of the laboratory and the clinic; in the German-speaking countries of Central Europe, it also mediated concerns about the costs of living in a too-rapidly urbanizing environment, the causes of the national devastation of a lost world war; the price of lost faith and lost values in a modern world. Architects of the holistic mind and brain sciences both fed off the energy generated by these larger concerns, and contributed authoritatively to them. By the end of the 1920s, much of this process of mutual resonance was becoming more and more politicized, and one of the most salient political arenas in which it began to play out was National Socialism.

Tracking this story across time did not leave me unmoved. I realized, with some initial dismay, that I could resonate to many of the discontents of the 1920s, and I was pretty sure that many of my generation could as well. The rejection of the "machine", the call to "wholeness" still worked on the imagination of the 1990s no less than on the imagination of the 1920s. Against the background of that conclusion, I felt called to develop a perspective on the implications of the German holistic project for our own time; to get some clarity on the question of what kind of responsibility we may have to be aware of the partially bad consciences of projects and ideals that may attract our own energy and allegiances.[7] In the course of this second-order reflecting on an historical project, I found myself involved in a series of sometimes difficult conversations with people ranging from scientist colleagues, to writers for New Age "holistic" journals, to family members of one or another now-deceased scientist, whose legacy I was accused of clouding. Sometimes those conversations led to published reflections by my conversation partners in their own right.[8]

Mind-Body Medicine and "QI": New Forays in Engaged Scholarship

My early work on the brain sciences provided an opportunity for using historical understandings as a platform for engagement with the logic of contemporary science. My later work on German holistic science provided an opportunity for engagement with the ambiguous ethical and political potential I saw in certain past and current intellectual projects. A still-ongoing project on the making of contemporary mind-body medicine has added to the portfolio of the "varieties of engagement" I seem to be prepared to take on board as a scholar, and raised new questions for me about ways of doing this style of work with intellectual integrity.

The origins of this newest project is relevant to understanding why it has taken the shape it has. In 1994, I was invited to join an interdisciplinary network of research scientists, sponsored by the MacArthur Foundation, that had been set up to develop strategies and models for elucidating the processes and mechanisms underlying the so-called "mind-body" interactions associated with changes in health and illness. As the only non-laboratory scientist in the group, my task was to find ways to bring the traditional concerns of the humanistic social sciences – concerns about culture, context, and meaning – into the framework of the group's thinking and research. This was critical-conceptual work from inside the frame of science, rather than from outside; and it slowly had an effect on my sense of my identity as a scholar. In the course of my work with the Mind-Body Network, I came to two conclusions: (1) that culture is not simply the stage on which bodies run their internal scripts, but rather is something that is itself constitutive of somatic experience, literally gets "under the skin;" and (2) that narrative, scripts, or stories are at once the patterning mechanisms that organize human activity into meaningful units, and the primary means by which meaning gets "under the skin" and influences somatic functioning.

Stories under the Skin is my tentative name for a new book project that will attempt to develop this argument. As it does so, it will also attempt to embed it in a more conventional historical-cultural analysis of the way in which the sciences of mind-body medicine participate, no less than any other cultural activity, in those acts of story-telling that organize human experience, both outside and under the skin. The book's argument takes its starting point from the claim that, over the past century, both the moral and the somatic experiences of healing and health have been shaped and infused by a range of stories we had learned to tell about the effects of our minds on our bodies. These stories–whether they are about "the power of belief" to cure our cancer; about how our sister's asthma "really" reflects the fact that she is "suffocating" inside an unhappy marriage; or about how the "stress" of modern life may kill us all – are intellectually persuasive in no small part because they appear to be grounded in the authority of science. At the same time, they *matter* with an existential, almost religious urgency because they appear to have something to tell us about how we might enhance our wellbeing, what meaning there might be in our suffering, and what kind of control we may be able to gain over our own mortality.

How to do the research on a project like this for which, arguably, some of the most important insights lay in the realm, not of text and words, but of bodily experience? My answer was that, to do this project well, I would have to engage with its material at a level below words as well. I would need to become an anthropologist as well as an historian of the arenas of mind-body medicine; and my methods would need to combine participatory understanding with observational detachment. So, over a period of several years, I spent time conducting interviews with cancer patients about how they think about the science that studies them; participated in a ten week program that teaches meditation to chronic patients at a medical clinic; submitted myself to various physiological laboratory tests to get an "under the skin" understanding of that experience; and visited various faith healing ceremonies in charismatic churches in the Boston area.

In the summer of 1997, I also went to China to carry out research for a particular part of this book project concerned with so-called *qi* and *qigong*, that had been a source of growing attention in the Western mind-body medicine scene. *Qigong* is a form of moving meditation associated with the revival of traditional Chinese medicine in modern times, and, more traditionally, with various forms of Chinese martial arts and Taoist spiritual cultivation practices. The goal of *qigong*, especially as it is taught in medical contexts, is to help people achieve balance within the supposed vital energy of their body (*qi*, pronounced "chi"). The idea is to find ways to overcome barriers to the free flow of *qi*, to keep it balanced through the body, and (in some understandings) to increase its quantity or intensity, sometimes to the point where it allegedly can emanate out of one's own body and affect the bodies of other people.[9]

All of this was relatively opaque to me when I first began to look into this part of the story. When I went to China, therefore, I thought my goals were relatively modest. I wanted to be able to write well about a phenomenon of what was clearly an important cultural and practical strain in current Western mind-body medicine practices. However, I had become convinced that, unless I had some "felt" sense of this world, I would just be writing about words – and mighty obscure words at that. I felt I needed some experiential grounding to write about this phenomenon with any integrity.

And when I went, this is what I thought I knew. I knew that organizing the current preoccupation in the West with phenomena like *qigong* was a narrative of modernist discontent. The narrative tells us that we in the "West" have fallen out of touch with our bodies' deeper rhythms and wisdom, gotten out of balance within ourselves. It urges us to look to the wisdom of the "East" that it suggests – in contrast to us – never forgot the ancient practices of mind-body integration and may be in a position to help us complete ourselves.

I knew that, in 1993, PBS in the United States had aired a three-part special, *Healing and the Mind,* presented by Bill Moyers. The dramatic opening segment of this show was called "The Mystery of Chi". In this segment, the viewer was taken on an hour-long journey to China, a country described as a "mind/body culture". It was made clear that we were here in China to bear witness to understandings of healing, mind and body that were not to be found "back home". The presentation showed acupuncture clinics, herbal pharmacies, calligraphy sessions, but the central focus of the show was on *qigong*. It was then made clear that *qigong* is a series of meditative movements that the Chinese believe is good for the health, but then the question was asked: what was the basis on which it rested? What was this thing called "*qi*"?

Qi was identified in *Healing and the Mind* (I am quoting now from the companion book) as a "mysterious" force, a force that, "as a physical reality ... makes no sense at all". Moyers goes on: "The fact that the [*qigong*] exercise is thousands of years old, and a hallmark of Taoist, Buddhist, and imperial Chinese scholarship, does not necessarily mean that human beings hold within them rivers, streams, and pools of vital energy. Nonetheless", he says, "the mystery challenges ...". The message was clear: Was our scientifically-based skepticism – rooted in a mere couple of centuries of experience – any match for thousands of years of Asian wisdom?[10]

The Moyers television series was turned into a best-selling video and also into a companion book that reached number one on the Publishers Weekly Hardcover Nonfiction Bestseller List in 1993, and stayed on the bestseller list all told for 23 weeks. Doubleday, the publisher claimed that *Healing and the Mind* is "the most widely read and influential book of its kind".[11] Back in 1971, the *New York Times* journalist, James Reston effected a sea change in American perceptions of acupuncture by describing an emergency appendectomy carried out on him in China (he was there covering Nixon's visit) that used acupuncture needles to relieve his postoperative pain[12]

Moyer's *Healing and the Mind* did something similar for *qigong*. Since 1993, there has been an explosion of informational websites and correspondence courses on both Chinese medicine and the place of *qigong* within it, many of which explicitly credited Moyers with having helped awaken the American public to the value and interest of qi and *qigong*.[13] The Dana Farber Cancer Institute in Massachusetts is currently running a study of qigong as an adjunct treatment for cancer. Emory University recently concluded a project, sponsored by the National Institute of Aging, that found, that in a three-arm study comparing *Tai Chi* (a related practice) to other interventions taught to frail seniors (individuals over the age of 70), *Tai Chi* was the only practice that significantly led to a reduction in injuries from falling[14] The National Center for Complementary and Alternative Medicine at the NIH has targeted *qigong* as one of its categories for potential research, and compiled a public bibliography on current laboratory and clinical research on the topic.[15]

In the popular realm, a journal called *Qi*, devoted to "traditional Eastern health and fitness", operates under the slogan, "Don't let Ancient Knowledge Become a Thing of the Past". One typical article in that journal, from 1995, spins out the message at greater length:

In China there is profound treasure. Marco Polo brought back small portions of it. For centuries traders carried bits of it out along the Silk Road. Still today there is treasure that we can borrow from China to enhance our world. An aspect of China's tradition that the West has completely discounted is the health care system. Science has been so busy creating new technologies for treating disease that we in the West believe that health care and medicine are the same thing. While we in the West have a fantastic and very expensive system based on treating people after they are sick, China has a very inexpensive system of health care based on keeping people well.[16]

All this I knew when, in the summer of 1997, I arrived at the Qigong Institute of the Shanghai Academy of Traditional Chinese Medicine (a government-sponsored center) for my first "under the skin" exposure to the exotic world of Chinese medical and meditative practices. And on the first day, I found myself listening with growing discomfort to a series of claims in the realm of the vulgar paranormal: descriptions of parlor tricks involving "*qi*". Sitting with the Director of the Institute, Dr. Chai, I heard about chocolate boxes in which the candy vanished through cardboard, pill bottles that flew through the air, and blindfolded children who "read with their ears". There were also allusions to more serious paranormal investigations being carried out in China, but Dr. Chai could not provide me with any details, since this was classified work (he said) that was sponsored by Chinese defense funding.

All the paranormal events, however, from the most mundane to the most awesome, were referred by him back to the radiating energy, *qi*, that here was identified with what he said Western researchers had called "*psi*". When I asked for the evidence that any or all these things had happened, he explained that it was hard to document or catch them on film, because of certain "laws" of the "psi field" that cause cameras to fail to function and observers to become "distracted", just at the decisive moments.

And then Dr. Chai told me a wistful story, and I realized that I almost recognized it, because it was "my" – it was "our" – story too. He told me how, in ancient times, many people in China had these powers but now, with machines, with technology, people have lost touch with them. To retrieve apples from the top of a tree, we use modern reaping devices rather than our psychic powers of concentration. In this sense, Chai told me, the Faustian bargains exacted by modern life had alienated the Chinese from their ancient powers. He himself had dedicated his own life to recapturing and defining that greatness. And this is where his own journey to reconnect to lost powers had led him – to chocolate boxes and bottles of flying pills. I said at the end that this must be a hard career choice, and he looked sadly acquiescent. "One must be very committed".

A day later, I met the man whom I decided was my real teacher: Dr. Chu. He taught me some *qigong* practice, seemed thoughtful and sophisticated, and answered my questions. One of the things he tried to do was to give me a sense of the history of *qigong* as a rational process of increasing discovery. It was a tradition, he said, that went back thousands of years, a fundamental part of Chinese culture, with the earliest writings dating to the Yellow Emperor's Canon of Medicine in 400 or so BC, and even before (he showed me a picture that hung in the Qigong Research Institute of Stone Age people dancing in a kind of circle – and explained that the Chinese believed this to be evidence of a very early form of *qigong*).

In the past several decades, he went on, *qigong* had been brought into the modern age. A critical turning point came in 1978 when, he said, the material existence of *qi* was experimentally proven! Two *qigong* masters were studied by a prominent physicist, Gu Hansen, at the Shanghai Nuclear Research Center. A special sensor picked up the energy he emitted from his hands (called *waiqi*), and noted that it contained infrared and electromagnetic components, as well as components that the scientist could not yet clearly identify. This was published in an article in the Chinese journal *Nature*. The authors of the article concluded that there was nothing going on here contrary to natural law – in that sense, nothing "paranormal". Qi was a form of energy that was currently unknown, but that eventually would be wholly susceptible to explanation in the language of physics. Listening to this, I didn't know what to think of the claims, but the conclusions made a certain cultural sense to me. After all, I knew we were working here within a Marxist society that had denounced all religious-mystical traditions as feudalistic, subversive superstition.

Or did I really know as much as I thought? A day or two later, I was shown a video of a Japanese television documentary called "*Qigong* in China". In this video, a Japanese Bill Moyers goes to China to investigate the practices and claims associated with this phenomenon. Though the film covered less overtly sensational questions of *qi* cultivation through moving meditation exercises, the documentary focused much of its attention on the drama of external *qi*. In particular, my attention was caught by footage of a *qigong* master who stood on a public stage, looking like an Asian version of a Western televangelist. He raised his hands up and outward towards the audience, allegedly emitting energy directly at the crowd, and – indeed – masses of people twisted, wept, struggled and convulsed as a result of what was called "*qi* infection"

I was troubled by the sinister feel of the scene; and so I asked if it would be possible to talk about the larger sociology of *qi* in China. Dr. Chu indicated he would be willing to do this. He muttered as he said this something to the effect that at this stage of his career he did not fear the government, he could be frank.

But in the end, he never managed to clarify matters. Instead, around this same time, he told me about another piece of the modern history of *qigong* in China, in which he was more directly involved. It had of course been claimed for a long time that acupuncture can be used to create a temporary analgesia sufficient to allow patients to undergo certain kinds of pain-free surgery, without drugs and fully conscious. The question that had interested Dr. Chu was how far the *qi* allegedly emitted directly from a *qigong* master might have a similar effect, since it might be supposed that similar processes were evoked in both cases.

On May 9, 1980, a successful operation was carried out on a fully conscious patient at the Shanghai's People's Hospital No. 8, with no analgesia except what was supposedly provided by the *qi* of the attending *qigong* master. Dr. Chu did not perform that first surgery, but he did perform others soon afterwards. He showed me a video that had been made of one of these.[17] I watched a bloody tumor being cut out of the breast of the patient, as Dr. Chu hovered close by. At the end, everyone stopped to smile at the camera, including the patient, twisting around from the operating table so the camera would catch his beaming face.

I did not know what to think. I said: all these claims about emitting *qi* were hard to believe, and were going to be difficult to explain to my colleagues back home. Could I experience this form of *qi* myself? He said "yes". Three separate experiences followed, all distinctively but inconclusively disconcerting. I will open up just the last one of these, provided by a doctor on a certain clinical ward I was visiting during my final week in Shanghai:

I kept a journal throughout this China visit, and this is what it had to say about that experience. The entry is dated Monday, August 4, 1997:

I write this in my hotel room, and my right arm and especially palm are still strongly prickling as if there were energy, like electricity inside them. The feelings are unmistakably persistent, somatic, and intrusive. They are not something that I have to think about, imagine or wonder if I am "really" feeling. And they had their source in something else so unmistakably intrusive that, at least for that moment, I could no longer waver and wonder if what was going on was "just" a product of wishing and fantasy: namely, a demonstration, directed into the palm of my hand, of what the massage doctor, Guo Xiang Fu, said was a particularly strong form of external *qi* he worked with, something produced by a practice called Kongjing Qigong (from his business card, I saw he is Vice Head of the Committee in China concerned with this subspecialty of qigong).

Dr. Guo had begun by looking at my palm to diagnose my overall condition and immediately began asking about ulcers in my stomach. Indeed, I admitted that the most serious illness I'd ever had had involved ulcers all through my stomach. Well, this was still a vulnerable point, he said.

Then I was on the massage table, and for some twenty minutes or so, he worked on me: a massage of a sort unlike any I'd ever experienced (to become a doctor specializing in massage therapy in China, you study specialized techniques of treatment for five years on top of your more general training). It was complex, somewhat painful, with occasional quite piercing sensations of discomfort as he discovered, and targeted in on key acupuncture points. Towards the end, he said apologetically that he would not be able to give me extensive external *qi* therapy, because he had too many patients waiting. He would, however, just show me briefly how it felt by emitting it into my palm. And so I put out my hand, palm upwards; and at first there was nothing in particular. Then I thought, well, maybe I feel a kind of warmth, not really sure. ...but then suddenly there was an acute sensation, like something electric energetic erupting in my palm. A less than brilliant exclamation broke out of me: "*wow!*" I looked over at his fingers trembling over my hand. And then I said: "I think I believe in this".

Engaging "QI", Part Two

What are the ethical obligations associated with a bodily experience of "belief"? I had not lost touch with those questions that had first called me into the history of science: questions about the relationship between the maps that tell us what it means to be human, and the territories of experience that are enabled, affirmed, or negated by one or another of these maps. From the beginning of my scholarly career, I had felt a personal stake in the larger question of how to ensure that our maps are serving us as well as they possibly can. What did a moment's experience on a shabby treatment table in a clinic in Shanghai have to do with any of that? And what kinds of further styles of engagement might it challenge me to take on board?

I returned home to the United States, with a suitcase of notes and videotapes, and, all other things equal, I might have concluded that I had enough material to digest for the purposes of completing one chapter in a larger book project. But all other things were not equal. Several months after my return from China, an individual approached me from the foundation world who had been interested in *qi* and *qigong* practices for some 15 years. He had heard of my recent trip to China, heard something about me, and had for me a proposal. In the late 1980s, he had gone to China himself to investigate possibilities for collaborative cross-cultural research into *qigong*. At that time, he was accompanied by David Eisenberg, author of the classic memoir of his own youthful immersion in Chinese medicine, *Encounters with Qi*. Eisenberg had also been the charismatic guide for Bill Moyers in the 1993 *Healing and the Mind* series.[18] In the 1980s, through contacts cultivated by Eisenberg, there had been some tentative promising developments, this individual told me. Then political events had intervened – most drastically, the 1989 Tiannmen Square massacre had erupted – and that first effort had to be aborted. But now there was, he felt, an apparent growing relaxation of top-down control of intellectual and cultural life in China (this was before the repression of the so-called Falun Gong movement in 1999).[19] He wondered if the time might be right to try again, and whether I might be interested in helping him try again. Would I be interested in helping to develop a new strategy for cross-cultural exchange and research involving Chinese *qigong* practitioners and Western laboratory scientists?

For three months, I demurred: on the basis of the strangeness of the whole idea, on the basis that I was too ignorant, on the basis a project of this sort would too far outside my disciplinary ken. But even as I demurred, something also pulled. There *had* been something clearly destabilizing about my experiences in China in 1997, something I had not yet been able to digest. What *was* the relationship between different institutionalized ways of knowing and what they enable one to see or experience? Why *did* Chinese maps of human bodily functioning look so fundamentally different from Western medical and physiological maps; and was it so evident that one culture had completely missed the boat and we had caught it?[20] If not, was there any way to bring the understandings derived from two different conceptual maps of bodily functioning into some kind of dialogue, with the effect of making more transparent, less self-evident the structuring assumptions of both?

The opportunities for learning here seemed potentially quite great; but the project felt like it might entail considerable risks as well – and not only because I would have to master a great deal of new and unfamiliar material. The risks also had to do with my ongoing dilemma of engaged scholarship and how to do it well. This would be a collaboration involving, at least in part, empirical laboratory work, but one that would entail navigating a highly complex set of cultural arenas. What would it mean for me to take it on? Finally I said to myself, if I accept this project, I will be engaged, but I will also need to commit to bringing into the arena of this work the analytic attitude I have practiced over the years as an historian of science. That is to say, I will need to hold steady enough to be at once an engaged actor in a cross-cultural collaborative effort, and also function as an observer-analyst of that same effort. And on that basis, after three months of demurring, I finally accepted the challenge.

One of the first steps for me then was to clarify the scope and quality of research on *qigong* already ongoing both in the United States and in China; and to identify possible collaborator-practitioners for any new project a group of my colleagues and I might want to develop in the future. In pursuit of both those goals, in September 1998, I returned to China, accompanied by a Chinese-speaking graduate student named Grace, visiting her homeland for the first time. We attended two conferences: "The 7th International Conference on Qigong", in Shanghai, an almost exclusively Sino-Japanese affair (I was the only American, and there was one European present, a French acupuncturist); and then, in Beijing, "The Fourth World Conference on Medical Qigong", with a crowd that included a mix of Americans and Europeans.

In the opening and closing public rhetoric at both these conferences, there was a tremendous emphasis on *qigong* as a flower of Chinese civilization, a national treasure that would serve as a healing force for all of humanity. To quote the awkward English translation of one of the presentation titles from the Shanghai conference: "The China Qigong Culture is Worthy of the Name of the Fountainhead of the New Idea about Human Civilization".

The *qigong* culture in question came into slow, uncertain focus for both my graduate student and me. And as it did, it became clear that nothing was going to be straightforward here. For example, an expanded vision of human potential was dangled before us, one that connected our powers to larger cosmic powers; and yet the demonstrations of that alleged potential struck us both as, in many instances, just this side of vulgar.

A key moment came early in the Shanghai conference when a *qigong* master gave an evening performance that included tricks like swallowing large heavy balls and swords, and making strips of newspaper burst into flames. He had a beautiful girl assistant at his side to hand him his props, and he moved among the audience dressed in a swirling yellow sequined cape and made up with V-shaped eyebrows that caused him to resemble Mr. Spock from *Star Trek*. The most poignant memory I have of that evening concerns Grace. Visiting her homeland for the first time, she was so embarrassed, both by the spectacle of this man and by the appreciative reactions of the audience of Chinese scientists, that she began to cry.

Yet, side by side with this phenomenon of *qi*-as-spectacle we also heard, over and over at this conference, about the imperative need for *qi* and *qigong* to partner with science. Only integration of "East and West", we heard, would allow for a full realization of the next chapter of human civilization. Graphs of laboratory data, slides of brain imaging studies, clinical outcome work documented in "before" and "after" video clips tracked us over the days, driving the message home that *qi* was valuable, *qi* was measurable, and *qi* was emphatically of the material world

During the Beijing Conference, I met a woman who is recognized as a pioneering laboratory researcher in this area: an immunologist named Feng Lida. Feng Lida had begun her studies of *qigong* in the late 1970s with research that allegedly showed that the *qi* emitted by a *qigong* master can alter the growth of bacteria in a petri dish. After 20 years of work in this vein, she had arrived at a place of honor within the *qigong* research community in China. This fact was symbolized by her positioning in the obligatory group photograph of all the participants at the Beijing World Conference. The organizers chose to seat her in the first row, exactly in the center of the picture.

I introduced myself to her, and asked her if she would tell me her story. We discussed how she had come to study *qigong* and what it all added up to. She told me she had been thoroughly trained in the methods and conceptual approaches of Western laboratory science. She had even studied molecular biology for two years at UC Berkeley, and had undergone eight-years of further medical training in Leningrad. And yet she seemed wholly comfortable in her role as the matriarch of this Chinese world of radiating energies to which we were bearing witness.

So I asked her: as she looked back on her life and career, was there anything that surprised her, anything she had learned about *qi* that might seem to challenge the materialistic, rationalist view of the world to which she clearly was committed as a scientist. And she shook her head, no, not at all. It was like with fax machines, she said then. At first blush, they too seem very mysterious. You put a piece of paper into a machine, press a button, and somewhere else in the world – where ever one wants – out comes an image of whatever was on the original paper. This strikes the neophyte as amazing, but that fact doesn't mean that the technology underlying facsimile transmission contradicts scientific principles. *Qi* and *qigong*, she said, were like fax machines in that sense. She then added as a kind of casual aside: did I know that Jesus Christ had actually been a *qigong* master, and that *qi* had been the actual force behind the various miracles attributed to him? There were several books in Chinese that explained all this. It was clear that, in her cosmology, *qi* was a marvelous thing, but it was no miracle.

This kind of intellectual negotiation may have worked for her, but Grace and I were increasingly restless. Was *qi* in the petri dishes, in the imaging slides, in the graphs, in the videos? Maybe something important was buried inside those visualizing technologies – I had not forgotten my destabilizing moment back in the Qigong Research Institute in Shanghai. But the arena for exploring collaboration between East and West all seemed tawdry and forced. As Grace put it: "Elvis has left the stage".

In fact, part of what also was going on – as I later came to understand – was that my own ideas about what I wanted from an East-West exchange were crashing up against the realities of China's modern history. Earlier in this essay, I described how, during my first China visit, Dr. Chu had told me a history of *qigong* in China: one that positioned *qigong* as a national treasure sanctioned by millennia of Asian experience, and now poised to be brought into the modern era through its validation and illumination by science. This, of course, was also the history that Bill Moyers had affirmed to millions of American viewers in *Healing and the Mind.*

It turns out that this is a newly written history, constructed since the 1970s to provide a sense of noble lineage and unbroken continuity to an entity that had never clearly been a single "thing" before – a thing called *qigong* – but instead had been many different practices and schools identified with everything from Taoist health and longevity practices to Buddhist meditative practices to martial arts. In the modern era, a certain kind of whole had been cobbled together out of varied selected parts. To be fitted into the whole, each of these parts had first to be lifted out of the local context in which it had originally developed – especially if it was a context that was now identified with "feudal superstition". It then had to be dusted off and reworked in ways appropriate for the goals of modern China.[21]

It is a striking fact that, when Westerners are told the history of *qigong* in China, there is rarely any mention of the fact that the modern *qigong* era did not only begin in a physics laboratory in Shanghai; it also emerged out of an attempt to manage and redivert what was called a "*qigong* craze" within China that began to spread across the country in the late 1970s, immediately in the wake of the Cultural Revolution. The *qigong* craze was a movement with a clear populist-religious thrust, connected in intention and spirit to old Taoist and other mystical beliefs and practices. As a mass movement, this form of *qigong* expressed itself in undisciplined, ecstatic forms of movement that bore a much closer family relationship to the world of the *qigong* master who could bring a mass audience of Chinese people to a state of convulsions, than it did to the sedate exercises practiced by elderly Chinese in parks across the country.

By the end of the 1980s, it was estimated that the *qigong* craze had "infected" some 60 million people, and a Hong Kong-based journal, *Contemporary*, claimed that *qigong* culture had become the strongest power beside the Communist Party.[22] Obviously something had to be done. You could not just extinguish the whole movement, but you could forbid certain expressions of it, and repackage other aspects of it into less threatening and more useful forms that might serve practical and ideological goals of the State. It was an only partially successful venture, pockmarked with a range of ongoing resistant elements driven in part by religious hungers,[23] and in part by an even more recent desire to capitalize off the potential market-value of *qigong* as more and more Westerners were seduced by its aura of exotic mystery, and were willing to pay to experience its secrets. The strange amalgams and graftings of different kinds of agendas that Grace and I had witnessed in these conferences had at least some of their roots in this stormy history.

Final Reflections

The story does not end here. Two days before Grace and I were due to return to the United States, we met a young American man who had been working as a translator at the Beijing conference. Paul had lived in China for almost four years, was studying at the Traditional Chinese Medicine College in Beijing, and seemed to be a sophisticated and deeply committed student of Chinese medicine. We shared with him our sense of unhappiness with the proceedings of the conference, how shallow and intellectually hand-waving we found so much of it. And then we asked him – was his own experience of Chinese medicine, including so-called *qigong*, all of this order? Was this all there was? No, he said firmly. No, behind the circus of public *qigong* demonstrations, there certainly was "another" face both of Chinese medicine, and of *qigong*. He talked about the face of that world as he had experienced it in the clinics of China, but the most important face he ultimately described turned out to be that of his own teacher of a certain form of meditative martial arts, Teacher Bai. This was the man he said, whom he "respected more than anyone else he had met so far in China".

Maybe, he said then, you can meet Teacher Bai. After a bit of scurrying around, Paul made it happen. Teacher Bai came to Paul's modest one-room living quarters in the apartment complex for foreign students at the Beijing Chinese Medical School. Tea was poured and for four hours we talked. About the nature of mind, about ways of disciplining its energies within the body, about the reality or not of *qi*. He was rather disinclined to use the word, he indicated. What was more important to him were the experiences people had, and their effects. Explanation would be a longer process.

I remembered again my own commitment to holding a space of distance between the institutionalized maps of our humanness and the territories of our experience, and found myself warming to this quiet man. Before the conversation was over, he had told me that he did not feel I would be able to make much more progress on my project of cross-cultural collaboration if I was not willing to get closer to the experience as it presented itself. As he put it, before I could have any idea how to engage science with these practices, "you have to come inside this world yourself". And then he said that, if I returned to China, he would be willing to be my teacher.

I went back to the United States and thought for two months about Teacher Bai's offer. Accepting would certainly mean complicating the balance of engagement and analysis this project demanded of me still further. This is how I saw the matter: if I accepted, I would be committing myself to a deeper form of participatory learning that might yield significant insights. At the same time, I would need to participate in a way that remained sufficiently self-aware of the cultural "strangeness" of what I was learning so as to be able to translate it into understandings that would serve the cross-cultural collaborative project I had also taken on board. I must then be committed to helping to create the conditions for a cross-culturally-informed empirical research project, but I must do so in a way that maintained a certain level of detachment from any specific outcome associated with that effort. Only then, I thought, would I be able also to see the whole effort with the eyes of an historically-informed ethnographer of the mind-body sciences in our time ... and in this way bring the project "back home".

I took a deep breath and finally accepted. In the summer of 1999, I returned to China for a third time, to spend five weeks as a student of Teacher Bai. Shortly thereafter, I was joined by a small contingent of laboratory scientists I had organized from the United States looking to explore possibilities for common ground with this practitioner and others. My "engagement with *qi*" is not over. This essay is merely a first interim report "from the field", so to speak.

More, my purpose here has really not been to offer any digested or comprehensive analysis of the "*qi* project". Rather, I have wanted to use some vignettes and choice moments associated with that project as a way of drawing out more fully some of the intellectual balancing acts that seem to be part and parcel of trying to live a life of engaged scholarship. There is definitely a knife-edge one walks here, and it is a shifting one, as projects evolve and morph. Everett Mendelsohn did not plant in me the specific seeds of my own restlessness that would result in the career of engaged scholarship that I find myself to be leading. He did, however, offer himself as an inspiring model of a scholar for whom engagement was not a matter of choice, but of obligation. He found his own formulas for interfacing scholarly projects in the academy with ethical and intellectual projects in the larger world, and walked his own knife-edge in the process. In this sense, while Everett was one of my first teachers in the history of science, it may be that one of the most important lessons he taught me was that the only real question one faces as an engaged scholar is, not whether you are or not, but how to do it well.

Harvard University

Notes

[1] Cf. my autobiographical piece, "Truth and Consequences" [*Section*: "Reflections from the Field"] *The Radcliffe Quarterly,* 83(2) (Summer, 1997), p. 17.

[2] Anne Harrington, *Medicine, Mind and the Double Brain: A Study in Nineteenth Century Thought* (Princeton, N.J.: Princeton University Press, 1987) [Paperback edition, 1988; Japanese edition (1991); Italian edition (1994)].

[3] The work of Roger Smith, especially his *Inhibition: History and Meaning in the Sciences of Mind and Brain,* (Berkeley: University of California Press, 1992) has taught me a great deal about ways of thinking well about this theme.

[4] See "Roger Sperry", The Nobel Prize Internet Archive. <http://almaz.com/nobel/medicine/1981a.html>

[5] "Nineteenth Century Ideas on Hemisphere Differences and 'Duality of Mind'" *The Behavioral and Brain Sciences: An International Journal of Current Research and Theory with Open Peer Commentary,* 8(4) (1985), pp. 617–659.

[6] A. Harrington, *Reenchanted Science: Holism in German Culture from Wilhelm II to Hitler* (Princeton, NJ: Princeton University Press, 1996).

[7] I have most directly addressed the ethical challenges raised for me by this project in two separately published essays: "Metaphoric Connections: Holistic Science in the Shadow of the Third

Reich", *Social Research: An International Quarterly of the Social Sciences* 62/3 (1995): 357–385; and "Unmasking Suffering's Masks: Reflections on Old and New Memories of Nazi Medicine", *Daedelus: Journal of the American Academy of Arts and Sciences,* 125 (1996), pp. 181–205.

[8] See, for example, W.M. Runyan (1998), "The Changing Meanings of Holism: From Humanist Synthesis to Nazi Ideology (Review of *Reenchanted science: Holism in German culture from Wilhelm II to Hitler*). *Contemporary Psychology*, 43, pp. 389–392.

[9] A Chinese-speaking undergraduate at Harvard, Ying Liu, took on the task of reviewing and familiarizing me with the basic Chinese philosophical interpretations of *qigong* and its different schools, both traditionally and more recently. Two basic texts (in Chinese) that proved helpful as a basis for our discussions were: W. Zhang and J. Chang, *The Dictionary of Chinese Traditional Qigong* (Taiyuan, Shanxi Province, China: Shanxi People's Publishing House, 1989); and *The Complete Book of Chinese Contemporary Qigong,* H. Wu,. Liang, K., Chen, W. and Zhang, P., eds. (Beijing, China: People's Sports Publishing House, 1997). A graceful and accessible introduction (in English) to the broader medical philosophy underlying such practices is Ted J. Kapchuk's *The Web that has no Weaver*: *Understanding Chinese Medicine* (New York: Congdon & Weed, 1983). And of course the work of professional historians of Chinese medicine like Nathan Sivin, Paul Unschuld, and Bridie Andrews offer richly nuanced understandings of these practices in both their traditional and modern historical contexts – work that is very different in both focus and style from what I, as an historian of the mind sciences who has stumbled onto Asian territory, am able to contribute.

[10] Bill Moyers, *Healing and the Mind* (New York: Doubleday, 1993).

[11] This comment is taken from a "publisher's note" on the back copy of the paperback edition of *Healing and the Mind* (1995). It speaks of the book as "the monumental best-seller (almost half a million copies in print!) that has changed the way Americans think about sickness and health – the companion volume to the landmark PBS series of the same name. In a remarkably short period of time, Bill Moyers's *Healing and The Mind* has become a touchstone, shaping the debate over alternative medical treatments and the role of the mind in illness and recovery in a way that few books have in recent memory".

[12] There has been a widespread misunderstanding that Reston's surgery was actually carried out using no anaesthesia other than acupuncture needles; in fact, the needles were only used to alleviate the pain following surgery, but that seemed remarkable enough to Reston. See his syndicated column describing the experience in *The New York Times* (July 26, 1971) pp. 1, 6.

[13] The Moyers presentation of *qigong* has also been criticized for offering "an embarrassingly credulous" presentation of the more flamboyant aspects of *qigong* performance in China. See the report by two members of the Committee for Scientific Investigation of Claims of the Paranormal (CSICOP), that publishes the debunking journal, *Skeptical Inquirer.*: Barry L. Beyerstein & Wallace Sampson (1995), "Traditional Medicine and Pseudoscience in China: A Report of the Second CSICOP Delegation". <http://www.csicop.org/si/9607/china.html>

[14] S.L. Wolf, Barnhart, H.X., Kutner, N.G., McNeely, E., Coogler C. and Xu, T., "Reducing frailty and falls in older persons: an investigation of Tai Chi and computerized balance training. Atlanta FICSIT Group. Frailty and Injuries: Cooperative Studies of Intervention Techniques". *Journal of the American Geriatric Society*, 44(5) (1996, May), pp. 489–97.

[15] For a general overview of the various kinds of clinical outcome studies in this area – of mixed quality – see the computerized "medical qigong database" that has been compiled by a non-profit group in California and which contains some 1300 abstracts from the literature, going back to 1986. This is available for purchase at <http://www.qigonginstitute.org>.

[16] Roger Jahnke, "Health-Care: China's Ancient Solution to the Crisis in Modern Medicine", *Qi: The Journal of Traditional Eastern Health & Fitness*, 1995; <www.qi-journal.com/jahnke.html#anchor585346>

[17] I have since stood with colleagues in an operating theater in Shanghai and watched Dr. Chu participate in such a surgical procedure live; I then watched a medical colleague of mine question the fully conscious patient immediately afterwards.

[18] David Eisenberg, *Encounters with Qi: Exploring Chinese Medicine*, written with Thomas Wright (New York: Penguin, 1987; 1985).

[19] This is a complicated story that has been widely followed in the press. Falun gong is a form of qigong practice that was created by Li Hongzhi in 1984 and began to see wider public following in the 1990s. Part of the appeal of the practice was the relative simplicity of the movements it taught. Part also came from the unabashed spiritual and ethical ethos within which the whole practice was packaged. After great numbers of followers materialized unexpectedly in silent protest outside a government building in Beijing in 1999, the No. 1 People's Court in Beijing sentenced four top organizers of the Falun Dafa (or Falun gong) movement to jail terms of 7 to 18 years. Significantly, all were members of the Communist party. They were charged with using a cult to "obstruct justice,

causing human deaths in the process of organizing a cult and illegally obtaining state secrets". The state also claimed that 1,400 followers of the Falun gong movement died because they rejected needed medical attention or committed suicide. For more details, see the Amnesty International Report (ASA 17/11/00) of March 23, 200, entitled " People's Republic of China. The Crack-Down on Falun Gong and Other So-Called 'Heretical' Organizations", This document exists online at: <www.amnesty.org/ailib/aipub/2000/ASA/31701100.htm>.

[20] See on this theme the brilliant recent work of Shegehisa Kuriyama, *The Expressiveness of the Body and the Divergence of Greek and Chinese Medicine* (Zone Books, 1999).

[21] Kunio Miura, "The Revival of Qi: Qigong in Contemporary China", in: *Taoist Meditation and Longevity Techniques,* K. Livia and S.Yoshinobu, eds. (Ann Arbor: Center for Chinese Studies, The University of Michigan, 1989).

[22] Thomas Ots, "The Silenced Body – The Expressive Leib: On the Dialectic of Mind and Life in Chinese Cathartic Healing", in: *Embodiment and Experience: The Existential Ground of Culture and Self,* Thomas Csordas, ed., (New York: Cambridge University Press, 1994), pp. 116–136.

[23] In 1996, the year before my own first trip to China, National Public Radio broadcast a report entitled "Spiritual Revival Sweeps China," as part of its series, "All Things Considered (May 27, 1996), with Mary Kay Magistad. (<http://search.npr.org/cf/cmn/cmnpd01fm.cfm?PrgDate=05/27/1996&PrgID=2>)

The text of this broadcast (that, as of this writing, seems no longer to be available on-line) ran, in part, as follows:

MARY KAY MAGISTAD: The Communist Party's discomfort with alternative belief systems stretches well beyond Christianity. In recent months, it has blasted what it calls superstitious Communist officials who visit Buddhist temples, and it has ordered restaurants to stop displaying shrines to the traditional Chinese god of wealth. The party is especially concerned about the recent surge of interest in Chi Gong, a blend of meditation and physical movement that is supposed to develop supernatural powers. Victor Yuan says the government's distrust seems to stem from the fact that Chi Gong is much harder to control than organized religions.

VICTOR YUAN: Unlike traditional religions, which are confined to churches or temples for their activities, Chi Gong as a new phenomenon is not yet restrained in these ways. People can go out and proselytize, convert, perform, and a lot of their first converts are authoritative figures from the Academy of Social Sciences, from the Communist Youth League, and other official figures.

MARY KAY MAGISTAD: A few months ago, the government banned the huge public meetings of Chi Gong enthusiasts and shut down many of the places that used to spread the word. But interest in Chi Gong remains. It's just become more discreet.

HELGA NOWOTNY

HOW TO TELL A STORY IN THE SCIENCES: SETTINGS AND LESSONS

Abstract

This paper is about story-telling in biography, autobiography and in the sciences. It examines the differences in which stories are told, depending on the teller, the intended audience and the socio-political and scientific setting. The main illustrations are taken from the discovery of high-temperature superconductivity. Although there is basic agreement about what happened, the stories told differ as seen through the perspective of the media, the history of science or through an interest in what the discovery reveals about the current transformation of the science system.

Life as Narrative

The occasion itself is telling, and story-telling will be my theme. The celebration of Everett's 70th birthday – a life rich in intellectual and scholarly achievements, lived with intense generosity, lasting friendships, and with the unabiding willingness and ability to engage with the betterment of the world through small human steps and vigorous action whenever and wherever called for – is perhaps the appropriate moment to resume a theme which Jerome Bruner fortuitously called "life as narrative".[1] Bruner's interest in different modes of thought brought him inevitably from what he called logical thought to another mode, quite different in form of reasoning, that is mainly embodied in stories and narratives. And nothing is perhaps more appropriate to illustrate this kind of thought than an analysis of the stories we tell about our own lives: autobiographies. A constructivist stance, based on the premise of some kind of 'world making' develops into a set of procedures for 'life making'. Stories may indeed happen, as Henry James once remarked, to people who know how to tell them – but then, how could we ever share with others what happens to us without knowing how to tell it? In this sense, we are all born story-tellers.

Describing lived time in the form of narrative is, of course, also an artful as well as a powerful means of constructing narrated (self) time – sequences need

G.E. Allen and R.M. MacLeod (eds.), Science, History and Social Activism: A Tribute to Everett Mendelsohn, 123–135.

to be set up and chronological orderings attended to, while the threads of meaning are being woven and unwoven until they emerge as colorful patterns, as intended or unintended patchwork. The weaver and the woven, like the dancer and the dance, then become indistinguishable while eye and ear adapt to different levels of depth and shades, to different scales of time, tones and melodies. Narratives then thrive on the pluralistic perspectives which they have to offer. Yet, there is also a deep paradox involved in the process of life-telling. In translating lived and experienced time into remembered and told time, one becomes conscious of time which has irrevocably passed, while simultaneously a pathway opens into the future which is still to be experienced. Past and future thus become linked in an extended present, that includes the narrating present.[2] Narratives confer to time its deeply human dimension and make room for its articulation, while creating the conditions for shaping and sharing human existence.

But narratives are never a one-way affair, nor do they ever stand alone. If art imitates life, according to Aristotle, life also imitates art. In the end, to return to Bruner, "we become the autobiographical narratives by which we tell about our lives".[3] Note that even in this sense of "becoming", narratives as lives are an ever unfinished process. Temporal experience is never encoded directly. Instead we are dealing with representations that encode the dynamic properties of events, experiences and episodes that take place in the real world, in such a way that we can remember them or act upon them.[4] Our way of telling about ourselves therefore may change continuously, since it is based upon indirect codings and yet, the accounts thus produced come to take control of our ways of life. The way how we tell what is changing becomes itself part of the story. Inevitably, it throws its shadow onto how our life will be lived. Other lives, that may have been possible but actually were not lived, like doors we could have opened but did not, continue to lurk behind what has just been assigned to the narrated past. The weaving patterns could have turned out otherwise, if we had chosen to change color and thread while weaving. This is the contingency of history, a message which is carried by each story being told. It is so – but it could have been otherwise.[5]

Contrary to a wide-spread belief, there is no room for naive realism in story-telling. Its constructivism is a doubly constrained one. It is culturally constrained by admitting certain forms of narratives and not others. The ones considered to be legitimate are admissible, since they are believed to achieve the kind of credibility which art imitating life and life imitating art warrant. In this sense, authenticity, the hall-mark of credibility which a narrator can convey to an audience, is itself culturally and historically contingent and therefore varying in form, content and outcome. The second constraint comes from the social bonds that bind narrators and listeners together. Narratives, including self-narratives, never stand alone. Stories are told to an audience that may well have its own version or multiple versions of what is being told. We are not only part of a *Denkkollektiv* (thought collective) in Ludwik Fleck's sense, but also part of an *Erzählkollektiv*, a story-telling collective. When Heinz von Foerster provocatively celebrated the narrative turn, to which he himself had so much contributed, he claimed that it is the listener, not the speaker, who assigns meaning to what is being said. A story-telling collective is therefore always also a listening collective that answers back, since each member has his or her own story also about what is being heard and what is being told.

Which is what we are doing here collectively. Starting from the premises sketched above, I want to continue by reflecting on story-telling in the sciences through the examination of three different settings in which stories are being told. The sciences are thought collectives with a powerful dynamic of their own, but they are also story-telling collectives: story telling is an integral way of scientific life. Different settings and audiences let the stories assume different social functions, but they also reach out into the actual world. Maybe this is what Nelson Goodman meant when he said that people make visions, and true visions make worlds. 'True" stories are the ones that have been rendered robust and story-telling in the sciences has achieved a remarkable degree of robustness by virtue of making and shaping "worlds".

Tribal Story-Telling

Story-telling presumably originated in a tribal setting. It displays a marked tendency to reinforce the bonds that bind together by celebrating the importance of past achievements and by putting the distance covered in the past into the perspective required by the present. The "tribal" component in scientists' lives, so aptly described and analyzed by Sharon Traweek and others, puts story-telling on center stage. It becomes a means to forge a tribal identity and communal solidarity, to initiate and socialize the young by upholding a canonical image of the achievements of the elders. The elaboration of tribal cosmologies and their enmeshing with the exemplary lives which are to be imitated form part of scientific world-making.[6] In such world-images, genealogies play an important part, since the search for precursors and the filiation of powerful ideas inevitably lead to the present. There they serve to position living scientists and their work. In this sense, the history of science is deeply entangled with story-telling. This implies taking a stand on current debates or controversies, summoning powerful allies on all sides, even if some may no longer be able to intervene. On a more general level, the narrative context emerges as being structured in ways that allows past work to be continuously reinterpreted and to project significant possibilities for further work. Thus, the narratives display a constant tension between the need for a coherent and shared understanding, while the thrust towards exploring new directions implicitly threatens or challenges the coherence just attained.[7] In this sense, story-telling and history-telling become a crucial force in stabilizating knowledge, treading a thin line between scientific conservatism, which attempts to preserve the consensus and shared understanding already achieved, and its intellectual radicalism, the search for creating novelty and ever new understanding, even if it entails the risk of upsetting what has already been settled.

Within a tribal setting trouble and confrontations may, however, arise when outsiders intrude, with accounts that differ from the ones privileged by the tribal authority. This is what Dominique Pestre and his colleagues were to discover during their work as professional historians charged with writing the official history of CERN, le Centre Européen de la Recherche Nucléaire.[8] Pestre recounts the ordinary experience of a stranger who arrives in a place to study the natives, only to discover that they do not exactly speak his language, organize their narratives following different rules, and listen to him only when it pleases and makes sense for them. In the case of CERN, the natives were quite powerful and could ignore or by-pass, rather than discuss with the stranger/historian, whenever contradictory stories emerged. While historians would readily acknowledge that all narratives are socially, temporally and spatially situated, to admit the possibility that there might be many narratives causes great uneasiness to those – in this case, physicists – who, as members of a tribe that strongly believes truth must exist and that it has to be the reference point against which all claims are calibrated.

Such a confrontation raises a number of other interesting issues, especially when, from a professional historian's point of view, the physicists' accounts have no concern for what historians would call the need to avoid anachronism. The CERN physicists do not hesitate to select and judge episodes which are chosen retrospectively, in accordance with current standards of proof, knowledge and categories rather than those that existed at the time of the discoveries or when the experiments were first conducted. Contrary to the story-line of a professional historian, physicists speak in a homogenous – or better, a homogenized – historic time, in which validating procedures and practices used in the past are equated without hesitation with those of today. The result is an image of an everlasting and never changing Science. As told by the CERN physicists, the stories they tell about their past achievements, universalize today's practices. They make current moral and behavioral rules as well as criteria of proof and procedures, appear as immutable. The crucial function such universalizing stories have in maintaining the values of the scientific institution become even more pronounced when external circumstances change dramatically. The interest in history and, more specifically, the self-history of an institution like CERN, has noticeably grown over the last decades. There is a clear link between the increase in ceremonies as highly-codified and publicized public events as well as in the greater interest in historical narratives and in resorting to public relation specialists in order to be recognized politically and by the mass media, and the altered relationships between physicists and the state, based on rising demands for accountability in social and economic terms. These changes are subtly reflected in different generational autobiographies written by physicists. The 'savant-philosophes' who sought to share their philosophical reflections with the educated public have ceded to the "empiricist temper", as Sam Schweber called it, tainted with a hedonistic or opportunistic streak.[9] In line with the rise in numbers and growing specialization, closer relations were established to industry - and to war. This is reflected in the autobiographical writings in the sense that it is no longer only about 'science' and the human values involved, written for a public that is curious to know about the science-dominated era in which it lives.

Rather, 'physics is fun' appears as a new theme, while the connections with the military never get mentioned.

Stories in Science vs. Stories about Nature

Much attention has been devoted to how scientists communicate - both within and between different disciplines and with the wider public.[10] Ongoing efforts, brandished under the banner of enhancing the 'public understanding of science', often proceed by trying to eclipse technical language, theoretical concepts and modes of thinking within disciplines in order to heighten the intelligibility of scientific content and insights. This form of popularization does not communicate any arguments, however. At best, it describes more or less coherent statements and findings that are to be believed by the lay audience. In contrast, argumentation requires to make explicit the assumptions that underlie the object of study and the process of researching it. In his analysis of scientific and popular texts written in and on biology, Greg Myers argues that articles written for the general public and for scientific colleagues are strikingly different. They represent two different versions of what scientists do and what science is about. Even very sophisticated popularizations tend to promote a view of science that focuses on the objects of study rather than on the procedures by which they are studied. The differences emerge in the narrative structure, in syntax, and in the contrasting views of science. Professional articles create a narrative of science; they follow what scientists do in their actual practices, arranging time into a parallel series of events that are made to appear simultaneously and supporting their claims. In syntax and vocabulary, in stylistic conventions like titles and introductions, the conceptual structure of the research field is emphasized. Popularizing articles, on the other hand, present a sequential narrative of nature in which the plant or organism, not the scientific practice, is the subject of the narrative. Hence the 'natural' chronology dominates. The narrative is organized around the externality of nature to scientific practices. Experiments may still be reported within popular narratives, but they function to make a story coherent, rather than to subordinate the chronology to an argument of simultaneity which takes it outside such chronology by reaching coherence in an argumentative way. Even when two articles seem to be about the same research, it may turn out that one is about garter snakes and the other about the chemical isolation of a pheromone.[11]

How to Tell Three Stories and Draw Different Lessons

The third setting is that of a fascinating case: the remarkable discovery of high-temperature superconductivity (HTS) by Müller and Bednorz in 1986. While accounts of the story's outline and of its narrative follows similar lines, different lessons can be drawn from telling it. The moral of a story tends to vary in accordance with the interests of those who tell it, but even more as a function of the audience to which the lesson-drawing is addressed. The discovery of HTS was unanimously hailed as a scientific breakthrough. Its fascination partly stemmed from the fact that it was the story of a discovery in which scientific ingenuity appears to win against all odds, both against conventional scientific wisdom and against the carefully upheld tenets of science policy-making, with its emphasis on priority setting, predictions of hot paths to be pursued, and promotion of scientific advances in conjunction with technological innovation. Moreover, it is a case that triggered an intense excitement, gripping scientists and the public alike. In the following, I will present three different versions of lessons that were drawn from the discovery of HTS. The first is the media story, which follows the rules of media narratives, in which the over-all lesson, especially in the U.S. media, strongly emphasizes what future research might mean in an intense climate of international economic competitiveness. The second lesson seeks to uncover the hidden roots of scientific creativity, its links with the history of science, and to spell out the resources upon such a discovery is based - and what needs to be done to preserve them. The third lesson moves in a similar direction, but emphasizes much more strongly what the case reveals about the current state of transformation of the science system. It examines "preparedness" for unexpected scientific breakthroughs and how individual scientific creativity is implicated by the "innovation machinery" of science with its selective filters. All three accounts tell "how a scientific discovery is made", but the moral of each story differs.

The Media Story

After so many dismal stories about potential risks, endless controversies, and cases of scientific fraud, the media were finally in a position to present HTS as a clear-cut scientific "success story". Readers were introduced to scientists who had sufficient individual freedom and creativity to follow their intuitions. A new kind of researcher was portrayed, who was neither of the old unworldly type, nor of the new greedy type brought forth by academic capitalism, but one able to adapt to the changed conditions of the research environment. A glimpse of a research area was offered, which was still basic science, but opening up towards the promises of technological application. These ranged in the short-term from applications in the defense and space sector, electronics and communication to the medium-term of medical and industrial ones. In the long term applications were expected in the electric power sector and in transportation. The "science in the making", thus told, was not only celebrating the triumphs of science; its "scientific realism" emphasized the close interactions and negotiations between science, politics, industry and a wider public. The gist of this portrait was the image of science as the driving force behind (individual and national) competitiveness.

The close involvement of the print media with HTS also assigned a new function to the press. At key moments in the rapidly developing new research field, newspapers reported discoveries even before scientific publication, a phenomenon that became known as "science through press conference". Even if this was a passing episode, the press, and with it the broader public, assumed the role of spectators and arbitrators in a mounting effort to turn a discovery into a major national and international research effort, sustained by the enthusiastic promises of future technological applications. The media stories of HTS, followed two years later by cold fusion, show that the wider contexts surrounding science have changed irreversibly. Science meets the public under altered conditions: trust and authority, support and cultural meaning no longer seem what they used to be. The media have become the most visible arena for the negotiations that now are conducted between science and the public. Lesson-drawing for public policies now passes through the media.[12]

The first, and clearest thread running through the whole HTS story was the interaction between science and its economic and political implications. The leitmotif of economic competition dominated from the beginning and was further enhanced when it appeared that initial technological optimism exceeded research realities, and that commercial applications were not "just around the corner". Seemingly in opposition to this dominant theme, was a second dimension. In the initial phase, the HTS discovery story held particular appeal because it was believed to require virtually no funding and few infrastructure investments. HTS arrived at the moment when U.S. politicians, scientists and citizens were discussing whether to invest billions of dollars in the construction of a superconducting supercollider – a machine that would allow the U.S. to regain the lead in the field of elementary particle physics – and when many fields of science were suffering shortages in funding. HTS could be presented romantically, evoking the "golden age" of science. Individual scientists, their creativity and perseverance, were portrayed as the embodiment of 'little science' that seemed to have returned with a huge promise. Even if this moral of the story could not be sustained, its appeal was strong and emotional.

Heroes are indispensable in any kind of story-telling. The U.S. media selected a number of researchers, such as Paul Chu, to accompany the public throughout the story, dramatizing the narrative's appeal by the tendency to personify research and equate the worker with the work. This was in marked contrast to press coverage in European countries, where, with the exception of the "local heroes" Müller and Bednorz, who had been awarded the Nobel Prize for their discovery in 1987, scientists were cast into impersonal roles. Despite the national differences, journalists came to accept that the "noble" activities of researchers had a real-world background of intense competitive pressure. They could no longer be cast as altruists. Chu's portrayal in the U.S. media is a good example of the new image of scientists from the 1990's: the affirmation of competitiveness, full cognizance and willingness to enter a race whose stakes are no longer set solely by personal prestige and scientific honor, but also partake fully in the commercial and technological global game. The hero scientist no longer stood 'outside', but has become a good and valuable member of the national team.

Looking back at the media story of HTS, it becomes clear that media reports not only were made to document priority claims of new results, but also provided scientists with three new negotiation assets to be used in their dealings with politicians and funding agencies: a means to stage priority claims publicly and thereby gain competitive advantage individually; a way of gaining increased public attention and interest for a new research field in the making; and a way their claims affected national economic goals, thereby enhancing national economic competitiveness. In these and similar respects, "selling" basic science has become a new feature in the transformation of the science system.

How to tell a story through the media has become an indispensable part of "selling" basic science, even if what was being "sold", namely potential technological applications, were still far away in the future. It was obvious to all who took part that what they hoped to engage in, fund, or set up and manage would remain the realm of basic science for years to come. No serious researcher or science administrator, no journalist adhering to professional standards, could or did claim otherwise. And yet every public representation of the new research field, every speech or interview, every international conference, resonated with the hopes and expectations that technological benefits would come, sooner or later. The lesson was clear: with the help of the media a hybrid space emerged in which negotiations for funding and mobilizing political support for a new research field could and did take place.[13]

The Hidden Roots of Scientific Creativity

But other lessons can be drawn from telling the story of HTS as a discovery. Gerald Holton and his colleagues have been intrigued to reconstruct the case as one of laying bare the hidden roots of creativity and their reliance on the historical filiation of concepts and theories.[14] But their ultimate interest lies in extracting the lessons the discovery holds for science policy. The authors distinguish different levels on which the story can be told. The first level is that of sketching the outline of events as they appear in chronological order. The second level of analysis goes deeper and searches for the historical resources that had enabled the success of the discovery team. These resources take into account the personal research trajectories of Müller and Bednorz, the influences that other researchers and their ideas exerted, and include the possible assets which institutions, their universities and the corporate institutions that employed them, put at their disposal. This leads the authors to identify four sets of resources on which the discoverers had drawn: the theoretical framework and ideas (schema) which motivated and guided them initially; experimental techniques and material resources (production); the means of gathering and analyzing data (observation); and theoretical concepts for interpreting the results (interpretation). While the genealogy of a scientific discovery can relatively easily be identified, the public literature can only set the stage for the emergence of a new scientific achievement. In addition, it is much more difficult to retrace the private dimension which rarely survives in public records, that degenerates into mere anecdotes or rumors hard to prove. The case of the discovery of HTS is unusual also in giving insight into the private thoughts of Karl Müller himself. He speaks openly about the persistence of his recurrent dreams and their symbolism, undoubtedly influenced in this by Wolfgang Pauli who was one of Müller's professors at ETH, the Eidgenössische Technische Hochschule, in Zurich. Pauli had published an essay with the psychoanalyst Carl Jung on the archetpypical conception in the work of Kepler, which induced Müller to become interested in Kepler's commitment to structures of high symmetry. In several interviews, and in an introspective article illustrated by the figure of a mandala, and even in his Nobel Prize speech, Müller publicly acknowledged this private dimension of the "personal thematic presuppositions" that guided his work.

Holton and his colleagues skillfully put together the picture that emerges from the different resources that a genealogy of a scientific break-through reveals. They rightly emphasize the unpredictability of events, rife with unintentionality. Most of the resources were "borrowed" either from other disciplines or from applied research. They were brought to bear on the discovery through equally unintended interactions or applications. Thus, the Müller-Bednorz story the authors tell unfolds very much in line with Müller's intuitions and guiding images: the perovskite structure that inspired Müller initially to investigate oxides in the first place (a highly unusual and risky move); their decision to turn to a barium cuprate compound that contained well-separated planes of copper and oxygen atoms, and other twists and unintended consequences. Müller is cast into a genealogical role: identifying with the work of Kepler, who tried to decompose planetary orbits into perfect circles, and was led to ellipses instead, but thereby helped prepare the way for Newton's *Principia*.

Holton and his colleagues are clear about the lessons to be drawn for science policy: major accomplishment in science depends upon education and research systems and their administration that supports a mixture of basic, applied and instrument-oriented developments. The laboratory in Rüschlikon near Zurich is described as an environment well endowed with such resources. Moreover, it is praised as one with a culture that gives good people the freedom to pursue projects with long gestation periods. The science policy lessons emphasize the rewards that come from having an administration that is sufficiently flexible in promoting the "borrowing" of resources across disciplines and between basic and applied research. Its culture is one that allows an unpressured pace of work and a sizeable degree of individual freedom in setting goals and directions of research. It encourages the use of all resources, including the sources of private inspiration on which administrative rule books and text books in science are silent. "If we wish to achieve noteworthy science, even if noteworthy is defined to mean only science with an economic payoff, the nation has no alternative but to support the seamless web of research".[15]

Unexpectedness of Discoveries and Preparedness

In the work I conducted together with Ulrike Felt, we reconstructed the historical outline of the case, following in somewhat less historical depth the genealogical filiation of theories, ideas and the material and instrumental resources that were brought into play. We were equally fascinated by the "private dimension" of Müller's insights and the frank public acknowledgment of the influences that he attributed to them. We also sought to search for the hidden roots of scientific creativity which the discovery of HTS displays in such a striking fashion, in an age where science policy is believed to hold the keys for promoting the unforeseeable, with the result that, when it happens, no funds are available to support the research that grow out of such a discovery. But we took a somewhat different tack in telling the story, guided by a different moral we saw in it.

First, we noted that the discovery of HTS was unexpected in at least three terms: its discoverers, the site of the discovery, and the ideas involved. In our view, Müller and Bednorz were outsiders; Müller, a specialist in perovscites who had only come into contact with superconductivity during a 20 months stay at IBM's laboratory in Yorktown that had been granted to him as a senior researcher before retirement. But to be an outsider is not such an unfamiliar pattern in the history of creativity. It has been termed the "novice effect" by Root-Bernstein. "Novices" are able to see things in a new light. The practical wisdom of this is simple, but far-reaching: to encourage scientists to change fields ever so often, to foster intellectual, if not also geographical mobility in scientists and to encourage the interdisciplinary "travel" of concepts, of tools, instruments and methods, and of research practices that go with it. Moreover, Müller broke out of the social hierarchy of his lab, looking for a new place in the cognitive hierarchy. This dual hierarchy provides different degrees of freedom. Only after Müller was appointed as an IBM research fellow did he enjoy the autonomy to set up his own research agenda, and it was his desire to pursue what he called his "own work" that led him to resign as a manager.

The unexpected site of the discovery of HTS is tied to social conditions that provide more creative environments. The IBM Rüschlikon lab, though part of a large-scale research enterprise, was relatively modestly equipped. In this respect, Müller and Bednorz had plenty of company; and many breakthroughs, especially in "little science", have been made with inexpensive equipment, and in inadequately funded labs with seemingly amateurish techniques. But this can in no way justify cutting research funds, since exploratory research aims to show that a phenomenon exists. If successful, it will be the starting point for another line of research that will need considerably greater funding to yield profitable applications. HTS provides a good example. A decade and a half after the discovery of HTS, work on technological applications proceeds with vigor, while theoretical understanding, although exciting, lags behind.[16] But Müller and Bednorz confirm another important lesson, one with strategic relevance to the degree of autonomy that researchers should enjoy in their work. Discoveries are surprises: all that can be planned is work that might lead to them. Autonomy for the individual scientist is a prerequisite, since he or she must be free to follow his or her own, perhaps idiosyncratic, strategies, combining public and private knowledge, and utilizing whatever appears promising from the collective reservoir of theories and techniques. Sufficient time is indispensable in exploring and following hunches. Such autonomy requires institutional niches deliberately carved out by researchers themselves or by their administrators. More is needed than fund raising.

Conclusion

What are the ideal conditions to foster creativity? Policy discourse is full of generalizations that hardly fit the concrete case of HTS or are outrightly contradicted by it. Any discussion of whether the Müller and Bednorz discovery can be interpreted as "mere chance" or "programmed success" must start with carefully defined premises. Even then, we are still faced with the fact that the process of discovery is at the same time intensely individual and collective, with an inherent tension between individual scientific creativity and the collective nature of science.[17] What can be learned from the HTS discovery and can the social organization of research enhance scientific creativity? Müller's and Bednorz's working conditions hardly fit the conventional image of the ideal. Their group was the smallest possible, below anything that policy-makers believe is the critical mass. Their interdisciplinarity consisted in Müller finding an assistant with skills complementing his own. The IBM lab in Zurich, while of good repute, was never declared a "center of excellence" and, as a pre-retirement Fellow, Müller worked in relative isolation. Except for his decisive stay at the Yorktown Heights lab, his mobility had been unusually low. So the IBM lab at Rüschlikon, widely envied, imitated, and considered symbolic after producing two Nobel prizes, does not provide a model for generalization.

What can be learned from the HTS discovery, despite its unpredictability? Müller and Bednorz persisted for more than four years in working toward a goal they set themselves, in the face of a discouragingly recalcitrant phenomenon. They defined their situation as one in which they could risk failure, a luxury for others with untenured positions, who are forced to keep up a steady flow of publications. Müller and Bednorz could set their own pace. Isolated and untouched by the ordinary pressures of work and successes, Müller's theoretical interest was aroused by his visit to the U.S. laboratory, where colleagues were working on the very practical problems of superconductivity.

Why can't such conditions be created on a broader basis? Why not create more working conditions resembling the Zurich model? Of course, this has been tried. But all attempts to collectively organize the individualistic mode of producing creative scientific work meet with an inherent constraint: the risk of failure is real. The question is: Who will assume the risk? If the risk of failure is restricted to a few individuals whose idiosyncrasies are tolerated by colluding managers as long as they fulfill their other duties, creativity can survive in unofficial niches in official institutions. If they are successful, the organization shares the glory. If not, failure remains inconspicuous, since the risk was never officially acknowledged. But if a research organization or funding agency officially decides to take risks on a larger scale, it is accountable. Institutions are charged with limiting the occurrence of failures, which means limiting risks.

Taken together, these are hardly the conditions to be created on a broader basis as a guideline to future breakthroughs. Policy-makers are keen to know, but not all insights and generalizations can be translated into policy. The process of translation is hindered by the inherent tension between the social organization of research – which is set up as a stable framework for reasonable expectations and predictable, reproducible results – and the unexpected, unpredictable discovery. By definition, the innovative breakthrough escapes predictability. Since it cannot be foreseen, it cannot be controlled. It eludes administrative competence, though the administrators of science seek to facilitate its occurrence. With hindsight, we can retrace the rugged thin line that connects the interface between private experience, intuition, persistence, individual drive and the social organization of public scientific knowledge. The latter is far more than the organization of a research lab. It pertains to a field and its traditions, its history of successful and failed concepts, of technological applications sought and sometimes found. On the private side of the equation, Pasteur's dictum that "in the field of observation, chance favors only the prepared mind", still holds. But preparedness also has a public side.

In our story two other main factors emerge next to that of unpredictability of individual scientific creativity. One is the preparedness of institutions, as well as of individuals. How well prepared were funding agencies and research councils whose task it was to foster the unforeseen, when it actually occurred? And how well prepared were different research groups to exploit the new opportunities? Pasteur's dictum of chance favoring the prepared mind also holds for institutions. Success meets those who are prepared for it: a sobering lesson for all those who had celebrated what was believed to be a return to "little science" and the self-sufficiency of individual creativity. The ultimate lesson for us therefore is one of the paradoxes that so deeply characterizes the present condition: while individual scientific creativity has never been so prevalent, the relative importance of the individual creative act shrinks in proportion to the numerous conditions that must still be met to integrate it successfully into the processes of innovation. Science and the social, economic and political conditions that promote and support it, have become a seamless "innovation machinery"; relentlessly and powerfully at work in exploiting the outcome of individual scientific creativity through a multitude of ever more refined selective filters. The enthusiasm with which individual scientific creativity in the discovery of HTS was greeted and received is poignant in the face of the profound changes that today's science system faces. As indispensable as individual scientific creativity remains, it has become increasingly contingent upon what the innovation machinery of research will make of it – after each new breakthrough.

If it is true that we become the autobiographical narratives we tell about our lives, maybe science also becomes the multiple stories it tells about its working. But let us not forget the settings, nor the lessons that we may wish to draw from it.

Swiss Federal Institute of Technology

Notes

[1] Jerome Bruner, "Life as Narrative", *Social Research* 54 (1987), pp. 11–32.
[2] Helga Nowotny, *Time. The Modern and Postmodern Experience* (Cambridge: Polity Press, 1994).
[3] Bruner, "Life as Narrative" (cit. n. 1), p. 15.
[4] John Michon, "On the Modularity of Time", Ms (1997).
[5] Helga Nowotny, *Es ist so. Es könnte auch anders sein* (Frankfurt a. M.: Suhrkamp Verlag, 1999).
[6] Sharon Traweek, *Beamtimes and Lifetimes: The World of High Energy Physicists* (Cambridge, MA: Harvard University Press, 1988).
[7] Rouse, Joseph. "The Narrative Reconstruction of Science", *Inquiry* 33 (1990), pp. 179–96.
[8] Dominique Pestre, "Historical Narratives at CERN in the 1980s: Between Political Usage and Legitimization, between Physicists' and Historians' Narratives". Ms (1997).
[9] Sam S. Schweber, "The Empiricist Temper Regnant: Theoretical physics in the United States, 1920–1950". *Historical Studies in the Physical and Biological Sciences* 17 (1986), pp. 55–98.
[10] Stephen Hilgartner, "The Dominant View of Popularization: Conceptual Problems, Political Uses", *Social Studies of Science* 20 (1990), pp. 519–539; *Between Understanding and Trust: The Public, Science and Technology*, Meinolf Dierkes and Claudia von Grote, eds. (Berks.: Harwood Academic, 1999).
[11] Greg Myers, *Writing Biology: Texts in the Social Construction of Scientific Knowledge* (Madison: The University of Wisconsin Press, 1990).
[12] Peter Weingart, "Science and the Media", *Research Policy* 27 (1998), pp. 869–879.
[13] Helga Nowotny, and Ulrike Felt, *After the Breakthrough: The Emergence of High-temperature Superconductivity as a Research Field* (Cambridge: Cambridge University Press, 1997).
[14] Gerald Holton, Hasok Chang and Edward Jurkowitz. "How a Scientific Discovery is Made: A Case History", *American Scientist* 84 (1996), pp. 364–375.
[15] Holton, et al. "Scientific Discovery" (cit. n. 14), p. 375.
[16] Maurice Rice, "Explaining High-Temperature Superconductors", *Physics World* (December, 1999), p. 55.
[17] Helga Nowotny, "Individual Autonomy and Autonomy of Science: The place of the Individual in the Research System", in *The Research System in Transition*, S. Cozzens *et al.*, eds. (Dordrecht: Kluwer Academic Publishers, 1990), pp. 331–44.

PETER BUCK

A HEALTHY REGARD FOR THE FACTS

Abstract

This paper examines the history of 1920s hospital surveys and statistics, the agencies and individuals producing them, the uses to which they were supposed to be put, and the grand irrelevance into which they fell, almost as soon as they were generated.

Introduction

Everett Mendelsohn's classic 1963 article, "The Emergence of Science as a Profession in Nineteenth-Century Europe", is best read as a cautionary tale of the *plus ca change, plus c'est la meme chose* variety. That interpretation is true to the text, both in letter – the professionalization of science is characterized as a "second scientific revolution" that advanced the less than revolutionary cause of "providing aid to science without disturbing the structure of existing institutions" – and in spirit. As evidence of how and how far professional scientists, not to mention scientific professionals, accommodated themselves to the nineteenth-century institutional *status quo,* Mendelsohn cited the phenomenon of "national 'styles' in the support of science, 'styles' which often governed the very form that the scientific establishment took". He might have gone on to note (but he did not) that his own essay left the nations and nationalism tradition undisturbed too. Although the British, French, and German cases that form the body of the paper were clearly meant to be understood as adding up to one large story, the tale was told one country at a time, and in a way that underscored the stylistic contrast between France, where centralizing government reformers and metropolitan elites were the stars of the show, and Britain and Germany, where the scientific action was mainly in the provinces.[1]

The stories I have to tell in this paper have to do with hospitals rather than the laboratories, universities, and scientific associations whose development looms large in "The Emergence of Science as a Profession in Nineteenth-Century Europe". They date from the 1920s, not the 1800s; are notably short on references to anything approximating a scientific revolution; and come entirely

G.E. Allen and R.M. MacLeod (eds.), Science, History and Social Activism: A Tribute to Everett Mendelsohn, 137–151.

from one country, the United States, which (of course) did not figure at all in the European history Mendelsohn was concerned to write. But the line separating metropolitan professionals from provincial ones carries over more or less intact from his slice of the past to mine, where it turns up in the guise of a stylistic feature of medical life in the United States that post-World War I commentators on the past, present, and future of the country's hospitals noted: while national organizations located in big cities were the hall-mark of emergent professions, the venues where professional work actually got done were overwhelmingly small town and local.

What follows is an account of how that division of professional labor both "governed the very form" the American hospital establishment took, to use Mendelsohn's phrase, and kept other potentially interested parties – including, most notably, the Rockefeller Foundation – from taking either it or its nominal affiliates out in the countryside seriously – except as cases in point of how minor menaces could become major nuisances, if too much was made of their failings. It is telling in that regard that Rockefeller studiously refused to be drawn into a discussion of all the things that might be done for small town hospitals, given the right mixture of urban wealth, professional expertise, and rural experience. The foundation's officers had money to burn, but a review of the situation persuaded them that neither they nor local leaders rich in experience were likely to learn anything of day-to-day use from the most professional sounding experts in sight, a trio of Chicago based agencies – the American Medical Association, the American Hospital Association, and the American College of Surgeons – that purported to speak for the hospital industry *tout court*. It did not help that those organizations returned the compliment by offering to take Rockefeller's money and run with it, while urging the great majority of small town hospital operators to find other work, on the theory that America was over-supplied with minor medical facilities, most of which were so poorly managed, badly equipped, and underfinanced as to be beyond redemption.

"The Emergence of Science as a Profession in Nineteenth-Century Europe" describes a world where the professionally ambitious participated knowledgeably and to good effect in the construction (and reconstruction) of institutions that, in turn, served them well. As will be apparent, I have a more jaundiced view of professional empire-building in general, and no great regard in particular for images of the sort the hospital establishment's leaders sought to project when they represented themselves as experts who knew what they were doing and had a decent grasp of the setting in which they were doing it. That is to say, my sympathies are with those contemporary observers of the 1920s health and medical scene who thought otherwise, and were disinclined to believe that anything close to a credible or usable fact about hospitals was available from the AMA-AMC-AHA consortium, or from anyone else for that matter.

Lies, Damned Lies, and Statistics

Facts of other sorts were never in short supply. Most of them turned up as by-products of schemes for straightening out the political economy of medicine, so responsible providers of health care could make more money by delivering higher quality services at prices that were competitive with those of cut-rate rivals – quacks, patent medicine venders, and drug stores – who were suspected of making off with "huge sums of money" ($500 million annually by conservative contemporary guess). No serious observer underestimated the difficulty of pulling off all those tricks at once, or failed to note how the nation's estimated $2–3 billion capital investment in hospitals complicated the problem: the late 1920s Committee on the Costs of Medical Care was only repeating a post-World War I commonplace about medicine when it noted that larger expenditures on plant and equipment had produced higher rather than lower costs, "contrary to the trend in most other human activities". Nor was it hard for industry analysts to lay out the vicious circle that had to be broken – rising hospital prices meant more empty beds, more lost revenues, more unrecovered overhead and maintenance costs, and hence still higher charges – or appreciate the importance of caution when taking steps to reverse that spiral, lest the interests of valued constituents be unwittingly compromised.[2]

Getting a reliable fix on either those constituencies or their interests turned out to be no more – and no less – tricky than figuring out from which of its many excesses the hospital business most needed rescuing. In both connections the ideas and data on offer were contaminated by the petty professional jealousies that energized all organizations seeking to monitor post-World War I health and medicine. For once, the rivalries on display went together with potentially consequential disagreements about what was worth knowing and why, as the industry's major trade associations were heavily invested in different bodies of information. With a view to letting hospitals know they could no longer economize by exploiting interns while pretending to educate them, the American Medical Association opted for maintaining a comprehensive annual census of all the institutions, beds, patients, and doctors with hospital affiliations in the United States. The American Hospital Association and its house organ, *The Modern Hospital,* dealt in ostensibly representative examples of emerging trends in finance and administration – which was only fitting for an organization dominated by hospital superintendents who knew that the "wholesale" delivery of medical services and the "retail" practice of medicine required different skills. At the American College of Surgeons, meanwhile, it also seemed that "millions of dollars may be saved" – but by collecting "dependable facts" about actual medical practices, then setting minimum standards designed to trip up institutions that suffered from "lack of organization and effort rather than lack of financial resources".[3]

Those exercises in data management drew less than rave responses from the enterprises being whipped into shape for their own good. The ACS's first try at turning facts into standards, a 1917 survey of 692 hospitals, was a minor disaster that had to be suppressed when just eighty-nine of the institutions canvassed managed to pass muster. The AMA fared only slightly better with its efforts: less than twenty percent of some 6,400 facilities queried in 1919 felt any need for "the essentials of an approved internship": organized staffs, clinical and pathological laboratories, X-ray facilities, working libraries, and adequate patient records. As for the American Hospital Association, its leaders described themselves as in charge of an organization of organizations, and therefore committed to speaking for hospitals "in the same sense and to the same degree as various other national associations represent and aid their particular industry and their constituent members". But in 1919, when the AMA was sending out circulars to over 6,000 hospitals, the AHA had just 208 institutional subscribers. It also offered only limited services. Apart from arranging for annual conventions, collecting statistics, and drafting informational bulletins, the Association functioned mainly as a mail drop; its president was neither the first nor the last to acknowledge the meagerness of its contribution, as of the mid-1920s, to the "idea of a completely organized hospital field".[4]

Post-World War I philanthropists ought to have been well-disposed towards even hopelessly stalled schemes for rationalizing the American hospital's internal structure and functioning, since the commitments involved – to standardized administrative procedures, professional practices, and the like – echoed much lauded pre-war foundation approaches to the revamping of medical education. But on first inspection the hospital establishment's habit of building its reformist castles on factual sand put off even observers who found much to admire in the blueprints up for adoption. While endorsing "standardization" as obviously "desirable", and praising the ACS's "splendid" efforts to show larger medical facilities how the "efficiency of [their] work may be very materially increased", a dyspeptic Rockefeller Foundation review of the situation, as of 1919, warned against investing in a field whose experts seemed informed about everything except what services which hospitals were providing to whom. Reading 169 contributions to knowledge persuaded Rockefeller's Director of Surveys and Exhibits, Ernst C. Meyer, that America had more medical facilities than it quite literally knew what it was doing with. He found no "summary statistics" whatever on the "types of cases of sickness cared for", except in a very few very large institutions, and "almost no collected and organized knowledge" about anything connected with "the vast majority" of the nation's hospitals and "their needs, particularly the fundamental needs concerned with their relation to the community".[5]

The knowledge that health and medical facts were not what they should be was ordinarily no excuse for not getting on with good works. More than mere statistical fastidiousness had to come into play before prospective benefactors wondered how the work being done could be all that good, or even worth doing, if no one could say for sure who it was good for. The hospital industry's informational lapses seemed disabling mainly to observers who suspected that making efficient use of a medical facility involved more than just making it efficient. In a September 1919 speech, the Rockefeller Foundation's Secretary, Edwin Embree, issued a not very oblique warning to the AMA-AHA sponsored American Hospital Conference that modern philanthropists would have little patience with experts who so consistently "overlooked the broader community aspect of hospital service". Having sat still for two days while everyone "talked much", and then mostly "of hospital standardization, of hospital organization", he raised the inevitable rhetorical questions: "Is there a possibility of our studying too much existing institutions rather than the needs of communities for the institutions? In making studies of hospital facilities, do we often enough study them from the standpoint of the community?"[6]

Keeping that catechism simple at Rockefeller in 1919 meant limiting it to urban America, and stipulating that interested parties elsewhere in the United States were decidedly not to be encouraged to waste their time studying hospitals from community standpoints. Otherwise, the call for allocating resources on the basis of information concerning needs promised to be self-defeating, because of all the attention it would divert from "numerous important life-saving and health-improving" programs on offer to the countryside that were sure to "outrank hospital service" in effectiveness. Directors of surveys and exhibits who understood their craft took it as given that any fact-gatherer who set out to show rural communities what they did not need risked fueling a movement that would be "far easier to start ... than to stop". If urban experiences with appraisal work on other sides of the health and welfare street were any guide, a small town that had a "survey made with a view to the possible establishment of a hospital" was liable to find its "local health forces" overwhelmed by a tide of "favorable public sentiment" on behalf of such a facility, more or less regardless of what the evidence itself showed. As the head of the Russell Sage Foundation's Pittsburgh Survey, Paul Kellogg, suggested, twenty years after that pioneering 1907 event, communities that did not wish to "engage springs of action" were ill advised to "box the compass of their opportunity for constructive development".[7]

1919 America was rife with experts looking to help communities gauge their health and welfare requirements, so big city philanthropists did not have to look far to see that local forces were neither the only, nor even the most important, candidates likely to spring rural fact-finding traps on themselves or on others. Meyer's literature search turned up an especially alarming number of public health professionals eager to show how they could penetrate the informational blackout by turning small town hospitals into collectors and generators of useful data. Accordingly, the one "suggestion" Rockefeller made an explicit point of rejecting was a proposal, widely touted by New York state's health commissioner, Hermann Biggs, for combining rural medical facilities and county health offices into complete "health centers" – under the supervision, it emerged, of Biggs himself – that were supposed to be the "mechanism" of the future for "surveying community conditions and needs in matters of health and sanitation".[8]

Specialists nearer the center of the hospital business had their own problems with the informational riddle on their small town flank. No one connected with the AMA-AHA-ACS combine needed to be told that rural enthusiasms were apt to be both misguided and difficult to squelch when aroused. It was likewise hardly news that nothing good would ever come from too much intercourse with state health departments. But history and logic conspired to shelter the industry's professionals from the chill wind of Rockefeller's warning not to confuse small towns with too many facts about local needs. Worries along those lines could hardly be expected to resonate with experience for trade association leaders who had just spent the better part of a decade failing to get the attention of more than a minuscule – and mainly urban – fraction of their nominal followers. Moreover, elementary economic reasoning of the sort specialists on hospital finance specialized in offered a clear warrant for taking that slice of the past as a guide to the ebb and flow of the only wave of appraisals likely to crest in the future. While big cities considering new medical projects could "afford to employ a hospital expert, or corps of experts, to make preliminary surveys" before proceeding, that was "obviously out of the question for rural communities".[9]

It was, however, equally out of the question for all self-respecting purveyors of facts and figures to go quietly into the night. The same fiscal calculus that explained why America's small towns would "never be able to command the services of high-priced hospital experts" also triggered predictable calls for actions that would keep those "cross-roads communities" from seeking assistance on the cheap elsewhere. Aware of competitors waiting to claim a share of the appraisal market, but mindful that the supply of genuine experts was almost as limited as the demand for their good offices, industry analysts sized up the opportunities for "missionary work" among the small hospital heathen, and announced that the "vast majority" could save their own souls, provided the right proselytizing agencies pointed a representative sample of sinners in the right direction. That modest proposal removed any semblance of paradox from Rockefeller's remarks about small town medical facilities and the lack of "collected and organized knowledge of their needs, particularly the fundamental needs concerned with their relation to the community". Repeated without attribution or quotation marks, those well chosen foundation words anchored a 1919 *Modern Hospital* editorial on how "the crying need" of the day was for "an investigation of the hospital needs of various rural communities, with a view to discovering some general principles which may be applied in any specific instance".[10]

Crying needs seldom made their way into hospital debates until someone with access to a podium had hit on a way to claim credit for taking the problem in hand. The ink was still wet on *The Modern Hospital's* version of Rockefeller's call for new knowledge when the magazine's managing editor, Mary Katherine Chapin, returned from a whirlwind tour of three upper Mid-West towns, armed with "faithful reports of personal observations" – and a three-part essay on "Some Rural Communities and What Their Hospitals Mean to Them" – that showed the universal local need for data could be uniquely well met by a trade journal willing and able to serve as a simple clearing house for information. Her AMA-ACS-AHA colleagues had, she thought, "underestimated" the level of small town interest in defining "the proper place of the rural hospital in the rural health program". So many districts were already "working on one or another phase" of the question, as of late 1919, that nothing appeared to be holding back the emergence of a "rural hospital policy" with a "definite shape", except the absence of "some means for coordinating the scattered efforts of various communities" and drawing "guiding principles" from their "collective experience".[11]

Unlike most of the square pegs and round holes hospital professionals spent the idle hour dreaming up and proposing to match with each other, Ms. Chapin's problem and the solution she had in mind for it were close enough to being real for their disqualifying theoretical flaws to be consequential: her forecast of coming attractions on the rural center stage simply did not fit with prevailing views of how and where small town hospital policy was to be made. Industry analysts were more enthusiastic about coordinating other people's work than being guided by it; consequently, the only principles they looked forward to extracting from experience were ones they anticipated imposing on it beforehand. Stock accounts of urban progress via standardization framed expectations of glad tidings from small towns served by not-for-profit hospitals with organized medical staffs, professional superintendents, and lay boards; outfits so structured appeared sure to have the technical resources quality medical care required: well-equipped laboratories, X-ray facilities, and the personnel to run them. Conversely and by the same token, any reported scattering of effort across the land could only be a measure of how hard put institutions out there were to bring or keep themselves up to snuff. The AHA set aside large blocks of time at its annual conventions for big city standard bearers to meet their country cousins on that assumption: "without proper education and supervision", rural hospitals seemed certain to either fall "prey" to pernicious local influences— "the unsupervised work of the undereducated physician" being the prime example – or "vegetate" from lack of "proper and frequent contacts with larger hospitals and bigger minds".[12]

Nowadays, professional journals connected with the hospital trade are almost always well advised to speak for the establishment they report to, year in and year out, rather than the hinterlands they report on only intermittently, if at all, because the world beyond today's organized pale is a genuine wasteland where undocumented health and medical facilities are unlikely to be good, and vice-versa. But *The Modern Hospital* was neither joining an obviously lost cause, nor just spoiling for a fight, when its managing editor embarked on a mission at odds with the operating agenda of her magazine's parent organization. No poor relation of a low budget trade association could afford to disregard a hint, however ambiguous, from so well heeled a potential patron as Rockefeller that a practiced reporter might be well rewarded for taking an interest in a few communities and what their hospitals meant to them – as distinct from offering to discuss what those communities might mean to their hospitals. With that prospect clearly in mind, and in anticipation of collegial objections that all the "guiding principles" worth having were already known, Ms. Chapin introduced her findings with a pointed foreword, whose import neither Embree nor Meyer would have missed, about how "the technique of hospital administration [had] become so absorbing" that specialists in the field often approached the subject as if it were "a self-centered world in itself".[13]

As at Rockefeller, so around *The Modern Hospital*, self-absorbed experts and the self-centered worlds they invented for themselves to be expert about were easier to skewer than improve on. Like Embree and Meyer, who never got beyond urging otherwise underoccupied professionals to study hospitals from "the standpoint of the community", Mary Chapin returned from her travels with no overarching conclusions about the American countryside, except that there were "facts" out there from which individual small towns might draw "any lessons which may be useful to them". Nonetheless, the "sketches from life" she drew cut hard against professional obsessions with organization charts and institutional standards: on her showing, any number of idiosyncratic success stories, and more than a few tales of instructive failures, could be told about even "a few representative communities" and their hospitals, even – or especially – by someone who knew all the preferred tricks of the trade and gave every evidence of having done her best to apply them to the situations at hand.[14]

Nor were the anomalies involved limited to secondary features of the conventional wisdom. The primary categorical questions establishment experts were learning to ask about rural and small town medical facilities – whether they were proprietary or not-for-profit, professionally administered or not, and up to or short of the accepted mark *vis à vis* technical resources – provided Ms. Chapin with the most grist for her mill. By almost all her accounts, no composite picture followed from those distinctions that corresponded with direct observations of how different institutions served different communities well or poorly. She found "resentful and suspicious" owner-operators of private shops running frankly commercial enterprises that nonetheless provided "a high grade of medical as well as hospital service", along with non-profit hospitals filled with doctors who had no interest in improved operating rooms and better laboratories. Examples likewise abounded of community institutions that had tried to organize their medical staffs, but wound up only adding to the mutual "distrust and jealousy" animating rural physicians, just as contrary instances showed untrained nurse-superintendents of proprietary facilities bringing employers to heel by exploiting the fact that the doctor "needed the hospital more than the hospital needed the physician".[15]

The possibility that such a seemingly inexplicable mix of good, bad, and indifferent outcomes might actually be inexplicable was not to be contemplated. But facts with so much promise to stimulate circulation and foster discussion were also not to be ignored by editors of trade publications who knew that small town hospitals were likely to outnumber big city ones for a long time. Mindful of that important home truth, and hewing to the program Rockefeller appeared to have laid down in 1919, *The Modern Hospital* spent the next several years encouraging local correspondents to pick up where Ms. Chapin had left off, and submit reports on how needs were being met in their districts. As of September 1924, the idea that the varieties of rural American medical experience could be traced back to the diversity of rural American society seemed sufficiently compelling for the magazine to round off its offer of a prize for the best essay on "hospital-community interrelationships" with a special call for entries from small towns where, it was anticipated, "the competition should arouse considerable interest".[16]

Apart from confirming establishment suspicions that there was no accounting for taste when it came to rural estimates of what qualified as a good hospital, the contributions from the field, both to the contest and more generally, made no noticeable impression on the larger minds in and around Chicago. Establishment attention remained mired where Rockefeller had found it in 1919, in the congenial debate about what it would take to build the kind of hospital an American could be proud of, or at least prosper in, which amounted to the same thing. Nothing in the competing mixes of fact and fancy available from the modestly expanded mid-1920s *Modern Hospital* warehouse aroused any sustained interest in the inverse question: how the nation's medical facilities could be adapted to the needs of the United States as it happened to exist at the moment.

It would be wrong, however, to accuse the trade associations of resting on their oars. Although unable to win more than a few small town converts to the campaign for an organized industry, and unmoved by *The Modern Hospital's* efforts to keep tabs on America's rural medical facilities in all their individual glory, AMA-AHA experts were busy trying to count the house. The statistics they gathered added an important dimension to the professional consensus that the minor institutions slipping through the establishment cracks could not be safely left to follow their own paths to perdition. The decade of the Great War had, it seemed, brought the United States too many medical facilities for its own good – 5,037 hospitals with 532,481 beds, according to a 1918 AMA survey, up from 421,065 beds and 4,359 hospitals only ten years earlier. Contemporaries reviewing that "staggering mass of figures" saw a zero-sum game impending in which resources tied up in one community's bad, small medical facility would be lost to larger and better institutions elsewhere. As a 1919 AHA spokesmen put it, the nation's "vast hospital investment" showed all the earmarks of becoming "a dead weight on the development of the future".[17]

Five more years of steady expansion failed to assuage those fears; the still greater totals for 1923 – 6,380 facilities, 755,722 beds – proved just one thing: "notwithstanding recent progress in many branches of hospital work, hospitals, all too frequently, just happen". It did not help that many of those happenings appeared to be occurring on very small scales. A 1923 Census Bureau survey found fifty-six percent of America's hospitals had forty beds or less, against twenty-three percent with over 100. The same report also placed half the nation's general purpose hospitals in rural districts or towns with fewer than 10,000 people. Worse yet, places in that demographic category seemed to be gaining instead of losing ground. AMA data for 1926 showed fifty-six percent of the counties in the United States with community hospitals, compared to forty-four percent in 1920; 383 counties had entered the lists in half a decade.[18]

Despite those impressive figures, trade association spokesmen were poorly positioned to say whether 1920s America ran to a shortage or a surplus of hospitals. Then as now, no one had a convincing formula for calculating how many institutions and beds the United States needed. Moreover, changes in coverage from pre- to post-war surveys accounted for much of the growth that seemed so alarming. The 1909 AMA census registered only non-government hospitals, and excluded institutions for nervous disorders and tuberculosis, as well as outfits "known to harbor quacks or to be in other respects unreliable". The figures for 1918 and 1923 were less discriminating: besides "facilities for the general care of the sick", they included the "special" TB hospitals, mental institutions, and government facilities left out of the earlier survey. Those additions made a difference. Government institutions contained sixty percent of the nation's hospital beds, in part because mental hospitals – mostly state or county owned – accounted for forty percent of them.

Knowledgeable commentators ordinarily had no difficulty distinguishing "special" facilities from "general" ones. But the inflated totals and exaggerated rates of expansion that resulted from lumping the two classes together suited AHA and AMA purposes. Dramatizing the post-war proliferation of institutions and beds helped explain away the failure of more than a few hospitals to follow establishment prescriptions for progress toward the ideal of an organized field. Pre-war comparisons revealed the laggard majority's shortcomings for what they were: consequences of that "astounding growth" no one wished to claim credit or accept blame for.[19]

Moreover, a demonstrated ability to gather data from as many enterprises as possible could be counted on to enhance a surveying organization's image of being in close enough contact with a large enough number of hospitals to speak for the industry as a whole. The rewards in heightened self-assurance, prestige, and sheer visibility that such canvasses yielded to their sponsors were all out of proportion to the resources invested in them, since they were usually done on the cheap, i.e. by mail. That meant even less heavy lifting than was involved in *The Modern Hospital's* self-strengthening program. The trade associations sent out questionnaires, the outfits being polled filled in the forms, returned them, then went back to work, while the record-keeping agencies counted responses, added up beds, patients, and dollars, performed the obvious calculations, and issued reports.

With no one checking the raw data, spokesmen for the hospital establishment sensed that their statistics left something to be desired, especially since so many of the institutions reporting in belonged to that class of small operations which the trade associations did not trust to do anything right.[20] But despite being of suspicious origin, the facts on offer looked incomplete rather than imperfect; better data were expected to show nothing very different, just more of the same. Nor was it hard for experts to spot premonitory indicators of motion toward the much desired stable state where institutions more likely to respond to authoritative guidelines characterized the field. The majority of America's medical facilities might be quite small, but "the average general hospital" seemed rather large. "That much talked of institution" turned up over and over again. Summarizing the Census Bureau's 1923 findings, *Hospital Management* editor, Matthew Foley, reported "Uncle Sam's idea" of the norm: seventy-one beds, 1,292 patients a year, daily costs of $4.81 per patient, and an annual deficit of $2,348.92. It did not matter that Census Bureau statisticians were skeptical of their financial data and surrounded them with responsible cautions – only seventy-five percent of the canvassed hospitals reported incomes and expenses; unreliable accounting practices meant "defective" returns "in many cases". Foley quoted the disclaimers, then repeated his characterization of the survey as "more comprehensive" than previous ones, and so a good guide to "average conditions".[21]

Finding average conditions corresponding to hoped for ones sharpened urban worries that the increasing number of rural institutions clustered around the small hospital median would turn into dead weights on the field's progress toward the larger hospital mean. 1920s authorities issued periodic bulletins disparaging a pattern of development that the AMA described as "greater in quantity and in capacity than ... in effectiveness". Experts warned small towns against saddling themselves with facilities built out of misplaced community pride or in response to the impulsive wishes of wealthy benefactors, the self-interest of local surgeons, or the calculated desires of local businessmen looking to evade their responsibilities for employee medical services.[22]

That was where *The Modern Hospital* came out too. February 1925 found the magazine awarding first prize in its essay competition on "the interrelationships of hospital and community" to a sometime New York City welfare administrator, former secretary of the Wisconsin State Board of Education, and newly installed Dean of Marquette University's Graduate School and director of its hospital college, Edward A. Fitzpatrick, for a paper that opened with a scathing account of the "social loss" due to the misdirection of money into medical facilities and services for which "no social need" existed. Communities stuck with such unwise investments were to blame, it seemed, for allowing architects "to build monuments to their architectural genius"; for letting heads of medical staffs "over-emphasize or under-emphasize phases of a complete essential program", according to their "personal predilections"; for permitting "wealthy donors" to dictate the "character and extent" of service "merely on the basis of their whim, or desire, or presumption"; and so on through ten paragraphs of very fine print documenting "the chaos of the present situation, the accidental or whimsical determination of major aspects of hospital service, the drifting, which is characteristic of much of our social life".[23]

Readers who got past that extended bill of particulars may well have thought they were also back where *The Modern Hospital* came in. Although based on little more than a year's scouting out of the hospital business, Fitzpatrick's indictment of the national "failure to realize the fundamental character of community need" showed a firm grasp of the key lesson the journal's editors had learned half a decade earlier: no account of drift, whim, accident, and chaos was complete until the origins of the mess had been traced to a shortage of data. As in 1919, so circa 1925, it seemed that America was off the hospital track because no one had assembled "verifiable facts" on the basis of which the nation's cities and towns could make informed estimates of their hospital requirements.[24]

There was another sense in which it was *deja vu* all over again at *The Modern Hospital:* the fact-gathering device its award-winning essayist proposed as the antidote to chaos was the same social and community survey procedure Rockefeller had tried to sell the hospital establishment on half a decade earlier. But the second time around the message had a harder edge, since it came with a sharp comment attached on the folly of amateur tinkering with the basic apparatus, *á la* the magazine's own recent trimming of the survey sail to fit the cut of its journalistic cloth. Five years after Mary Katherine Chapin confidently forecast that the "scattered efforts" of America's small towns would converge toward a rural hospital policy with a "definite shape", once a catalog of instructive stories was available for inspection, Fitzpatrick dismissed that whole idea with a one sentence remark about "the experiences of other people in other places" being no substitute for "the actual facts in the local situation". Rockefeller's emissaries to the field would have appreciated both the distinction and the comparison he drew to support it. Americans were building too many medical facilities of the wrong sort in the wrong places, it seemed, because no one was even beginning to do for hospitals what experts on other subjects had long since done for schools, churches, banks, stores, and various other "community factors": namely use social surveys as bases "for comprehensive planning of community needs".[25]

The Future of An Illusion

It is just as well that 1920s hospital facts and figures were widely understood to be so flawed as to be wholly unusable; otherwise, even more data might have been collected, in which case there would be no avoiding the temptation to see whether, in retrospect and with the usual benefits of hindsight, the various worlds hinted at in the literature can be made to add up to something. As it is, although a more or less plausible hospital past can be reconstructed from the information on offer, the argument for resisting the urge to get on with that job pretty much makes itself, once the task is recognized for what it entails – i.e. picking and choosing among contradictory findings that no one on the spot put much stock in – and what it leaves undisturbed, viz., the image which the hospital establishment sought to project of its information gatherers as experts doing their best to meet demands for information about the industry's present state and future prospects.

That said, there is or ought to be no doubt that keeping everyone uninformed about goings on in the hospital field without encouraging too much misguided foraging about for local data took some expertise. Giving credit where credit is due to that part of the past is mainly a matter of stopping where most newcomers preparing to strain the evidential soup started, and reflecting on the lessons for the future that they were invited to learn – or see to it that others learned – from the one indisputable fact around: namely, that the professionals monitoring post-World War I America's hospitals always seemed to have their data structured precisely so they could specialize in professing ignorance about the contexts in which the institutions under their purview had to function.

Contemporaries who wrapped the hospital statistics mystery in that enigma tended to think of both riddles as just waiting to be cracked with the hammer Fitzpatrick wanted his new-found colleagues to try. This was not surprising, since social and community surveys impressed almost everyone as the model of choice for "applying social research to community progress" in 1920s America.[26] By 1925 industry analysts were themselves near the head of the parade when it came to explaining why an "orderly and scientific survey" was preferable to "the higgledy-piggledy, inefficient way" cities and towns were in the habit of appraising their hospital requirements. As with the weather in general, however, the fog obscuring those needs and opportunities made for good conversation mainly because no one was doing anything about it. A slice of the reality behind the rhetoric surfaced when just thirteen reports on "hospitals and sanitoria" – and none of the AMA-ACS-AHA conglomerate's multiple contributions to knowledge – qualified for inclusion in the standard guide to post-World War I surveys: a 1928 Russell Sage Foundation compendium that, informed reviewers were quick to note, documented how many of the genre's "most notable" examples (nearly 500 of the 2,775 items on the list) concerned health and medicine.[27]

It would be nice to think the wholesale omission of the hospital industry's self-appraisals from a catalog of "fact-finding studies made as a basis for social action" showed that someone, somewhere had tumbled to the two enduring truths about agencies claiming professional expertise in the field: their statistical brews contained little in the way of facts, and were seldom concocted with a view to promoting social action. But it is sufficient to note the gulf – which Russell Sage's catalogers would have been more comfortable discussing – between keeping track of America's hospitals and mapping the rest of the health and medical landscape according to the social survey rule of thumb whereby all useful charts were local and keyed to "problems and conditions having definite geographical limits and bearings".[28] That aspect of the division of expert labor reveals more about the post-World War I hospital industry than all the reams of statistics and other odd facts that spokesmen for the trade tried to share with an unresponsive public, but eventually had to settle for handing on to future generations of gullible social scientists.

Harvard University

Notes

[1] Everett Mendelsohn, "The Emergence of Science as a Profession in Nineteenth-Century Europe", in *The Management of Scientists,* Karl Hill, ed. (Boston: Beacon Press, 1963), p. 4, p. 32, and *passim.*

[2] Ernst C. Meyer, "Relative Value of Hospitals and Dispensaries as Public Health Agencies and as Fields of Activity for the Rockefeller Foundation", Rockefeller Foundation Archives; *Medical*

Care for the American People: The Final Report of the Committee on the Costs of Medical Care (Chicago: University of Chicago Press, 1932), p. 13.

3 Rosemary Stevens, *American Medicine and the Public Interest* (New Haven: Yale University Press, 1971), pp. 118–31; "Seventeenth Annual Session of the American Medical Association: 'Victory Meeting' at Atlantic City, June 9–13, Celebrates Advances in Medicine and Surgery Made During the War", *The Modern Hospital,* XIII (1919), p. 53; Andrew R. Warner, "Development and Progress in the Field of Hospital Administration", *The Modern Hospital*, XIV, p. 177; Pliny O. Clark, "Community Funds for Maintenance and Capital Expenditures", *The Modern Hospital* XV, p. 413; Meyer, "Relative Value of Hospitals and Dispensaries as Public Health Agencies" (cit. n. 2).

4 A.R. Warner, "The Future Tasks and Problems of the American Hospital Association", *The Modern Hospita* XIII (1919), p. 255; John E. Ransom, "Review of the Work of the A.H.A. during 1924", *The Modern Hospital*, XXIV (1925), p. 9.

5 Meyer, "Relative Value of Hospitals and Dispensaries as Public Health Agencies" (cit. n. 2).

6 "Remarks of Mr. Embree at American Hospital Conference, September 1919", Rockefeller Archives.

7 Ernst C. Meyer, "Hospital Service in Rural Communities", *JAMA*, 72 (1919), p. 1136, p. 1462; Paul U. Kellogg and Neva R. Deardorff, "Social Research as Applied to Community Progress", *First International Conference of Social Work* (Paris, 1929), I, pp. 794–5.

8 Meyer, "Hospital Service in Rural Communities" (cit. n. 7), p. 1135; "Memorandum on the New York Health Center Bill", *The Modern Hospital*, XIV (1919), pp. 382–3.

9 "The Growth of the Hospital Idea", *The Modern Hospital,* XIII (1919), p. 125.

10 *Ibid.*, p. 125.

11 Mary Katherine Chapin, "Some Rural Communities and What Their Hospitals Mean to Them", *The Modern Hospital,* XIII (1919), pp. 374–78.

12 Margaret J. Robinson, "Increased Efficiency is the Keynote of Small Hospital Section", *The Modern Hospital*, XXV (1925), p. 425.

13 Chapin, "Some Rural Communities" (cit. n. 11), p. 374.

14 *Ibid.*

15 *Ibid.*

16 "The General Program for a Prize Essay Competition on 'The Interrelationships of Hospital and Community' Being Held by *The Modern Hospital* and *The Nation's Health*", *The Modern Hospital* XXIII (1924), 44–46; "Registration for Prize Essay Competition Closes September 15", *The Modern Hospital* XXIII (1924), p. 237.

17 "The Growth of the Hospital Idea" (cit. n. 9), p. 124.

18 "The General Program for a Prize Essay Competition on 'The Interrelationships of Hospital and Community' Being Held by *The Modern Hospital* and *The Nation's Health, The Modern Hospital,* XXIII (1924), pp. 44ff. Martha J. Megee, "Application of Social Service to the Problems of the Small Hospital", *The Modern Hospital,* XXIV (1925), p. 105.

19 "The Growth of the Hospital Idea" (cit. n. 9), p. 124.

20 As late as 1935, a Public Health Service study in fact found major discrepancies in the accounts that hospitals gave of themselves. One out of every five institutions reporting bed capacities twice in one year sent in figures differing by ten percent or worse. One in ten showed deviations over twenty percent. Hospitals with twenty-five beds or less produced especially marked inconsistencies. One-third could not or would not tell within ten percent how many beds they had. Reports of average daily occupancy rates varied even more. Over a third of all the hospitals surveyed returned figures off by at least ten percent from one report to the next. In almost half those cases, the margin of error exceeded twenty percent. As for the small facilities, three out of five did not know within ten percent how many people per day they treated. See, Joseph W. Mountin, Elliott H. Pennell, and Emily Hankla, "A Study of the Variations in Reports on Hospital Facilities and their Use", *Public Health Reports*, 53(1) (1938), pp. 17–25.

21 Matthew O. Foley, "Finish Hospital, Dispensary Census", *Hospital Management*, XIX (1925), pp. 30–33.

22 Homer F. Sanger, "Hospital Facilities and the Medical Profession in the United States", *The Modern Hospital*, XXIV (1925), p. 465.

23 Edward A. Fitzpatrick, "Interrelationships of Hospital and Community", *The Modern Hospital*, XXIV (1925), p. 133.

24 *Ibid.*

25 *Ibid*, pp. 133–4.

26 Paul U. Kellogg and Neva R. Deardorff, "Social Research as Applied to Community Progress", *Proceedings of the First International Conference of Social Work* (Paris, 1928), I, p. 793.

[27] Joseph J. Weber, *First Steps in Organizing a Hospital* (New York: Macmillan, 1924), p. 18; Allen Eaton and Shelby M. Harrison, *A Bibliography of Social Surveys: Reports of Fact-finding Studies Made as a Basis for Social Action; Arranged by Subjects and Localities* (New York: Russell Sage, 1930); Kellogg and Deardorff, "Social Research as Applied to Community Progress" (cit. n. 26), p. 793, pp. 795–6.

[28] Eaton and Harrison, *A Bibliography of Social Surveys* (cit. n. 27), p. xxiv.

JEAN-JACQUES SALOMON

THE DARK SIDE OF PROGRESS

Abstract

The issue of progress as a feature of human history has come under question in the west since the Enlightenment, but at no time more profoundly than toward the close of the second millennium. The "end of History" was prophesized and a *malaise*, increasingly associated with the growth of science and technology came to replace the unquestioned faith in progress that loomed large in the nineteenth and first half of the twentieth centuries. While there is no reason to reject the possibility of progress, especially in the moral and social realm, it is no longer possible to assume it will happen, as post-Enlightenment thinkers so often did, as an inevitable product of developments in science and technology. The link between scientific and technological progress and the sort of social/ moral progress envisioned in the Enlightenment, is clearly no longer realistic. However, history is still an open book and the idea of progress especially social progress not dead. But clearly for such progress to occur the same amount of time and attention must be paid to understanding the principles of social science that has previously been lavished on the natural sciences.

Introduction

As for any man or woman with a real culture and a deep interest in what is going on in the world, there are several simultaneous or overlapping individuals in Everett Mendelsohn. This is not meant to diminish his importance as an historian of science, but just to insist on the many facets of the man beyond his professional aura and achievements. I do not remember exactly when and how we met, most likely before I was teaching in Cambridge for the first time at MIT (1968). We met indeed during many other occasions between Harvard where I lectured and CNAM (Conservatoiare National des Arteset Méaatoers) where he lectured, and during the International Congresses of History of Science which took place in Tokyo and Moscow. At that time, I had already created with Derek de Solla Price the International Council for Science Policy Studies (ICSPS), then attached to the International Council of History of

G.E. Allen and R.M. MacLeod (eds.), Science, History and Social Activism: A Tribute to Everett Mendelsohn, 153–166.

Science, which in turn was part of the International Council of Scientific Unions (ICSU) network.

During the Kyoto Congress, Derek and I fought to obtain for the ICSPS an autonomous status within the ICSU, the "science policy studies people" appearing as rivals to the formal "historians of science" (for instance, Thomas Kuhn argued that we might more easily, because of our policy orientation, be supported by administrations, government or industry and thus appear as competitors to the historians in terms of raising resources, which in a sense was not wrong). The ICSPS indeed became an autonomous organization, whose members were economists, sociologists, historians and policy-makers as well, but I personally always insisted that historians of science should be well represented, if only to defend the cause – felt to be threatened – of history of science (my own first training) before the powers-that-be who invest money in scientific research activities. This is how Everett became one of the first obvious American representatives in the ICSPS and later on its President for several years.

The ICSPS used to meet every two years in a capital city, whether in the East or the West, which was not meaningless from a political standpoint during the Cold War: we were certainly less instrumental in peace-making or arms control than the Pugwash Conferences (of which, in fact, many of us were also members), but we were able at least to bring people together from very different horizons, regimes and ideologies which, in spite of some tensions, led to open discussions and often helped specialists to breathe fresh air (and some to escape). I remember, for instance, a meeting as early as the late 70s in a country house belonging to the Polish Academy of Sciences during which we felt for the first time that the communist system was breaking down as we witnessed the differences of opinion expressed in exchanges among the representatives from the East. And it was during one of these meetings that I discovered how deeply Everett was involved, as an associate of the Quakers and a citizen, not only in protest against the Vietnam war, but also in the attempts to facilitate peaceful exchanges between Palestinians and Israelis – and thus how close we were not only from an intellectual but also a social point of view.

In brief, I discovered, behind or beyond the brilliant academic, whose prestige as an historian was already great in France (especially through his friendly relationship with one of my teachers at the Sorbonne, Georges Canguilhem), that Everett was committed to see a world whose social and political posture was not doomed to be subjected to injustice, wars and massacres – a man of progress, who believed that history can evolve in directions that do not simply imply a negative balance-sheet. Another of my Sorbonne teachers was Raymond Aron, who was of the same generation as Canguilhem. They were close to each other for many reasons: they came from the same "grande Ecole" in the rue d'Ulm, both had fought for a better world against fascism and Nazism (during the war, Aron in London and Canguilhem in the French Resistance), but both were rather sceptical as to the view so often shared (and often with arrogance) by scientists that the road to new knowledge is necessarily the road to social and political progress. We often talked, Everett and I, about this "European scepticism" whose roots are deeply entrenched in culture as much as in history, as illustrated, for instance, by Husserl's *Kritik* or Musil's *Man without Quality*, so basically different from the average American's optimism about technical change. It is in remembering our discussions on these topics that I present this paper, part of thoughts I developed more in a new book on science and technology in this century – and beyond.[1]

To start with, it may seem paradoxical that the most recent (and devastating) denunciation of the notion of progress in fact came from the United States. This could be a repeat of the debate of the 60s on the "end of ideologies" that I evoked then in these terms: "Already at the first trumpet blasts which were to sound the knell of the 'ideological age', Raymond Aron noted that this controversy, however different its sense in the United States, could boast of a long history in Europe: In turning back from ideology, he wrote, the American anti-ideologists did not turn very far; some of them merely turned back from Europe".[2] Did the end of the cold war, on one side, and on the other the proximity of the new millennium suffice to make an American view so akin to the traditional scepticism and criticism of progress as illustrated in Europe by a longstanding tradition of poets and philosophers?

The fact is that, in 1996, the French journal *Commentaire* published an article by William Pfaff, a commentator on the *New York Herald Tribune* and a philosopher in his spare time, under the title: "Du progrès: réflexions sur une idée morte" [On progress: reflections on a dead notion]. As if echoing the pessimism then prevalent in France (and in Europe) – a pessimism which fortunately seems overcome today – the newspaper *Le Monde* opened its columns during the month of August 1996 to a debate on this theme. "What if there was no progress in the past?" wondered William Pfaff. "And what if there was no reason to suppose that the future would be better than the present or, worse still, better than the past? This is a crucial question, and particularly pertinent on the eve of the third millennium. The end of the first millennium disappointed the general expectation that it would be the Last Judgment (and the end of time). The second millennium ended with disarray in the secular optimism of modern times, and with the affirmation that the course of History is unpredictable".[3] But even if the hymn to progress is no longer intoned with the same conviction, and if the intellectual models used to analyze it are clearly discredited, should we go so far as to argue that it is a myth and an obsolete idea?

The Ideology of Progress

We inherited from the 18th century the idea that advances in the natural sciences would inevitably lead, as the physical conditions of life improved, to the perfection of human beings and society. The ideology of progress that sustained the Enlightenment basically affirms that progress in knowledge and techniques has a moral dimension such that human beings will one day be better than they are now, both individually and collectively. In other words, this means seeing History as a transition from an inferior to superior state not only of mind, but also of social organization and moral behavior. From Kant to

Marx via Hegel and Auguste Comte, but also for many liberal thinkers, there was throughout the 19th century an obsession with History conceived as political progress inextricably linked with advances in knowledge and techniques, as scientific discoveries and rationality would gradually overcome poverty and injustice, and ultimately guarantee universal democracy.

Further back than the Enlightenment, we should remember that this vision of History as Progress, which is so typical of the West, goes back to Judaeo-Christian origins. Most other cultures have developed without any sense of history, let alone of progress. On the contrary, some religions reject any belief in a historical sequence, with stages or changes in the course of existence conceived as a "story" with a beginning and an end. For Hindus, existence is circular, a perpetual return; Buddhists try as far as possible to free themselves from events, and therefore from history. By contrast, the distinctive feature of the biblical story is that it has a beginning and a direction, that it is teleological, leading somewhere, and once the end is reached humanity is supposed to give an account of itself. Our life on Earth has a sense, even more so if the promises about the after-life influence, perhaps even determine, the events in which human beings take part. Ultimately, human destiny is intelligible, shaped in the same way that human beings will have prepared for their own salvation. Thus the whole history of the West is imbued with the idea of progress, especially since the Middle Ages, which represent, as Jean Gimpel has rightly argued, the first industrial revolution, thanks to the mills, monasteries, abbeys and castles.[4] This is then the eschatological vision of Judaism, Christianity and Islam, driving their followers towards mastery of matter and energy with a determination that the mathematization of nature after Galileo was to make increasingly effective.

Following the scientific revolution in the 17th century, and even more the industrial revolution in the 19th century, this teleological vision of history lost its religious basis and, in the most extreme form of positivism, arrived at the notion that science is the only means of understanding the world and saving humanity. I hope I may be forgiven for rushing through the transition from the religious approach to progress to the secular version of the 19th century. But I must also mention the American and French Revolutions, which insisted – each in its own way – that the steady removal of constraints on human freedom was what History was all about. To this the debates about progress later added references to the theory of evolution, the source of social darwinism and eugenics, which were to inspire the pseudo-progressive ideology of Nazism, and to historical materialism, the source of the radical revolutionary passion that inspired the pseudo-progressive ideology of the communist regimes. During the two centuries that I have compressed so drastically, it is clear that the idea of progress was constantly presented in debates about the economic, social and political future of societies, first in the West and then in the developing countries that were colonized by them, as if it were a matter of the simplest possible equations: progress equals civilization, and civilization equals progress.

Thus the first frontier to the idea of progress was drawn by this confusion between the conquests of scientific rationality and the course of the historic destiny of humanity. Many of the illusions (and hence also the disillusionment) of the twentieth century, marked by so many massacres and disasters, arose from this confusion which, it has to be said, was deliberately maintained by philosophers and political leaders. Just as Hegel saw Napoleon as the ultimate embodiment of the "universal soul", how many 20th century intellectuals have hailed Lenin, Stalin, Hitler or Mussolini as accomplishing the historic destiny of either the proletariat or the nation, and in both cases as the very model of progress?

In fact, it is not the idea of progress as such that makes it a myth, but the claim based on it of control over the laws of history – the great mystery being, as François Furet said, why so many intellectuals succumbed to the myth. A pathology with universal dimensions in the case of communism, nationalist dimensions in the case of fascism, the two ideologies that have dominated the history of the 20th century constantly "heightened their effects by making their partisans more fanatical: the test of power, rather than softening them, merely increased their misdeeds and crimes. Stalin exterminated millions of human beings in the name of the struggle against the bourgeoisie, while Hitler killed millions of Jews in the name of the purity of the Aryan race". If, as François Furet argues, there was a mysterious evil shaping political thought in the 20th century, the contribution of the idea of progress should not be minimized.[5]

The second frontier of progress – and it is well to remember this, because it brings out the bias in our attitudes as Europeans or Westerners – is drawn by the difference between universality defined in logical terms and in geographical terms. This point, made some time ago by someone very close to me, is essentially that the universality claimed by Western science only exists and works within the framework of institutions based on Western rationality.[6] From a geographical standpoint, the idea of Progress is not accepted universally in so far as the application of Western rationality is unknown, unrecognized or often even attacked in many developing countries. It is clear that the irruption of the idea of progress in these countries, through imperialism and colonization, did real violence, and the memory of it still underlies the misunderstandings, resentment or fascination that these countries feel toward the West; the theme of progress spilled over into a vocation to "bring civilization" to the unenlightened, since by definition the opposite of progress was stagnation, corruption, at worst barbarousness. Yet the gap between the universality of progress in logical as against geographical terms was so great that to close it required recourse to gunboats, slavery or even, in the case of China, the deliberate, massive spread of opium.

The Two Approaches to Progress

This view of progress is, of course, not in the least objective; it is based on the passions and biases inherent in intellectual and political-ideological debates. It is hardly surprising that, after the defeat of Nazism and the implosion of communism, some think that the triumph of the market means that we have come to the end of History: liberalism having won a knock-out victory, we can expect nothing further of the future, and the end of the 20th century marked the end of both history and progress. I return to this notion, put forward by Francis Fukuyama immediately after the fall of the Berlin Wall, if only to demonstrate that it was merely another kind of illusion or ideology generated like other fantasies by the approach of the third millennium.[7] It is nevertheless important to remember that while in the 19th century, Marxists and liberals were convinced that History was heading toward the Good, there were nevertheless some individuals, mainly poets like Baudelaire, Milton or Keats and a few philosophers, who, far from singing the praises of progress, distrusted and criticized it. Baudelaire, for example, wrote: "Mechanization will so thoroughly americanize us, progress will cause our spiritual sides to atrophy such that nothing, in the bloody dreams ... of utopians, will be comparable to these positive results".[8]

One can in fact go back to Rousseau to find the source of the most profound critique of progress, and it is not a coincidence if, in all the contemporary criticisms of industrial society, from Heidegger to Lyotard or Debord or Baudrillard, one can still identify an echo of the themes dear to Rousseau: nostalgia for primitive Man, respect for nature, an indictment of "advances in arts and sciences" that devour human beings and society, denunciation of the damaging effects of urban living, and so on. All thinking about the alienation of human beings in society inevitably starts and returns here: progress is artificial, it distances us from the natural world and therefore from "real" life. And the more that progress takes the form of an increasing number and range of manufactured consumer goods, the more the attack on progress as the source of artefacts turns into an indictment of the loss of control over our environment suffered by humankind.

Nonetheless, objective views of progress do exist. They underlie, for example, the economists' calculations of the growth of output and productivity. Indisputable conclusions can then be drawn as to the rise in the standard of living and quality of life: the material improvements cannot be challenged. In this sense, the "great hope of the 20th century", to quote Jean Fourastié, has in fact been fulfilled in *economic* terms in so far as we take account of "the facts, but only the facts about production, consumption, hours worked, health and life expectancy".[9] The Marxist concern with worsening poverty lost its force in the period of unprecedented economic growth after the war, and while it is true that since the 1970s the industrialized countries have again had to cope with moments of intense crisis as fresh tensions were generated by unemployment and the gap between rich and poor widened, no one can challenge the undoubted material improvements that have increased in number and become accessible to many more people thanks to science and technology. Criticisms of progress from the Left do not attack the notion as such, but rather the misallocation of its benefits and the distortions of a system that maintains (and often increases) pockets of worsening poverty in wealthy societies.

If we look at the long term, the most impressive indicator of technical progress is the almost tenfold increase in per capita income in the last 200 years. Yet this purely economic indicator gives no hint of the improvements in individual and collective well-being that have come as a result of this substantial rise in incomes: longer life expectancy, falling infant mortality rates, eradication of certain diseases, higher standards of education, faster communications, improvements in living conditions, guaranteed social protection, greater opportunities for leisure activities, mass access to tourism and culture, etc. Whatever the inequalities that remain and that have indeed increased since the 1970s, and however large the pockets of poverty in wealthy countries, the overall outcome in *material* terms is clearly impressive. In fact, if one focuses narrowly on economic indicators, the same can be said for many developing countries, where living conditions have undoubtedly improved, at least for certain sections of their populations: one has only to recall the results of the "green revolution" in India or the economic growth of Brazil since the 1960s.

Another objective view of progress, again measured purely statistically, is given by the comparative indicators of investments and output in science and technology. I am referring in particular to the studies undertaken since the 1960s, based on numbers of articles in scientific journals and, more recently, comparative measures of technical innovation derived from patents and licences. In this area, progress appears to be an incontrovertible reality, not in the least obsolete in that it is reflected in an unprecedented and constantly growing number of articles, journals and scientific meetings which must surely generate *an increase in knowledge and power*. In addition to the combination of new knowledge and techniques, this objective view is clearly manifest in the updating of the techno-economic paradigms on which the industrial revolution has been based since the beginning. It is now obvious that we are witnessing a new phase in the history of industrialization, which is shifting us out of a technical system based exclusively on energy and natural resources, and into one that is intangible, dominated by information flowing through communications networks, biotechnologies and new materials. Mankind has never before had access to such efficient, intelligent and multipurpose tools as those that the revolutions in information technologies, genetic engineering and new materials provide and are likely to generate in increasing numbers.

Yet it is here in the ideological framework of progress that we encounter another frontier, which gives rise to controversy all the more readily because the indicators available to us can do no more than record statistics. For example, it is not enough just to count the number of articles published in order to assess the output of a scientist, a laboratory or a nation; it is important also to take account of the content of the publications, their contribution in terms of fresh information or even discoveries. In the same way, quantitative data on output and productivity will continue to be "black boxes" as long as they are interpreted using no other methods or approaches besides economics in its narrowest form, for instance by examining the structure of businesses, how they are organized and managed and, at national level, undertaking studies of systems of education and training, or the use and distribution of skills and élites, etc. The truth is that once one leaves behind the kinds of fact measured by economists, the evaluation of progress is less straightforward; it becomes a matter of subjective judgment and describes only particular situations within specific social categories, since the whole social question comes down to asking who benefits from progress.

Furthermore, our economic indicators are incapable of assessing the social costs and drawbacks of economic growth and technical progress (for example, for the environment). Nor can they take account of the enormous problems created by unemployment or the need to adapt to the different skills required by the new technologies. The economists' quantitative data cannot measure the gains whereby new knowledge and techniques broaden and deepen our understanding of nature and ourselves – yet this is what above all should be reckoned as progress. But they are also unable to evaluate the losses, in the transformations forcibly brought about by technical change, that correspond to the disappearance of a heritage of skills, culture or amenities which may have as high a cost for the future of humanity as (in the sphere of natural phenomena at a global level) the loss of biodiversity.

Progress in a Suspicious Age

For these reasons, the notion of progress is henceforth liable to be treated with caution: not only has the straight road leading to greater knowledge and material improvements deviated from the even less straight roads to "happiness" and "greater morality", but it also clearly involves costs, upheavals and dangers out of all proportion to the supposed (or vaunted) advantages resulting from the transformations and innovations. People now prefer to talk about technical change and economic growth, as if to eliminate from the notion of progress any hint of concomitant ideological presuppositions. The idea that the more science and material progress there is, the more humankind must advance inexorably towards a better life is more in line with the illusions of the 19th century than the disillusionments of the 20th.

Yet, among the disillusionments of the twentieth century, the great increase in awareness of the harm caused by the growth process has had as its corollary a questioning not only of technology but even of science itself: both are now perceived as equally implicated in the nature and extent of the harm done. In the past, the ideology of progress coincided so neatly with the outlook of the scientific establishment that whenever there were real threats of accidents or catastrophes, the blame was laid on technology, engineers, managers or their underlings, but never on the scientists themselves. In the eyes of the fundamental research worker, indeed, scientists merely produce knowledge, and knowledge is as neutral as Aesop's tongue, i.e. whether it turns out to be good or bad depends on the use made of it. For example, Fritz Haber, Nobel laureate in chemistry, who was responsible for introducing poison gas as a weapon, was never accused of war crimes (of course such a concept did not exist at that time); on the contrary, even though he was Jewish, he was showered with favors by the Nazi régime.

It is indeed this image of the "pure scientist" and with it attitudes to scientific progress, that have changed. In the past, scientific activities could flourish and take the risk of being shown to be the cause of, or a contributor to, terrible consequences. Now, science is virtually expected to prove its innocence before going any further. This is the real meaning of what the Americans were the first to call *technology assessment*: an evaluation of technology, perhaps, but with a sense of being on trial. At this trial, science is now as much in the dock as technology. None of the changes that have occurred since the 19th century seems to me as significant as this challenging of science resulting from the wartime destructions and massacres, the harm caused by growth and the dangers connected with major technological risks. From the poison gas of World War I to the gas chambers of World War II, from Seveso and Bhopal to cloning, from oil slicks to risks of nuclear contamination, from the mercury at Minimata to the fall-out from Chernobyl, it is clear that in addition to the industrial massacres and the long-term effects on the environment there is now another factor: the possible consequences for the genetic capital of plants and animals, above all of the human species – consequences that are irreversible and transnational, if not global. And how can one avoid seeing links between the anxieties arising from modern scientific advances and the shadow of uncertainty and ultimate catastrophe cast by the nuclear arsenals?

The painter Goya said that the sleep of reason creates monsters, but now we know that reason also creates monsters when it is awake. As well as the natural cataclysms that human beings have always known they could do nothing to stop, there are now technological catastrophes of mankind's own making, using its most sophisticated knowledge. To quote Robert Oppenheimer, "physics has known sin" after Hiroshima and Nagasaki. At that time, this admission of an uneasy conscience shocked most scientists, who felt that since it had not been their decision to use the bombs, they were not in the least responsible for the consequences: for instance, in Ernst O. Lawrence's words, it was quite simply "unprofessional" to express scruples.[10] Since then, the problems raised by the long-term management of nuclear waste that nobody knows how to deal with, or by the nuclear reactors built on the same model as Chernobyl that are still functioning in the ex-USSR, the worries about global warming or about techniques of cloning mammals that clearly can (and one day surely will) be applied to humans – all force us to reflect upon the limits that society can or should place on the conduct of scientific research.[11]

On the one hand, the institution that is the most striking embodiment of rational investigation – and hence of progress – is now wondering "how far is too far". On the other, the concern to have some social control over research has burst into the deliberations previously reserved exclusively for scientists, as is clear from the creation of various Committees on Ethics at the national and international levels to which non-scientists and sometimes laymen and laywomen contribute. Science the endless frontier, as it was defined by Vannevar Bush at the end of World War II, is now confronted with frontiers built up by human reactions, even resistance, and various social institutions. So the idea of boundless progress that underlay the rise and the success of science as an institution is turning into an idea that limits should be imposed on scientific activities. And this not only comes from outside, but also from within the scientific institution. Take, for instance, the remarks of the great physicist,

Freeman Dyson, of the Institute for Advanced Studies at Princeton, showing how an uneasy conscience has gradually developed within the scientific community: "My scientist friends may justly protest that the calamities of American society are caused by drugs, or by guns, or by racial intolerance, or by illiteracy, or by bad schools, or by broken families, rather than by science. It is true that the immediate causes of social disintegration are moral and economic rather than technical. But science must bear a larger share of responsibility for these evils than the majority of scientists are willing to admit. When we look at historical processes on a time-scale of fifty or a hundred years, science is the most powerful driving force of change ...".[12]

The Never-Ending Story

Perhaps we should eliminate from this notion of progress all the elements that appear not to be related to science and technology, as Paul Valéry did in his lifelong interest in the subject: "I used to try to take a positive view of what is called Progress. Once any moral, political or aesthetic consideration had been removed, Progress seemed to me to be reduced to the rapid and highly noticeable growth of (mechanical) force available to mankind, and to the accuracy of the forecasts that mankind can achieve".[13] For myself, I am far from sure that reducing progress to these two factors – mechanical force and mathematical precision – does justice to a notion that has been a real step forward in human development. But there is no need then to succumb to the siren voices of transcendency or faith: after all, the consequence of mammals' brains becoming increasingly complex was the human species, even if we are still part animal, as Freud tells us. When indeed we focus on ourselves, how is it possible to eliminate all moral, political or aesthetic considerations? It is from this angle that, in spite of all those who proclaim the end of history, or a process of constant renewal or decline, the idea of progress does not yet seem to me to be dead.

The end of the century is not the end of History in the sense argued by Francis Fukuyama, the American follower of Hegel and his French commentator Kojève, from his analysis of the collapse of communism and the triumph of capitalism. In this radical shift, in which both Left-wing and Right-wing dictatorships fell around the world and the numbers of free-market democracies increased, Fukuyama attributes a decisive role to science. On the one hand, science brought about the defeat of fascism by means of weapons and then of communism through economic means; on the other, the uniform impact that science now has on the future of all societies means that, with free-market democracy left alone on the stage, History must be at an end.

Nobody should underestimate the role played by science and technology (or, to be more precise, the use made of scientific knowledge and technical innovation) in these political upheavals. But one may question – as I do – whether the human story is really going to stop with this happy ending. Besides, Fukuyama himself was not entirely convinced, given that in the conclusion of his essay he wondered whether the coming together of freedom and prosperity would not condemn humankind to such boredom that people would be driven to start wars just in order to have something to do – or to allow scientists and engineers to pursue their research!

The paradox of the West, the cradle of modern science, with the accumulation of discoveries and industrial applications that science helped to multiply, is precisely that it is now having to think about the real costs of its achievements, the considerable risks involved, the gulf between the force now at our disposal and the wisdom (or lack of it) shown by human beings. It is true that the experience of the past century, which has been barbarous with the help of (and sometimes in the name of) science, raises doubts about the link between advances in knowledge and advances in human morality postulated during the Enlightenment. The "malaise of civilization" is obvious, and while it is nothing new to readers of Freud or Husserl, it has never before generated so much agonized debate. The end of the century also marked the twilight of the second millennium, which leads some to think that the end of the world is at hand, as happened at the end of the first millennium – admittedly in a completely different context – in large part because of the importance of science for all aspects of life today.

In the religious climate of 1000, the end of History was expected to be Judgment Day marking the start of either the kingdom of heaven or eternal damnation. In the secular climate of today, the issue is no longer the redemption of mankind, nor even our moral or metaphysical salvation, but simply our survival in the shadow of nuclear arsenals, threats to the environment, nuclear waste and the greenhouse effect. This is what the Last Judgment means for our times and it generates all manner of apocalyptic warnings. Moreover the great fear was that our computers could not recognize the date 2000 and would treat the start of the new century as if it were simply going back to the decades already gone ! The fear – and I take it to be a highly revealing sign of our times – did not deal with the end of the world and the urgent necessity to prepare one's own salute, but with the possible collapse of our computerized systems: a revealing fantasy of the Golem specialist of marketing and public relations selling the necessity to rejuvenate the old or buy new softwares! In brief, industrialized societies have simply substituted the notion of the vulnerability of technology to that of the vulnerability of human soul. No other age – whether we take as our point of comparison the century that is now ending or, even more striking, the whole of the last millennium – has seen such enormous upheavals resulting from progress in science and technology, but in spite of these advances and performances, the most sophisticated technical systems appear to be running the risk of breaking down for extremely silly reasons.

Nevertheless, the millenarian fears about the uncertainties of the 21st century should not create the illusion that science, any more than History, will stop here, and consequently the idea of human progress – however ramshackle it may now seem – will not stop either. Even Fukuyama in a more recent paper has discovered that, as long as science will continue to add new knowledge, his end of history will be postponed and mankind will not be doomed to face the reign of the Nietzschean "last man" forever purged of heroism and *megalothymia* [14] Clearly nobody now can say whether we are heading for a better world, even less whether human beings will behave more morally than before. We Europeans are hardly in a position to preach to others, after a century that started and ended in Sarajevo: what happened and what is still happening in the former Yugoslavia is enough to prevent one from believing that there is any likelihood of human beings becoming perfect. And yet, can one not argue that, nevertheless, in the (very) long term and in spite of – some would say because of – the huge cost in sacrifices and victims, there are some signs of progress in social evolution?

From One Renaissance to Another

If one takes a long-term view, it is very tempting to see a parallel between the upheavals during the Renaissance and those of today: the growing legitimacy of modern science in the course of the 16th and 17th centuries is echoed in the rise of technology in the 19th and 20th centuries, each of these long periods being marked by terrible wars and massacres, but also by intellectual revolutions. In the Renaissance, people had to cope with the advent of printing just as now we have to cope with computers; they fantasized about the exploration of the New World as we dream of extra-terrestrial space; they threw themselves into all kinds of different sects before and after the Reformation, just as we are now seeing a simultaneous resurgence of a longing for faith and of religious fanaticisms; they engaged in dreadful wars just as we, after two intra-continental civil wars, are confronted with tribal conflicts even within Europe; they sought unity under the domination of empires in the universality of the Church whereas we now seek unity under strong economic constraints to counter globalization and fear of domination by the United States.

This parallel is, I admit, daring. Nonetheless, if there has been progress since the wars of religion, lasting long after the French Revolution, it has been via a greater tolerance, an acknowledgment of the legitimacy of different forms of religious conviction. This does not mean that religious conflicts have disappeared even in our own societies, nor (as we see with every passing day) that this recognition of others' rights is universally shared. The spirit of tolerance is nevertheless demanded in our societies and included in the United Nations Charter, and it is significant that war between European democracies has become inconceivable since 1945. In the same way, it is impossible to deny the step forward represented by the almost universal recognition of human rights, the new procedures for consultation, arbitration and trial set up at the international level, the legal proceedings undertaken against some war criminals, the early stages of a law about intervention in another nation's affairs, or the efforts made between Rio de Janeiro, Tokyo and Buenos Aires to deal with the global threats to the climate.

We have had a splendid illustration of these developments in the arrest in London of General Pinochet, hounded down by the memory of his victims not just in Chile, but in Spain, Switzerland, France and elsewhere. The final outcome hardly matters, whether the senile general will or will not escape his trial – it was and it will ever appear as an opportunity to demonstrate that a universal conscience does exist and places the law above nations and their sovereignty. One of these days (but when?), all states will indeed be subject to a law that is above them: this is inherent in the treaty establishing an International Criminal Court as a permanent tribunal, with the whole planet under its jurisdiction. Admittedly, none of this is yet certain, and the treaty suffers from many loopholes that compromise its universality. And what is still going on in Kosovo, Rwanda or Congo shows that a lot remains to be done. However, even if the gains are still fragile, it is hard to imagine that they will now be overturned. And while it will certainly take a long time for the international community to accept the need to implement systematically the measures that the individual nation states have nevertheless agreed to, it remains the case – like an "idea of Reason" in the Kantian sense – that societies are moving toward a world order that is gradually, if very slowly, recognizing the need to defend human rights, the founding principle of the United Nations.

From this viewpoint, as from many others, History remains open-ended, which is why I find the notion of "the end of History" even more debatable than that of the end of progress: it means taking for granted that humanity has a destination for which it has the key, and if that leads somewhere, the direction taken will be guided objectively thanks to what can be revealed by science, reason or faith. It is hardly surprising if the massive changes since 1989 have given some people the impression that the future has become unpredictable – as if, prior to the collapse of communism and the spread of free-market democracies, it had ever been predictable! It is one thing to claim to give history a direction, to attribute a logic to events as if they signified the irreversible advance of humanity toward collective moral or political progress; it is another to be tempted to believe or to proclaim that the paradise of liberal democracy will remove all the difficulties, risks and setbacks, and will make it redundant to seek knowledge and power.

A page has been turned, but the book remains wide open to new chapters, and we cannot know their extent or their contents: the fact that the story is open-ended may also mean that the human adventure is not without hope. The dark side of progress teaches us some humility in spite of the considerable scientific and technological achievements of the past century. It is a lesson that justifies the extension of the principle of complementarity to human affairs, as Robert Oppenheimer, following Niels Bohr, has done in several of his lectures. This extension is not unwarranted in our years of pseudo "science wars", if one remembers how Emilio Segré, who heard in 1927 Bohr's famous lecture in Como as a young engineering student, explained later: "Two magnitudes are complementary when the measurement of one of them prevents the accurate simultaneous measurement of the other. Similarly, two concepts are complementary when one imposes limitations on the other".[15] And thus similarly, in the new forthcoming century, civilization will have to face the simple fact that endless progress implies also limitations on progress.

To be sure, the disappearance of totalitarian dictatorships has not meant the eradication of the violence, passions and barbarism of which individuals, groups and nations are capable. No more has it eliminated, despite the uncertainties, the ambiguities and disappointments of progress, the reasons – even if they are illusory – why some people strive ceaselessly for a better world. There is a kind of other-worldliness in this idea that liberal democracy might mark the end of the human adventure, and this other-worldliness seems to me no less debatable than the positivism revealed by some scientists, when they assert that everything is susceptible to reasoning or that only the reductionism of the scientific method can encompass humankind, nature and the world. The notion of progress may have been discredited, but we shall not avoid the unpredictable and equivocal advances of science and technology, any more than we shall avoid the uncertainties and surprises of history. Yet there is no reason not to suppose that, in the long term, the efforts of certain people, among them Everett Mendelsohn, will astonish future generations as not having been in vain.

Conservatoire National des Arts et Métiers

NOTES

[1] J.-J. Salomon, *Survivre à la science – Une certaine idée du futur* (Paris: Albin Michel, 1999).

[2] R. Aron, "Fin des idéologies, Renaissance des idées", in *Trois Essais sur l'âge industriel*, Paris: Plon, 1966), p. 198, quoted in J.-J. Salomon, *Science and Politics* (Cambridge: MIT Press, 1973), p. 160.

[3] W. Pfaff, "Progress", *World Policy Journal*, 12 No. 4 (Winter 1995/96) (translated from the French, *Commentaire*, Eté 1996, vol. 19, no. 74).

[4] J. Gimpel, *The Medieval Machine: The Industrial Revolution of the Middle Ages* (London: Penguin Books, 1977).

[5] F. Furet, *The Passing of an Illusion: The Idea of Communism in the Twentieth Century* (Chicago: University of Chicago Press, 1999).

[6] C. Salomon-Bayet, "Modern Science and Coexistence of Rationalities", *Diogenes*, Unesco, No. 126, April-June, 1984.

[7] F. Fukuyama, *The End of History and the Last Man* (New York: The Free Press, 1992).

[8] Baudelaire, *Fusées*, XXII.

[9] J. Fourastié, *Les trente Glorieuses* (Paris: Fayard-Pluriel, 1979).

[10] J.-J. Salomon, "Terror and Scruples", Introduction to *Science, War and Peace* (Paris: Economica, 1989).

[11] See J.-J. Salomon, "Le clônage humain: où est la limite?" *Futuribles*, 221 (June 1997) and chapters VIII and IX of *Survivre à la science* (cit. n. 2).

[12] F. Dyson, *Imagined Worlds* (Cambridge: Harvard University Press,1996), pp. 103–104. I have shortened the quotation from the passage where Dyson does not hesitate to blame science for the lay-offs resulting from lack of training in IT, the low pay scales for unskilled youth, the families with computers that become "hereditary castes", the children of families without computers becoming more and more marginalized, and hence – with no hope of jobs – encouraged to join gangs and be tempted to commit crimes.

[13] P. Valéry, *Regards sur le monde actuel*, Sur le progrès.

[14] F. Fukuyama, "Second Thought: The Man in a Bottle", *Los Angeles Times* (translated in *Le Monde*, "La post humanité est pour demain", n. 5, July–August 1999). We find the same questioning about the possible creation of a "new man" thanks to the progress of biotechnology in Peter Sloterdijk's lecture that provoked a scandal in Germany (French translation, "Règles pour le parc humain: Réponse à la lettre sur l'humanisme", *Le monde des débats*, Supplément, Octobre 1999).

[15] E. Segrè, *From X-Rays to Quarks* (New York: W. H. Freeman, 1980), p. 167.

PETER WEINGART

THE LOSS OF DISTANCE: SCIENCE IN TRANSITION[1]

Abstract

Distance is a prerequisite for the production of scientific, i.e. certified, trusted knowledge. The paper analyzes how the distance of science to politics and the media is eroded and instead a closer coupling is created by pressures to legitimate the scientific enterprise in a mass democracy. This loss of distance may have repercussions for the institution of science.

Introduction: The function of distance

The end of the century or even the millennium is the occasion for diagnoses about the fundamental change of society. Two of these seem to contradict each other. One postulates a transition of the industrial society to the *knowledge society*.[2] The other one claims a hardly less fundamental change of the *mode of knowledge production*, from the academic to a 'postnormal' or simply to a 'mode 2'.[3] The former suggests, among other things, a general scientification of virtually all sectors of society, a widespread diffusion of scientific knowledge replacing other kinds of knowledge, and the emergence of knowledge production as the primary productive force.[4] The latter claims the loss of institutional identity of academic science as the university loses its monopoly of knowledge production; the scientific disciplines lose control over the direction and quality of research, as much knowledge is produced in the context of application. Thus, political, economic and other criteria gain importance in determining the quality of research. As academic knowledge production comes under pressure of societal legitimation, institutional science becomes increasingly oriented to societal values, political goals, and media representation.[5]

These diagnoses agree in one respect. Scientific knowledge is central in the modern industrialized society. However, they disagree in another important respect. Propagandists of the knowledge society tend to regard science as a separate and superior type of knowledge, while proponents of the new modes of knowledge production stress the trivialization of academic knowledge and its ultimate dilution into general social practice. Dualisms of this kind are

G.E. Allen and R.M. MacLeod (eds.), Science, History and Social Activism: A Tribute to Everett Mendelsohn, 167–184.

rarely long lived and should not be taken too seriously but they do serve to sharpen the analytical perspective. In the following, I want to bring the two diagnoses together. My argument is that the implicit (sometimes even explicit) romanticism of the second view overlooks the crucial mechanism by which scientific knowledge develops: what will here be termed *distancing*. The first view, the thesis of the ongoing transition to a knowledge society, fails to take into consideration the apparent paradox of the success of science: the scientification of society does not leave the institutional identity of science untouched. My purpose, thus, is to trace significant recent institutional developments of science that are believed to occur *between* the extremes of the complete dissolution of science as the central institution of knowledge production on the one hand, and its institutional identity on the other. The focus of my analysis will be on the function of *social distance* as the crucial precondition for the production of superior and certified knowledge.

Distance, Differentiation and the Communication of Truth

Modern Western societies are functionally differentiated. Historically, the transition from their predecessors, the stratified societies, took place during the 17th and 18th centuries. The emergence of modern science as a differentiated specialized and functionally specific system was part of this process. The gist of this process was not, as the lore of science has it, that some mysterious new methodology was found that fulfilled the prerequisites of the discovery of nature and provided the key to reading the book of nature's secrets. Rather, it was the gradual emergence and social stabilization of a particular mode of communication that was devoted to the distinction between *true* and *false*. Science in its modern form, thus, constituted itself, as well as its distinction (and its boundaries) from other functionally differentiated systems.[6] Ever since then, political or juridical issues are not settled by science, nor are scientific issues settled by majority vote or the verdict of just or unjust.

The historically relevant question which has been a central theme for historians of science is: which particular social conditions made this development possible? This is also the question sociologist of science Robert K. Merton once asked, giving two kinds of answers. One was strictly sociological. By implication, it addressed the historical problem: how could communication be expanded socially and geographically and be stabilized over time, and it identified the social prerequisites of the production of 'certified knowledge'[7] The more historical-sociological answer tried to explain the contingent development, i.e. the emergence of the *new* science of the 17th century, by unveiling the underlying religious and economic motives that formed the social frame for the new mode of knowledge.[8] The particular motives of Merton's analysis, and the peculiar conditions of its reception, (especially the anti-functionalism of the 1960s) have diverted attention to questionable empirical claims and epistemological assumptions that were attributed to the norms of the scientific ethos.[9] A fresh look at this study reveals, apart from its theoretical complexity, its historical basis. Daston's *moral economies* of science are de facto a historical description of Merton's sociological, functionalist analysis.[10] Historians of science interested in the history of objectivity have reconstructed the emergence of scientific communication and identified the interplay between *collective* discourses on true and objective knowledge and trust in people and in institutions.[11] And they have described the process of differentiation of that communication from the everyday context, i.e. the differentiation of a systematic and focused critique of truth claims from the psychological and social attributes of their proponents. Porter has termed this type of social innovation the "technology of distance", and Fuchs has demonstrated how the historical accounts can be reconciled with sociological theory.[12]

Merton's formulation of the 'ethos of science' may thus be interpreted as the sociological condensation of the extended process of the differentiation of modern science. He has countered the criticism directed against it (mostly by sociologists) by identifying two "strategic research sites": 1) the analysis of apparent paradoxes in the behavior of scientists and the resulting psychic and social conflicts;[13] 2) the study of typical causes of attacks against science or conflicts between science and its social environment respectively.[14] The second is close to our present concern especially if one disregards the immediate occasion of the original writing, i.e. the threat of Fascism and Communism in the 1930s and 40s.

One conflict is implicit in utilitarian expectations of science. It arises, according to Merton, because of disappointed expectations of the utility of scientific knowledge, and because of the renunciation by scientists of any responsibility for social and/or economic consequences their knowledge may bring. The more pronounced the orientation to knowledge for its own sake, the more endangered is the social esteem and legitimacy of science.[15] This hypothesis has proved to be prescient: after decades of lavish funding of basic research under the 'social contract' forged immediately after the war, this arrangement has come under pressure. First, the demand on scientists to reflect on the social consequences, above all with respect to the impact of technology on the environment, and to assume responsibility at least in terms of widening the scope of analysis, has led to an entirely new arena of research, namely the assessment of science and technology. Second, in all Western industrialized countries, the appropriation of funds to science has increasingly become attached to utilitarian expectations, a development which has triggered the very diagnoses of a changing mode of knowledge production.

The second type of conflict that 'indicates' the existence of the ethos concerns the core operating principle of science, that Merton termed organized scepticism. This tends to bring science into opposition with political authority, religious symbols and other firmly established institutions. This conflict is by no means limited to the various forms of abuse and suppression under totalitarian political regimes such as Merton had in mind. The same conflicts occur in liberal-democratic contexts, though in less radical form. One pertinent example is the opposition to the values of universalism, objectivity and political neutrality and the propagation of particularistic criteria connected with the political movements of the late 1960s until today.[16] Postmodernism has been associated with the values of a ‘counterculture’ which gives preference to personality, authenticity and creativity without standards. Ezrahi has rightfully pointed to the irony that the science criticized by this movement once was inaugurated to “free men from the authority of the past, the authority of the transcendental and the authority of dogma”.[17]

For Merton, conflicts of this nature were indicators of the existence of the scientific ethos. For the purposes of my own analysis these observations indicate a partial *loss of distance* between science and different sectors of its social environment. Contrary to trendy loose language suggesting a ‘blurring of boundaries’, i.e. a highly unlikely reversal of the secular process of functional differentiation, I want to claim that something more complex and interesting is happening. The more than tenfold growth of the knowledge production sector in half a century, the resulting pervasive presence of science in every niche of society, the permeation of virtually all aspects of everyday life with scientific language, thinking, symbols, and derived artifacts characterize what is now called the knowledge society. Science as a mode of knowledge production and as an institutional arrangement has been so successful that it has become the main structuring force in modern societies. Thus, it is to be expected that the institutional boundaries will not remain identical over time. But this is not to be mistaken as a reversal of functional differentiation. Instead, what can be observed are particular *close couplings* between science and politics, the economy, the media, and law. These new arrangements have peculiar repercussions on knowledge production, on notions of true and false, certain and uncertain, and on the demarcations between science and non-science. In the following, I will focus on just two of these, the couplings between science and politics, and between science and the media.

The Science – Politics Coupling

Scientific Expertise and Political Legitimacy – A Paradoxical Relationship

On March 28, 1979, the “Three Mile Island” Nuclear Reactor at Harrisburg, Pa. suffered a major accident. Commentator Walter Cronkite opened that evening’s CBS news by saying that this had been the “most confused day in the history of the news media”, and then went on presenting the contradictory information from the site of the accident. In fact, this marked a fundamental change in the public recognition of the authority of scientific experts and the certainty of scientific knowledge. Seven years later, almost to the day, a much more severe accident occurred at a nuclear power plant in the Ukrainian city of Chernobyl which, in effect, dealt the death blow to nuclear power in most European countries. A similar spectacle unfolded. Politicians and scientific experts alike engaged in media debates which culminated in a politician’s ostentatiously eating salad in front of the national TV audience in the futile attempt to demonstrate that vegetables were not affected by the Cesium fallout. “Three Mile Island” and “Chernobyl”, became symbols not only of the vulnerability of nuclear power technology but also of the loss of authority based on reliable knowledge and unanimous expertise as well as a loss in credibility in politicians who relied on scientific authority.[18] Since then, many other incidents involving other areas of knowledge or other technologies have revealed the same pattern. They have been given the label of the *risk society.* The safety of recombinant DNA, the ethics of reproductive technologies, the application of biotechnology in agriculture and in food processing, the ethical, political and economic implications of sequencing the human genome, the depletion of the ozone layer, the implications of CO_2 emissions for anthropogenic climate change, the transferability of mad cow disease (BSE) to humans, are all cases around which major public controversies have evolved. Their common features continue to be the uncertainty of scientific knowledge, disagreement among experts, alignment with opposing political positions, and public distrust of scientific expertise. And more generally speaking, the controversies revealed the enhanced role of scientific knowledge in *legitimating* political decisions, as well as its potentially *de-legitimating* function, i.e. its threat to economic and political interests.

Reactions to the two nuclear accidents among the politicians and scientists reflected strongly held concepts about the nature of scientific knowledge and its role in policy-making. The vision of a linear relationship between a perceived problem, a set of proposed solutions based on certified scientific knowledge, and subsequent political decisions instructed by such knowledge, proved to be illusionary. The political landscape had changed and become more complex. Today, few remember that at the beginning of the modern era of *science for politics* (i.e. an institutionalized role of scientific advisors in politics) the most pressing issue was the unaccountable and thus illegitimate influence of scientific experts in democratic governments.[19] The fear of such an influence had some justification as long as scientists commanded an elite status and scientific knowledge an unchallenged authority. But 'Three Mile Island' and 'Chernobyl' symbolize the turning point. On the one hand, science-related policy issues had become civilian rather than military and were, thus, exposed to public attention. Perhaps even more important, however, was the crucial change initiated by the accidents: the general accessibility, in principle, and factual use of scientific expertise across the political spectrum. This 'democratization' of expert knowledge has calmed the fears that experts can undermine the democratic mandate.[20] At the same time, it has radically altered the conditions of the science-politics relationship.

This relationship has been conceived in two simplified models that represent the two sides of the legitimation dilemma with which policy-makers and their experts are confronted in modern mass democracies. In principle, decisionist policy-making (i.e. policy-making with little or no regard to established knowledge) suffers from a rationality deficit, whereas technocratic decisions (i.e. such that are virtually determined by knowledge) suffer from the lack of legitimating public consent. Current democracies operate on the basis of "*two* circuits of legitimation of public decisions".[21] Decisions have to be *rational* in light of existing knowledge, and they have to be made by representatives of *delegated* power. The legitimating authority for the former is science which brings experts into governments, for the latter it is public consent which is normally obtained in elections.

At least three basic assumptions underlying the technocratic and decisionist models do not stand up to empirical test. These are (1) the supposed linear sequence of (political) problem definition, (expert) advice and (political) decision, (2) the value free concept of scientific knowledge, and (3) the disinterestedness or political neutrality of scientists. Neither the technocratic notion of the separation of 'truth' from power can be upheld, nor the democratic or decisionist model carry sufficient legitimacy. A host of empirical case studies have corroborated the recursive nature of the advisory and decision-making processes. The model of scientific advice to politics that exemplifies these interdependent processes may be dubbed the *recursive model* of the science-policymaking relationship. Instead of the linear chain of 'problem perception, expert advice, political decision', the process has to be modeled as a recursive loop. The model, thus, entails the following stages: (1) The perception of the problem may come either from the scientific community or from policy-makers. (2) In the political process it is transformed according to political criteria of relevance. (3) As a political program funding research for further clarification of the initial problem, it is handed back to the scientific community. (4) The scientific community, in turn, executes the pertinent research whose results become the basis for continuous adaptation of the initial problem perception.[22] This is an idealized reconstruction, which means that the different stages are not necessarily neatly separated either institutionally or temporally.

The gist of the model is that it reflects the close coupling or, to put it in a dynamic perspective, the reciprocal nature of two processes: the *scientification of politics* that goes hand in hand with the *politicization of science.*[23] Mention of just one case, the 'Ozone-debate', must suffice as an illustration.

The problem of ozone depletion became an issue of international negotiation and agreement only after two scientists, M.J. Molina and F.S. Rowland, had advanced the hypothesis that Chlorofluorocarbons (CFC's) played a crucial role.[24] In fact, the discourse on *risk* which since the mid-70s has conspicuously been tied to key technologies, is in large measure driven by research probing into potential dangers coupled with rising expectations of certainty.[25] In other words: science plays an increasing role in defining the problems for which it is then called to give advice about once these problems are on the political agenda. This illustrates best the *scientification of politics.* Science, one actor among many in the political system, takes part in setting the political agenda. The politicization is inseparably linked to this. Immediately after publication of the Molina/Rowland hypothesis, two coalitions formed, one favoring rapid regulation of CFCs (scientists, environmentalists, consumers, politicians), another one opposing such regulation (CFC producers and consumers, lobbies for the chemical industry, scientists and politicians). The proponents of regulation, i.e. the few scientists who had published the crucial reports, became spokesmen and activists on behalf of their convictions even before they could prove their hypotheses. The ozone debate, among others, demonstrated the role of scientists as political actors, regardless of the differences in political systems or motives or interests that observers may attribute.[26]

The conspicuous *politicization* of the scientists in this case is by no means exceptional. The general point is that from the outside the scientists' positions in a controversy are identifiable in political terms and are sometimes (as in this case) even perceptibly determined by their politics rather than by available knowledge. Internally, this is inescapable. The scientific expert in the decision-making process inevitably assumes a *political* role. Or as Roqueplo put it: "What transforms scientific knowledge into an expert appraisal is its inscription within the dynamics of decision-making. Yet this inscription, at least in the case of scientifically and politically complex questions, immediately leads the scientist to express opinions or convictions which (however scientifically founded) cannot in any way be identified with knowledge in the strict sense which science generally affords this term".[27] It is as a *systematic* result of this coupling rather than the idiosyncratic behavior of individuals that experts provide expertise beyond their realm of knowledge.

The same applies to the role of scientific knowledge. Instrumental relevance cannot be separated from political relevance. The more scientific knowledge becomes an ingredient of policy-making, the more it becomes politicized. In essence, the reciprocity of scientification and politicization is tantamount to a *loss of distance* between scientific knowledge and political decision-making. Here, then, a *paradox* becomes apparent: Given the legitimating function of (authoritative scientific) knowledge in politics the general accessibility of that knowledge has led to a *competition for expertise* and to an inflationary use of expertise. However, that competition intensifies controversies in policy-making rather than alleviating them, thus inflating the legitimating role of expertise and diminishing its legitimating power.

The Close Coupling and its Impact

The increasing loss of distance in the relation between science and politics in contemporary society is most conspicuous at the institutional level. In fact, it is the result not only of an enhanced role of science and the increase of political issues with a scientific-technical component. Rather, the crucial force is *democratization*, i.e. the leveling of the legitimacy of socialstructural privileges and elite status. Although scientific knowledge still occupies a privileged status in Western society it is no longer the basis for privileged social status. Democratization of scientific expertise means here that advice is no longer given by a distanced elite of eminent scientists whose cooptation into this function is based more on social relations with the political elite but that scientific expertise is spread widely, and that it is available, in principle, to all political groups.[28]

In addition to the institutionalized expert advice in regulatory practices and legal procedures dating back to the 19th century the more recent general accessibility of scientific expertise has led to its generalized use in political controversies. Note that this by no means indicates a diminishing importance of scientific knowledge. On the contrary, recourse to scientific expertise is the most important resource in political controversies.

This accounts for the particular dynamics of the use of scientific expertise in policy-making and its inherently paradoxical nature. To the extent that political controversies are being carried out by the adversaries taking recourse to expert knowledge, competition for the latest knowledge arises. In principle, the competition for the latest, and therefore supposedly most compelling scientific knowledge drives the recruitment of expertise beyond the realm of consensual knowledge right up to the research frontier where knowledge claims are uncertain, contested, and open to challenge. Thus, the scientification of politics, regulatory and legal practice literally produces its opposite, the politicization of the expertise involved.

The pervasive discourse on risk and uncertainty must be understood in light of this mechanism. The closer the integration of scientific knowledge into policy-making the higher are the expectations of certainty, reliability and safety. But at the same time the political reliance on that knowledge systematically exceeds its potential thereby pushing beyond the limits of certainty. The *loss of distance* characterizing this new arrangement is best exemplified by the fact that many technological developments (nuclear energy, genetically altered plants and food, waste disposal etc.) are implemented experimentally. The hitherto institutionally separated process of experimentation in the laboratory had to achieve closure before implementation would begin. At least in cases where the complexity of the technology does not allow it, this separation is suspended and the respective technologies are implemented with considerable uncertainty – expressed in the probability of failure – as to their actual performance.[29]

In spite of the fact that the paradoxical outcome of the competition for scientific expertise, the production of uncertainty rather than certainty, is widely noted, the arrangement as such, and the corollary concepts of science, politics and the advisory process remain surprisingly stable. The reason for this is the lack of an alternative: there is no other system of more reliable knowledge and consequently none providing higher legitimacy for political decisions. But two reactions can be observed both on the side of institutional politics and science. One may be called *contraction.* It is the attempt to limit the access to scientific expertise and/or to 're-privatize' (i.e. close off from the public) controversies until closure is achieved. This is, in effect, an attempt to *regain* the *distance* which has been lost to democratization. Since the latter cannot easily be repealed, this distance has to be regained within its context, rendering chances of success very small. This may be attempted by introducing a new level in the hierarchy of scientific institutions (e.g. the establishment of scientific advisory councils) on which are conferred the role of ultimate authority in expert advice.[30] Governments and science have formed an alliance to keep controversy confidential until consensus is achieved – for example, in a body such as the Intergovernmental Panel on Climate Change (IPCC). IPCC is a consortium of climate research institutes and policy-makers from all over the world who, before reporting their research findings and recommendations to the media, reach a consensus among themselves in confidentiality. In this way the experience of the nuclear power debate, where contradicting experts neutralized each other, is avoided. Here again, the attempt to limit the amount of expert knowledge in order to augment its authority and utility is obvious, and up to this point, in the case of the IPCC this policy has been very successful.

Another strategy pursued by the scientific community may be called *expansion*. Although it appears to be contradicting the first it is actually complementary. It, too, is a reaction to the abundance of scientific expertise in policy-making. However, here the issue is how to capture attention. The answer is seen increasingly in overselling to the media. There is evidence that fields of science try to capture media attention, and thereby indirectly the attention of policy-makers, by establishing what could be called the 'catastrophe-cycle'.[31] This leads us right to the second example of close coupling, that between science and the media.

Science and the Media

Signs of Change

A young unknown scholar publishes a dissertation on a sensitive subject. The media announces that the book, marketed by a commercial publisher, will trigger a major debate within the scientific community. Although that prediction is only partially realized, the media themselves stage a debate among themselves and the leading scholars in the field. The author appears on national television talk shows and gains enormous publicity. The media criticize representatives of the discipline and hail the youthful author's appeal to the 'hearts of the public'. At his home university deliberations begin to appoint him to a chair for which he is a strong competitor. After the end of the media debate articles in scholarly journals severely criticize the author's research. To the extent that the media do report this 'peer review' at all they distance themselves from their previous enthusiasm. In the end, the chair remains unfilled[32]

This admittedly extreme example of a scholar receiving unusual media attention indicates a new proximity, a new close coupling between science and the media. Another aspect of this coupling appears in an incident that is clearly not exceptional. A Dutch AIDS-researcher announces the discovery of a vaccine. When pressured by colleagues to give proof of his discovery he has to admit having exaggerated his claims. When attacked for fraud he justifies his behavior by declaring that only with such exaggerations is one able to obtain the desired public attention and support.[33] This, too, indicates a new kind of relationship between science and the media. What are the characteristics of this new relationship, and what are the repercussions to science, if any?

The Independence of the Media and the 'Medialization' of Science

Just as in its relationship to politics there is an established mode of science's relation to the media that is experiencing rapid and fundamental change, that of popularization. It derives from a hierarchical concept of types of knowledge, according to which scientific knowledge is superior to popular, everyday-knowledge. Science has a monopoly on truth in society. The media merely transmit information. In this arrangement the control over the adequacy of this transmission lies with science. To most scientists, popularized knowledge is in the best case simplification, in the worst case pollution[34]

This concept of popularization implies a passive and generally unspecified public, and the process of communication from science to the public is conceived as being unidirectional. The public is excluded from the production and validation of knowledge and by implication is considered to be incompetent to judge the transferred knowledge.[35] The popularizers, whether they be authors or the modern media, are not granted an independent function in this asymmetrical model.

In media and communication studies, and more recently in sociology of the media and of science, it has long been recognized that the concept of the media being strictly dependent and reflecting a mirror image of reality is naive.[36] The core insight is that the media operate under their own parameters (profitability, public attention) and therefore, not surprisingly, have developed their particular strategies of representing the world. These have been identified as ways in which journalists judge the 'news value' of information and apply criteria of selection: actuality, sensation, personalization, and locality have been identified as such criteria.[37] Obviously, the parameters of operation and the selection criteria have an impact on the way in which news is selected and processed. In this sense it may be said that they construct their own reality in the same way as science does. They just use different instruments, different approaches to 'reality' and different forms of representation. Thus the complaint (especially frequent among scientists) that the media provide 'wrong' or 'distorted' reports is futile.

To the degree that the media gain more and more importance in shaping public opinion and even public perception of reality, the monopoly of science in judging representational adequacy may be weakened. Science's abstract criterion of truth is now being confronted by the media's criterion of public acclaim. The reliability of information (as represented by the prestige of a scientific journal) now competes with the level of diffusion (represented by the circulation figure of the newspaper or the number of viewers of a TV broadcast). The validation criteria of science are not necessarily replaced but are supplemented. In the case cited above, the media went even further: with the self-assurance supported by the command of mass attention, they posited their own criterion of 'emotional appeal' against that of science, i.e. 'truth'.[38] This is an obvious indication of a general democratization. Just as in the political arena, the elite status of science and its dominance in representing the world is challenged, not to say eliminated. Instead, the mass media control public attention according to their own criteria of relevance.

In view of the media's operational independence, it is increasingly questionable to hold on to the current model of popularization. The more important the media are for the structuring of public discourse, the more important it is for science to gain media attention. Or put in different terms: the stronger the dependence of science upon public consent, the more important is attention and consent of the media. This dependence is exacerbated in times of scarce funds as in mass democracies the allocation of resources is determined, albeit not directly, by public vote. In other words, the hierarchical asymmetry implied in the traditional popularization model may be replaced by a more symmetrical arrangement, and the communicative distance may be narrowed.

The recourse of science to a non-scientific public is not a novel phenomenon. In a sense, it is part and parcel of popularization. Originally unspecified and diffuse publics were envisioned by popularizing authors. But as time went on these non-scientific publics become more focused and their involvement assumes a more strategic function.[39] For scientists to take recourse to the public serves the purpose of mobilizing legitimacy with reference to two types of problems: the securing and expansion of the boundaries of science *vis à vis* its social environment, and the settlement of conflicts within science which cannot be solved internally. As long as these mobilization efforts are brief and diverse the risks involved for science in narrowing the distance between it and the media are few. It is for good reasons, namely to protect both the epistemological distinctness and institutional independence, that part of the unwritten ethos of science is the disdain for prominence by public acclaim rather than reputation by peer review, i.e. for media stardom, or for pre-publication in non-reviewed journals. Both on the level of sanctioned behavior, and of communication, this exemplifies the distance between science and the media. If, however, the strategic use of the media by science becomes a common feature, this will in all probability have repercussions for science. The phenomenon at issue may be termed the *science-media-coupling*. It is the basis for the thesis of the *medialization* of science. This phenomenon, which is typical for modern mass democracies, takes on different forms. But in all cases central mechanisms of the self-direction of science are affected. Three cases can be distinguished: *pre-publication* of results in the mass media, the attainment of *media prominence* by scientists, and the *strategic exaggeration* of research findings triggering 'catastrophe discourses' to attain public attention.

Pre-Publication

The most spectacular case of pre-publication in recent years has been the announcement of the successful achievement of cold fusion. The story has been well documented.[40] The gist of it was (1) that the announcement of cold fusion by the two researchers, B. Stanley Pons and Martin Fleischmann, was considered a sensation, given more than three decades of concerted efforts in the major industrial countries and a projected effort for another three decades to achieve (hot) nuclear fusion; and (2) that the supposed discovery was disclosed at a press conference before any internal review process had taken place, and without any detailed information given about the experiment that could be the basis for such a process. The result was hysteria among concerned research groups all over the world, and in the media.

The press served in part and for a limited period as a source of information and as information broker for the scientists. Lewenstein characterizes this situation as one of *instability* of information.[41] This phase was reflected in the actual information behavior of scientists, in the insecurity, and contradictions in evaluating Pons' and Fleischmann's results, as well as in their evaluation of its impact on science.

Another aspect of the science-media-coupling becoming apparent is the *amplification* of communication effects. As can be expected, reactions among scientists to the Pons/Fleischmann press conference were divided.[42] Critics warned that media hysteria would lead to 'bad science'. Wrong results published without control could mislead research laboratories into impasses. Proponents saw positive impacts, in that excitement would increase the speed of scientific debate and the exchange of opinions and results. The conventional view favors a slow communication system operating with the control mechanism of peer review; the more 'modern' favors a stronger and faster focussing of attention. However, the latter overlooks a consequence of amplification namely, the higher risk of misleading a much larger number of laboratories and scientists than was previously the case.[43] Indeed, research laboratories all over the world entered the race to confirm or refute the Pons/Fleischmann experiments. It is conspicuous that researchers involved directly or indirectly see the case of cold fusion as confirmation that the control system of science functions. At the same time, they systematically overlook or downplay the sheer breadth of reaction and the cost incurred by it due to the incomplete description of experimental setups in the press and its amplifying effect. The prevailing attitude is that, given the extremely high scientific and economic stakes, no one can afford to let an opportunity pass by. The costs of following a wrong path must be borne.

The lesson from this episode is that the scientific community, while continuing to proclaim the principal tenets of the scientific ethos pertaining to communication and peer review, in cases of particularly high economic expectations and correlative media attention succumbs to the temptations of direct communication to the media. This signals a breakdown of distance, and the rule of not communicating to the public before the internal communication process with peers has provided the requisite certification for the knowledge claims involved. The traditional controls may still operate, but at high cost.

Media Prominence and Scientific Reputation – the Dangerous Inversion

A second case of the close coupling of science and the media affects the internal steering mechanisms of science and in the long run may have a profound impact on the overall course of scientific development. *Prominence* as defined by the media, and often the result of media attention, is a specific media construct. It pertains to peoples' personalities and the assumed public interest in their activities. When scientists become prominent media stars – as in the case of Nobel Laureates – this is only partly due to their reputation as scientists. They may also be assigned media prominence because of their unusual personality traits and/or the popularity of the research topics they work on.

Reputation in science, on the other hand, is the result of past research and its evaluation by the community of peers. It serves to structure and focus attention in a communication process which is otherwise overburdened. Reputation is accumulated credit and trust in future achievement based on peer judgement. Thus, it is the medium by which decisions of quality and relevance of research are made. The function of distance is to delineate these decisions and evaluations from other standards such as the popularity of a scientist or the economic utility or political acceptability of his/her research topics. A breakdown of this distance would entail the risk of confounding these issues and, thus, of losing the efficacy of a differentiated communication.

With scientists increasingly seeking to attain public attention through the media, the question arises how scientific reputation and media prominence relate in directing the course of science and the policy decisions to support it. From the *perspective of science*, it is theoretically possible that political attention is directed to media-prominent scientists and their topics and that this attention is translated into decisions of resource allocations even though the scientists' prominence is not in line with their reputation. From the *perspective of individual scientists*, it is likewise imaginable that the path to success which has previously been blocked in internal evaluations may be found via prominence in the media. From the *perspective of science policy*, this creates the potential problem of focus directed to topics whose state of research does not justify the support suggested by media attention. In other words, the *science media coupling* may theoretically lead to a competition between media and scientific criteria of relevance and validation. In modern mass democracies such a development is increasingly probable.

So far the number of clear cut cases that show an inversion of scientific reputation and media prominence are probably few and far between, and none of those known are dramatic in their consequences.[44] Rather, they are indicators of what may lie ahead. Perhaps the most telling case in this respect was the German print media's orchestration of the Goldhagen Debate. It culminated in the media's explicit criticism of the historians whose 'dry research' did not 'reach the hearts' of people, thus placing the emotional appeal to the masses over academic standards of precision and representation of fact.[45]

Strategic Orientation to the Media and Credibility

A third case of science media coupling, and the correlate loss of distance, is the strategic adaptation of scientific discourses to the anticipated expectations of the media. It has become an observable pattern that scientists anticipate their media effects when addressing policymakers and the public. In fact, that particular kind of reflexive behavior is suggested and legitimized as part of a 'Public Understanding of Science's strategy of bringing science closer to the public. Information about the public, marketing, and manipulation are difficult to differentiate. Catastrophic discourses are a case in point. The threat of climate change, of a rapid decline of genetic diversity, of an irreversible destruction of the ozone layer are potential global catastrophes announced by scientists, as is the prognosis of a pending collision of a meteorite with earth. They are dramatized pronouncements and prognoses calling for immediate action that are taken up and amplified by the media. Often enough these dire warnings ignite politically effective discourses. At first, the strategic elements, if there are any, are not visible. Only when public attention wanes do dissidents and critics speak up and question the original catastrophic scenarios. In this phase the allegation may be formulated explicitly that the predicted threats were a conscious misleading or at least a public relations version of the truth.

This pattern is becoming routine in the relationship between science, politics and the media, and it reveals a dynamics which has the loss of distance as its underlying principle. Scientists, trying to convince politicians of the need to act rapidly in order to avert what they are convinced will be a catastrophe of global dimensions have to attract attention. Attention is a scarce resource and in politics can only be obtained if severe threats to legitimacy are launched that can mobilize public support. But the stakes in this are high for science. Threats have to be substantiated especially when the action to avert them involves sacrifices. And the suspicion that their proclamation is in the scientists' self-interest can be ignited easily. Also, it can be expected that the public gets used to scientific catastrophe scenarios. Once these effects take place and become part of the repertoire of public discourse it can no longer be judged reliably whether the drama of declarations by scientists is based on certified knowledge or if it is only construed with the view to their PR-effect. This has happened quite clearly in the scientific and public debate over climate change where allegations of strategic, media oriented catastrophism have been voiced both by sceptical scientists and journalists alike. Once a discourse has reached this point it is virtually impossible to restore the credibility of an authoritative source of truth.[46]

In the competition for attention all actors try to gain control but none of them ultimately control the game. The results are discourses of competing claims: the catastrophes claimed by science are becoming ever more global. There is an obvious paradox of the close coupling between science and the media: The more intensely scientists compete for public attention by initiating catastrophe discourses, even if they aim to mobilize support for the aversion of sincerely registered threats, the greater the danger of losing credibility. This, too, is an effect of the loss of distance between science and the media.

Conclusion

These two examples of close coupling, that of science and politics and science and the media illustrate a new complexity in the relationship between science as an institution and its social environment. Contrary to those analysts who interpret this as a 'blurring of boundaries' or as a de-differentiation, I argue for retaining the distinction drawn by systems theorists between the level of system codes (truth for science, power for politics, etc.) and the level of organizational structure. The kinds of blurrings of boundaries can, indeed, be observed on the level of concrete organizations. In order to keep this analytical distinction, I have spoken of *close couplings*. In discourses about science, the notion of truth does not disappear, nor is it replaced by the notion of power, or that of popularity, emotional appeal or others. The relevant sociological question is, however, if and in what ways the complex interrelations observed on the organizational level affect the differentiation between science and other social systems that has emerged over the last three centuries.

Porter's concept of *social distance* is useful in identifying instances where respective mechanisms operating in science (knowledge production), politics (securing legitimacy), and the media (securing attention) interfere with one another and lead to paradoxical repercussions. Distance is an expression of differentiation on the organizational and behavioral levels. The breakdown of distance is an outcome of a secular democratization of industrial societies that affects the three systems. It leads to a generalization of public discourses, and to the demise of privileged opinion-leaders, as well as of privileged sources of knowledge. Neither certified knowledge nor the authority of office or birth confers legitimacy. The ruling logic has become that of claim and counter-claim, selected by public attention and subject to the forgetfulness of the ongoing discourse. This development cannot be controlled or repealed by moral argument.

There is, thus, reason to beware of the enthusiasm apparent in the role of the 'concerned scientist' or of the seemingly unquestioned political correctness of a 'critical' reflection of science (and technology). The ambivalence of the delineation between science and non-science, between consensual knowledge and contested value judgments that has its roots in the very beginning of modern science cannot be eliminated. The social invention of 'distance' as a device to achieve consensus in communication over time and space may have had its day. We don't know what the world will look like.

University of Bielefeld

NOTES

1 This article presents a condensed version of an argument that is developed in more detail in a book, *Die Stunde der Warheit? Zum Verhältnis der Wissenschaft zu Politik, Wirtschaft und Medien in der Wissensgesellschaft,* (Weilerswist: Velbrück, 2000). In particular, it draws on two chapters that have appeared as articles P. Weingart, "Science and the Media", *Research Policy*, 27 (1998), pp. 869–879; P. Weingart, "Scientific Expertise and Political Accountability – Paradoxes of Science in Politics, *Science and Public Policy* , 26 (1999), pp. 151–162.

2 R.E. Lane, "The Decline of Politics and Ideology in A Knowledgeable Society", *American Sociological Review*, 31 (1966), pp. 649–662, p. 650; D. Bell, *The Coming of Post-Industrial Society,*

(New York: Basic Books, 1973); N. Stehr, *Arbeit, Eigentum und Wissen, Zur Theorie von Wissensgesellschaften,* (Frankfurt: Suhrkamp, 1994).

[3] M. Gibbons, *et al., The New Production of Knowledge,* (London: Sage, 1994); S.O. Funtowicz, J.R. Ravetz, "The Emergence of Post-Normal Science", in: *Science, Politics, and Morality. Scientific Uncertainty and Decision Making*, R. von Schomberg, ed. (Dordrecht: Kluwer Academic Publishers, 1993), pp. 85–123.

[4] Stehr, Arbeit ... (cit. n. 2), p. 36.

[5] Gibbons *et al., The New Production* ... pp. 5, 7, 8; Funtowicz/Ravetz, *The Emergence* ... pp. 90, 109, 117, 121 (both cit. n. 3).

[6] N. Luhmann, *Die Wissenschaft der Gesellschaft* (Frankfurt: Suhrkamp, 1990), p. 273.

[7] R.K. Merton, 'Science and Democratic Social Structure" in: R.K. Merton, *Social Structure and Social Theory,* (Glencoe: The Free Press, 1957, 1942) rev. ed., pp. 550–561.

[8] R.K. Merton, *Science, Technology and Society in Seventeenth-Century England,* (New York: Harper Torchbooks, 1970, 1938). The sequence of these steps was, of course, the other way around.

[9] Everett Mendelsohn, "Robert K. Merton: The Celebration and Defense of Science", *Science in Context,* 3 (1989), pp. 269–289.

[10] The difference between her historical analysis and Merton's is based on her epistemological reading of his 'ethos'. That is not necessarily the only possible reading. Lorraine Daston, "The Moral Economy of Science", *Osiris,* 10 (1995), pp. 3–24, 7.

[11] S. Shapin, *A Social History of Truth. Civility and Science in Seventeenth-Century England,* (Chicago und London: The University of Chicago Press, 1994).

[12] T.M. Porter, "Quantification and the Accounting Ideal in Science", *Social Studies of Science,* 22 (1992), pp. 633–652, p. 640; S. Fuchs, "A Sociological Theory of Objectivity", *Science Studies*, 11 (1997), pp. 4–26.

[13] R.K. Merton, "The Ambivalence of Scientists", in: *The Sociology of Science. Theoretical and Empirical Investigations*, N. Storer, ed. (Chicago: University of Chicago Press 1973), pp. 383–412; R.K. Merton, "Behavior Patterns of Scientists", in: N. Storer, *The Sociology of Science*, pp. 325–342.

[14] R.K. Merton, "Science and Democratic Social Structure ..." (cit. n. 7), pp. 550–561.

[15] R.K. Merton, "Science and Democratic Social Structure ..." (cit. n. 7), p. 545.

[16] Yaron Ezrahi, *The Descent of Icarus. Science and the Transformation of Contemporary Society,* (Cambridge: Harvard University Press, 1990), p. 53.

[17] Ezrahi, *Descent of Icarus* (cit. n. 16), p. 56.

[18] P. Weingart, "Das 'Harrisburg-Syndrom' oder die De-Professionalisierung der Experten", in: *Kernenergie: Gefahr oder Notwendigkeit,* Helga Nowotny, ed. (Frankfurt: Suhrkamp, 1979), pp. 9–17.; W. Krohn, P. Weingart, 'Tschernobyl – das grösste anzunehmende Experiment," in: *Kursbuch,* 85 (1986), pp. 1–25.

[19] A plethora of writings appeared featuring concepts like "scientific power elite", "new priesthood", "scientific estate", "new mandarins" etc. that reflected this concern. Cf. D.K. Price, 1967, *Scientists and National Policy Making,* R. Gilpin, R. Wright, eds. (New York: Columbia University Press 1965); R.E. Lapp, *The New Priesthood,* (New York: Harper & Row Publishers, 1965); *Knowledge and Power,* S.A. Lakoff , ed. (New York: The Free Press, 1966).

[20] Of course, this is not to say that the problem of accountability has disappeared altogether. There remains a fundamental tension between political power and its legitimation on the one hand and scientific knowledge on the other which becomes immediately apparent in goverments' persistent interest to keep expert advice confidential because it is considered a threat to the autonomy of political decision-making. P. Roqueplo, "Scientific Expertise among Political Powers, Administrations and Public Opinion", *Science and Public Policy* 22 (1995), pp. 175–182, 177.

[21] Roqueplo, "Scientific Expertise among Political Powers, Administrations and Public Opinion" (cit. n. 20), p. 176; Ezrahi, *Descent of Icarus* (cit. n. 16).

[22] Cf. the reconstruction of four funding programs in Germany (Environmental Research, Nuclear Fusion, Biotechnology, Electronic Data Processing) and the 'War on Cancer' in the US in: *Geplante Forschung: vergleichende studien überden Einfluss politische Programme auf den Wissenschaftsentwicklung,* W. Van den Daele, W. Krohn, P. Weingart, eds. (Frankfurt am Main: Suhrkamp, 1979). For a detailed account of the emergence of the 'Program for the Environment' *(Umweltprogramm)* cf. G. Küppers, P. Lundgreen, P. Weingart, *Umweltforschung – Gesteuerte Wissenschaft* (Frankfurt: Suhrkamp, 1978).

[23] P. Weingart, "Verwissenschaftlichung der Gesellschaft – Politisierung der Wissenschaft", *Zeitschrift für Soziologie*, 12 (1983), pp. 225–241.

[24] M.J. Molina, F.S. Rowland, "Stratospheric Sink for Chlorofluormethanes: Chlorine-atomcatalysed Destruction of Ozone", *Nature*, 249 (1974), pp. 810–812.

[25] P.M. Wiedemann, "Tabu, Sünde, Risiko: Veränderungen der gesellschaftlichen Wahrnehmung von Gefährdungen", *Risiko ist ein Konstrukt – Wahrnehmungen zur Risikowahrnehmung,* Bayerische Rück, ed. (München: Knesebeck, 1993), pp. 43-67, p. 57; P. Weingart, "Large Technical Systems, Real-Life Experiments, and the Legitimation Trap of Technology Assessment: The Contribution of Science and Technology to Constituting Risk Perception", in: *Social Responses to Large Technical Systems, Control or Anticipation,* T.R. LaPorte, ed. (Dordrecht: Kluwer, 1991), p. 14.

[26] R. Grundmann, *Transnationale Umweltpolitik zum Schutz der Ozonschicht: USA und Deutschland im Vergleich* (Frankfurt: Campus, 1999), Chapter 6; in this particular case, Grundmann argues, the *political style* in each of the countries, adversarial in the US, consensual in Germany, does not explain the outcome of the ozone debate because they practically inverted into their opposites during the 1980s (see p. 335).

[27] Roqueplo, Scientific Expertise ... (cit. n. 20), p. 176.

[28] The former was C.P. Snow's worry based on his experience with the Tizard-Lindeman rivalry during the war years in the British government. Snow characterized Churchill's relation to Lindeman as "court politics". Cf. C.P. Snow, *Science and Government,* (Cambridge, Mass.: Harvard University Press 1961), pp. 57, 63.

[29] W. Krohn, J. Weyer, "Society as a Laboratory. The Social Risks of Experimental Research", *Science and Public Policy,* 21 (1994), pp. 173–183.

[30] The idea of a 'Science Court' put forth in the 1970s had a similar objective. Its fate should be a warning. Cf. "Task Force of Presidential Advisory Group on Anticipated Advances in Science and Technology, The Science Court Experiment", *Science,* 193 (1976), pp. 653–656.

[31] Boehmer-Christiansen lists "fusion, many 'health' issues, cancer, limits to growth, star wars" as examples aside from the "environmental domain". "In each case, unsubstantiated promises or threats are made. Global warming is the threat through which Global Change research is funded". S. Boehmer-Christiansen, "Reflections on Scientific Advice and EC Transboundary Pollution Policy", *Science and Public Policy,* 22 (1995), pp. 195–203. pp. 202, 203 fn 17.

[32] P. Weingart, P. Pansegrau, "Reputation in science and prominence in the media: the Goldhagen Debate", *Public Understanding of Science,* 8 (1999), pp. 1–16.

[33] R. Hagendijk, J.Meeus, "Blind Faith: Fact, Fiction and Fraud in Public Controversy over Science", *Public Understanding of Science,* 2 (1993), pp. 391–415.

[34] S. Hilgartner, "The Dominant View of Popularization: Conceptual Problems, Political Uses", *Social Studies of Science,* 20 (1990), pp. 519–539, p. 519; J. Green, "Media Sensationalisation and Science: The Case of the Criminal Chromosome", T. Shinn, R. Whitley eds., *Expository Science: Forms and Functions of Popularisation,* Yearbook Sociology of the Sciences Yearbook IX (Dordrecht: Reidel, 1985), pp. 139–162.

[35] R. Whitley, "Knowledge Producers and Knowledge Acquirers: Popularization as a Relation Between Scientific Fields and Their Publics", *Expository Science: Forms and Functions of Popularization,* Sociology of the Sciences Yearbook IX, T. Shinn, R. Whitley, eds. (Dordrecht: Reidel, 1985), pp. 3–30, p. 4.

[36] W. Lippmann, *Public Opinion* (New York: Harcourt, Brace, 1954 [1922]), 317; N. Luhmann, *Die Realität der Massenmedien* (Opladen: Westeutscher Verlag, 1996).

[37] J.F. Staab, *Nachrichtenwert-Theorie: Formale Struktur und empirischer Gehalt,* (München: Alber, 1990).

[38] Weingart, Pansegrau, "Reputation ..." (cit. n. 32), pp. 9–10.

[39] M. Bucchi, *When Scientists Turn to the Public: Alternative Routes in Science Communication,* Thesis submitted to the European University Institute, Ms. (Florence, 1997).

[40] F. Close, *Too Hot to Handle: The Race for Cold Fusion* (Princeton: Princeton University Press, 1991); J. Huizinga, *Cold Fusion: The Scientific Fiasco of the Century* (Oxford: Oxford University Press, 1993), rev. ed.

[41] B. Lewenstein, "From Fax to Facts: Communication in the Cold Fusion Saga", *Social Studies of Science,* 25 (1995), pp. 403–436, pp. 415, 417.

[42] Reactions were obtained by interviews of US and German scientists involved in fusion research. Cf. P. Weingart, "Science and the Media", *Research Policy,* 27 (1998), pp. 869–879.

[43] *Newsweek* cited an American researcher as saying: So many scientists had been lured into cold fusion that it "probably brought the rest of science to a halt for the last months" (*Newsweek*, May 8, 1989, 44).

[44] An analysis of leading German print media in 1996 focusing on nine scientists showed three to fall in the category where media attention is followed post hoc by attention in the scientific community. Cf. Weingart, "Science and the Media" (cit. n. 42).

[45] The debate was about the book by Daniel Goldhagen, *Hitler's Willing Executioners* and confronted the media with professional historians who questioned Goldhangen's thesis about an

inherent selectionist Anti-Semitism among Germans. Weingart, Pansegrau, "Reputation ..." (cit. n. 32).

[46] Cf. The special issue of *Public Understanding of Science*, 9 (2000) on 'Global Climate Change and the Public', in particular P. Weingart, A. Engels, P. Pansegrau, "Risks of Communication: Discourses on Climate Change in Science, Politics and the Mass Media", pp. 261–283.

GARLAND E. ALLEN

RADICAL POLITICS AND MARXISM IN THE HISTORY OF SCIENCE

Abstract

The radical political activism of the 1960s and 1970s became the starting point for an exploration of Marxist historical materialism as it could be applied to the history of biology. In particular Marxism helped to point out the relationship between the development and accumulation of capital following the great burst of industrial growth in the United States after 1865, and its expansion into agriculture. The "industrialization" of agriculture provided a stimulus for the development not only of agricultural technology but also Mendelian theory and its associated social theory, eugenics.

Introduction

This paper grew out of a 1997 History of Science Society Symposium on the history of radical political thought in the 1960s, and its influence on our understanding of science in its economic and political context.* As I had been involved in a number of these discussions and activities, I focused this paper rather specifically on the applications of Marxist theory to my own area of academic interest: the history of science, especially the history of biology, genetics and evolutionary theory in the 20th century. Beginning in the turbulent period of 1968-70, the attempt to understand and apply Marx's writings have had a profound effect on my own thinking about history. Thus, it seemed that in the context of a volume honoring Everett Mendelsohn, such a topic might provide a way to explore how this approach can be useful in a field where it has been so underutilized. In the early-to-mid-1960s Everett began to emphasize the social relations of science in a contemporary context. His concern and enthusiasm for these issues must have seeped into my subconscious at the time I was doing graduate work under his direction between 1961 and 1966.

In exploring how Marxist methodology has informed my own views in history of science, I have two motivations. The first is that it provides the

G.E. Allen and R.M. MacLeod (eds.), Science, History and Social Activism: A Tribute to Everett Mendelsohn, 185–201.

opportunity for self-examination and clarification. As most of us who teach are well aware, there is no better way to understand what you really think about something than to try to present it cogently to others. The second, and hopefully more general reason, is that I would like to suggest how Marxism can be useful both in the pursuit of scholarship in the history of science as well as in providing a more comprehensive understanding of the world we live in today. Several considerations have molded my thinking on these issues. One of the reasons for studying history, is, as Marx said, "not merely to interpret the world, but to change it". From my perspective understanding without action is impotent; and action without understanding is chaotic. The main aim of action – in the present context to produce a more just, egalitarian, and humane world – cannot be accomplished without considering those facets of life to which Marx directed attention: economic, political, educational and social life and the institutions they spawn; the philosophical underpinnings of economics and society; technology and its relation to science; and the recognition that human beings are both a part of nature and transcend the objectified nature portrayed by the Western scientific tradition.

Marxist and radical analyses in general are not popular today. The reasons for this have little or nothing to do, in my mind, with the contents of either Marx' analysis of capitalism or his general view of history, but are contingent upon political biases and perspectives that have plagued Marx's writings from the beginning, and that have taken on their own distinctive characteristics in our time. The spread of global capital, the fall of the old communist regimes, and the new political and economic alliances emerging throughout the world have led to a chest-thumping optimism among world capitalist leaders that the free market system has overcome its problems, and will continue to outlive its socialist contenders. This view seems highly naive to me, and I can only agree with a financial analyst on the Op Ed page of the Sunday *New York Times* in October, 1997 who found himself arguing that Marx's analysis of the economics of capitalism is more correct than most economists are willing to acknowledge.

Let me define two terms that I have already introduced and will be using in this paper. The term *radical* I use to describe views deeply critical of any establishment or *status quo*. Radical analyses question social, political, and economic relations, but can come from either the political right or left. *Marxism* is a form of radical analysis using the theoretical and political concepts laid down by Marx and Engels, and expanded upon by their followers. In this paper I refer to Marxism in its up-dated form, what some scholars and activists call neo-Marxism, rather than Marxism as Marx himself understood it (indeed he even disavowed the label of "Marxist" in his own time).[1] Neo-Marxism is the form of Marxism to which I was introduced with friends and comrades in the radical science movement of the 1960s and 1970s.

This essay describes how I came face-to-face with Marx's theories and the impression they made in the early 1970s, and subsequently on my view of history and history of science. The fact that this process occurred largely in conjunction with activism will perhaps reinforce both social constructionist and Marxist claims, that ideas tend to follow from concrete experience, and only then come into play in altering understanding of that experience. At the end, I would like to suggest how this radicalization process influenced my research by giving one example from an area of the history of genetics: the emergence of genetic and eugenic theory in the context of industrial capital expansion into agriculture in the United States in the early 20th century.

Beginnings: Radicalization Without Marx

Before 1968, my involvement with the world of political activism or causes was at best, informed by a liberal, New Deal background, with its roots in organized labor in the early decades of the 20th century. My first political involvement of any kind was in the Civil Rights movement, when I went with a contingent from Boston to take part in Martin Luther King's Selma-to-Montgomery march in 1965. The two most enduring lessons I learned from the Civil Rights movement were: (1) power relationships and achieving justice were not quite what common middle-class ideology and school book presentations held them up to be; and (2) activism – direct activism – was absolutely crucial to achieve political and social progress. Civil Rights led directly into the anti-war movement which was building-up from 1966 as the war in Southeast Asia escalated. Participating in the anti-war movement brought the choice of joining activities of groups such as students for a Democratic Society (SDS), or becoming involved with Eugene McCarthy's presidential campaign in the spring and summer of 1968. I chose the latter since "radical" was hardly part of my vocabulary at that time, although its meaning soon did become clear. The disastrous Chicago Democratic convention in August of that year was the first major step toward my understanding of the relation between politics and power.

It seemed clear from the experience of that convention that even in the best liberal Democratic tradition politics-as-usual was establishmentarian and self-serving. Further involvement with direct action politics of the anti-war movement moved me steadily in a more leftist direction. Thus began a long standing association with a new group known as Scientists and Engineers for Social and Political Action (SESPA) and a five-month trip to harvest sugar cane in Cuba with the Vencermos Brigade in 1970. The latter was a particularly moving experience, both emotionally and intellectually as I worked and talked day by day with people trying to build a new society based on radically different concepts of social organization, responsibility, and democracy.[2] It seemed inescapable that oppression was everywhere, and that in the U.S. we lived in "the belly of the monster". Society needed a major overhaul, from the bottom up and not from the top down. Exactly how that overhaul was to take place was not clear, but it seemed to me to involve at the very least something more than individual action. What would be required was collective action and a drastic change in the way we do things, in including the distribution of wealth.

When I first met the SESPA crowd at the 1968 AAAS meeting in Chicago, I became aware that the same problems affecting society at large were mirrored in the practice of science itself. The SESPA group was loosely organized, but through its *Newsletter* and later its magazine, *Science for the People*, it offered a systematic, radical critique of science and its relation to technology and establishment politics, especially with respect to the conflict in Southeast Asia. The sense of a nation-wide network of active scientists speaking out and taking stands on economic and political issues was a great catalyst to my own somewhat halting efforts in this direction. I still knew virtually nothing of Marx or Marxism – indeed, in 1970 I had never even read *The Communist Manifesto*.

Enter Marx

Six months in Cuba were the turning point. Marx had much to say on all the issues in which I was becoming involved. Cuba underscored my belief that ideas and theories had to be matched by practice (the Cubans are eminently pragmatic and action-oriented people). Systematic analysis of the origins of current problems are essential in determining their solutions. While in Cuba I began to read Marxist literature that had previously escaped me, and found it compelling in terms of what I had experienced in the civil rights and anti-war movements, and in understanding what I had learned about Cuba's place in the history of colonial and economic imperialism. On returning from Cuba, I began to read much more seriously, in study groups with SESPA friends and colleagues in St. Louis, the works of some contemporary neo-Marxists: Harry Magdoff's *The Age of Imperialism,* (1969), Paul M. Sweezy's *A Theory of Capitalist Development* (1942), Ernest Mandel's two volumes *Marxist Economic Theory* (1962/68),[3] and the economically minded *Monthly Revie*w which, unlike some of its European contemporaries, had not emasculated Marxism by ignoring or repudiating its economic foundation. Neo-Marxism was particularly meaningful because it incorporated an understanding of the nature and the origins of imperialism, which was the key feature, as I came to see it, in understanding events in Southeast Asia. Nothing, perhaps, pointed more clearly to the importance of a Marxist analysis than seeing in 1965 the publication of a map in of *The Manchester Guardian*, showing the actual and potential oil leases by western companies in the South China Sea off the coasts of Vietnam, Cambodia and Thailand (Fig. 1). The American news had never mentioned anything like this. So, it seemed to say, that's why the U.S. was so tenaciously involved in Vietnam. Lots of things suddenly began to fall into place.

At about the same time (Autumn-Winter 1971) I became involved, through SESPA, with a newly organized group called the Committee Against Racism (later the International Committee Against Racism or INCAR) and its parent group Progressive Labor Party. INCAR was organizing to combat the recent claims by Berkeley psychologist Arthur Jensen, published tin 1969 in *The Harvard Educational Review* (1969), that the persistent differences between blacks and whites in I.Q. scores was due to genetic differences in intelligence between the two populations.[4] Jensen's claims had attracted an enormous

Oil interests in Southeast Asia.

Oil profits prolong war

By Barry Weisberg
Pacific News Service

The quest for oil appears to be assuming greater importance in the formation of overall war policy for Southeast Asia.

The Saigon government recently delayed, from February to March or even later, invitations to petroleum companies to bid on South Vietnam's offshore oil concessions. These concessions are located in the Gulf of Thailand and the southeast offshore region adjacent to the penal colony of Con Son. The Thieu-Ky government has not yet determined if it will offer all 18 offshore leases in a block or whether it will stretch out the leasing tially massive underwater oil deposit stretching some 3000 miles on a continental shelf which connects Japan, Indochina, Indonesia and Australia. Four major offshore finds in Southeast Asia to date have been made by American companies: Atlantic Richfield, Cities Service, the Union Oil Company and Natomas Oil Company of California. The Wall Street Journal reported rumors Sept. 22 that Standard Oil of New Jersey (ESSO) had discovered a huge reservoir of petroleum in its 28,000 square mile concession off the coast of Malaysia, directly adjacent to the South Vietnamese blocks. ESSO has made no public announcement to date, realizing news of the find would

The importance of Southeast Asia oil stems from predictions that within the next 10 years the industrial world will consume as much petroleum as was produced in the entire previous history of oil. The U.S. consumes one-third of the world's production and American petroleum companies control three-quarters of the non-communist supply. While only 3% of the U.S. supply comes from the Mideast, the 59% of Mideast production controlled by American petroleum companies is sold to the oil-hungry nations of Western Europe and Japan. The recent discord in the Mideast between the Organization of Petroleum Exporting Countries (OPEC) and the large petroleum

Figure 1. Newspaper showing various oil interests and leases in the South China Sea. It was articles such as this that showed a wholly different side to the Vietnam War. it was something more than supporting freedom in the abstract. [From *The Guardian* (New York, Sept. 19, 1973: p 19.]

amount of attention, much of it highly critical. What was most disturbing , however, was that despite the criticism, it was being used to undercut such existing programs as Headstart, school bussing, and even the early stages of affirmative action programs. In researching a talk for the main organizational meeting of INCAR in New York that autumn, I re-read Mark Haller's *Eugenics: Hereditarian Attitudes in American Thought.*[5] Haller had done a remarkable job in presenting eugenics as an intellectual and social movement, but he fell short, it seemed to me, in analyzing the genetics that eugenicists actually employed to support their theories, and had virtually ignored the larger economic context of the late 19th and early 20th centuries in which eugenics flourished. Haller's book made me see, however, that Jensen's claims were not new – they had been made in almost the same form (minus some statistics), a half-century earlier. Moreover, they seemed likely to be serving the same function today as they had served before to justify existing social policies and proposed reforms on seemingly scientific grounds. Eugenics, with its hierarchy of racial and ethnic groups, served the interests of the economically and politically powerful, giving the strategy of divide and rule a biological basis.

In several study groups organized by INCAR and Progressive Labor Party, I started reading Marx himself. Up to that point I had shied away from Marx because I was told that his style was difficult and dry, and I knew that *Das Kapital* and other works were long! As in so many other cases, however, direct experience proved the opposite. Reading Marx turned out to be quite an eye-opening experience. The book was long, to be sure, but it was not dry: Marx includes a wealth of specific and interesting details, and has an ever-present, if biting, sense of humor. Reading Marx was like reading Darwin for the first time: it opened up a whole new way of looking at the world. In fact, I found myself inadvertently comparing Marx to Darwin.[6] Most important, I found Marx's (and Engels') methods of historical and dialectical materialism, and the powerful ideas of base and superstructure, immediately applicable to understanding not only history, but the contemporary world as well.

Components of Marxist Theory

Marx was a synthesizer. The number of specific aspects of society that he tied together was compelling. The first was Marx's and Engels' *materialist conception of history.* Marx and Engels suggested that historical causality might be understood as something more than a sequence of changes propelled by the efforts of great men or chance circumstances. It meant simply that the things people do for most of their waking hours – i.e. their work and the way it is organized – ultimately colors the way they view the world and how they act on those visions. It is self-interest – one's own perspective – that provides the social vantage point from which events are seen and interpreted. This just seemed sensible – that a labor strike, for example, appears very different from the point of view of workers and owners. On a broader scale it meant that great historical developments take place not primarily for abstract, or ideological reasons, but for concrete and material reasons. For example, it had become clear that U.S. involvement in Vietnam was not primarily about securing "freedom" in the abstract to the people of Southeast Asia but was about control of resources and markets. Broadly conceived, economics seemed to be the *motor* that determines ultimately how far, how fast and in what direction the social vehicle can run. Material conditions are not the *only* forces operating in history, but they are the most persistent and relentless. Ultimately, directly or indirectly, they affect everything else. This meant to me, as an historian, that I could not hope to understand major historical changes without first knowing the economic and material conditions under which those changes took place.

The materialist view of history also made sense to me from a biological point of view. No ecologist, for example, would ever try to understand how an ecosystem functions, or how it develops over time, without first knowing in detail the food webs and energy relationships on which it is based. While other components and interactions within ecosystems alter the way food webs and relationships change (i.e. evolve), all come to a halt or are significantly changed if the food web is significantly disrupted. Knowing the materialist foundation is necessary, if not sufficient, for understanding how ecosystems or social systems function and evolve. Historiographically, this was a powerful idea, because it provided a single method from which to begin the analysis of the complex interactions that make-up any society at any point in history. I had a similar reaction to Darwin when I first read his own works (rather than textbook summaries), and began to comprehend his singular method of understanding the process of evolution. I am not a monist in the Haeckelian sense, but I do admit a special proclivity for synthetic and unitary explanations. Historians of science seem to have a long-standing aversion to seeking singular causes, while, ironically, in the sciences we study that is one of the main goals: to discover fundamental principles that unite disparate phenomena.

Another important feature of Marx that I began to appreciate, because it resonated with my personal experience, was his concept of *class analysis*. I had been subjected in both college and graduate school to a very ambiguous idea of the nature of class and its meaning in history. While few historians deny that classes exist, or that they have played some role in historical developments, the nature and influence of class was as frequently debated in the social sciences as was the nature of species in the biological sciences. Yet the more I read of Marx, the more I realized that: (1) social classes have existed and do exist, for example in contemporary United States; and (2) they have played and continue to play a critical role in historical change. To the claim that classes are nebulous and difficult to define, I can only say that if you have ever had the concrete experience of standing on a picket line you see class divisions drawn very quickly and very firmly (i.e. who will and will not cross the line, and the attitudes of those whom you encounter under such circumstances). Abstract arguments about the reality of classes or the historical role they play fade into insignificance (and seem ridiculously academic) in the face of direct and concrete confrontation in the real world.

Perhaps the most all-encompassing and ultimately most useful feature of Marxism, as I came to understand it, was the philosophy of *dialectical materialism*. Dialectical materialism as a philosophical method has attracted scientists from J.B.S. Haldane to J.D. Bernal and Joseph Needham in England in the 1920s and 1930s, to Richard Lewontin and Richard Levins in the United States in the 1970s and 1980s.[7] I became particularly fascinated with dialectics as a means of analyzing and understanding the *dynamics of changing systems*. It seemed to be a method of analysis and not simply, as was often claimed, a set of dictates imposed on the world. It suggested that complex systems are more than the sum of their parts, and that, as a corollary, to understand such systems it is necessary to find the means of studying the *interaction of the parts* as a unity. This approach stood in stark contrast to the analytical and reductionist method – implicit in much of our political and social life but especially prevalent in the sciences – that sought to isolate parts of a whole and study them separately. It also suggested how changes in systems could be understood by looking at the contradictory or opposing forces, or processes operating within them. Dialectical change was pre-eminently evolutionary, and I finally saw why Marx could be considered an evolutionist. Both he and Darwin saw changes in any system as resulting primarily from changes occurring *within* the system itself, rather than specifically imposed upon it from the outside. For Darwin, organisms evolve not primarily because the environment changes (although that does happen), but because of the interaction of heredity and variation (two opposing forces) in a competitive dynamic – where more successful variation leads to differential fertility. For Marx, societies evolve because of internal contradictions to which people react from their particular class perspectives.

To take one example: Marx's use of a dialectical approach in *Capital* comes in his analysis of the evolution of the capitalist system as it was developing in western Europe and to a limited degree the United States (in various places he provides a glimpse of how the same method could also be applied to other economic/social systems, such as chattel or feudal societies). To indicate both the usefulness of the dialectical method, and how Marx saw societies as inevitably evolving, consider his discussion of the "law" (originally enunciated by David Ricardo in 1817) of the tendency of the rate of profit to fall,[8] and how this affects the historical development of both the base and superstructure of a capitalist society. Marx's own explanation is intriguing but a little convoluted. It can be simplified in modern terms as follows: In the recurrent cycles of buying (or manufacturing) and selling commodities, capitalists extract surplus value as profit – i.e. that portion of the price above and beyond the cost of obtaining or producing the commodity. Thus if it costs $100 in wages (to simplify the model) to produce a commodity, and it is sold at a 10% profit, the price will be $110. However, since only $100 total was paid out in wages, there is not enough aggregate money in circulation to buy the commodity, and it ends up being unsold (it is unrealized return). This situation right away represents a primary contradiction (in Marx's terms between wages and prices). Workers are dissatisfied because they cannot buy the very commodities they produce, and capitalists are dissatisfied because they cannot sell their products and realize a profit. This is the process that tends to make the profit rate fall.

As Marx points out, there are several ways in which the system as a whole can (and does) respond to this set of contradictions: (1) reduce the price (hence the profit rate) as a means of capturing more of the existing market; (2) reduce costs by obtaining the commodity or the raw materials from which it is produced at a lower price; (3) reduce costs by lowering wages; (4) increase the price of the commodities that are sold to compensate for those that are not sold; and finally, (5) increase the market by expansion, bringing in new buyers. In reality, all of these function more or less simultaneously, though some will take precedence over others at one time or another. But as Marx points out, all of these responses lead to contradictions of their own. (1) reducing prices causes the profit rate to fall, the very outcome the response is meant to address; (2) obtaining the commodity or raw materials more cheaply goes against the interests of the seller and historically through colonialism and imperialism has involved capitalist countries in wars and other conflicts over control of raw materials or sources of commodities; (3) lowering wages means that even less money is available to purchase the commodities in the ensuing business cycle; (4) increasing the price of the commodity (inflation) has the same effect as lowering wages. Finally (5) the only real way out is expanding the system through increased markets, a process that historically has produced conflict between capitalist entrepeneurs, and at any event has ultimate limitations. The outcome of this cycle of responses, counter-responses and modifications is that capitalist systems are not capable of maintaining a steady-state, and must therefore constantly change (evolve), enacting new laws governing everything from import and export quotas, labor, regulations (i.e. no-strike or limited-strike laws), monetary policy, taxation , etc. The specific responses at any one time in history are not always predictable, but they are not usually random – they follow the logic imposed by the existing contradictions and the options available for resolving them. Marx gives many examples of how changes in the economic base such as the ones described above lead to changes in the social superstructure as any society evolves. But it all derives from and relates back to economics – not in some vulgar sense of economic determinism, but in the same sense that as food webs change in a natural ecosystem, so must the other ecological relationships that depend upon them.

The value of dialectics as a *process* of understanding complex systems has slowly gained adherents in the west since the 1960s, and has long been an important mode of analysis in Russian social and natural science. For example, Loren Graham informs me that despite everything, a survey of Russian physicists in 1996 showed that 49% thought that dialectic materialism was by far the best way to understand nature, while an additional 21% thought that it was at least as useful as any other philosophical orientation. Following Engels' *Dialectics of Nature,* Lewontin and Levins have argued that dialectics provides an extremely powerful method for analyzing biological systems, especially in ecology, population genetics and agriculture.[9] Steven Rose has made the same point, giving especially pertinent examples from the neurosciences.[10]

Yet another particularly illuminating feature of Marx's and Engels' viewpoint was the importance they gave to understanding the history of science and its relation to technology. Marx said of Darwin: "Darwin has interested us in history of nature's technology, i.e. the formation of the organs of plants and animals, which organs serve as instruments of production".[11] In animals, of course, the technology becomes an end in itself, while in human society the technology can lead to a quest for its theoretical underpinnings, namely, to science. Concomitant with the materialist theory of history, one would *expect* that technology would drive science more frequently than science would drive technology – although the two have always had reciprocal influences on one another, and have co-evolved throughout history.

As the relation between economics, technology, and science began to become a little more clear, I began reading earlier historians and sociologists who had attempted to work out some of these relationships in the context of specific histories of science – with, I might add, varying degrees of success. In graduate school I had read an abbreviated version of Boris Hessen's "The Economic Roots of Newton's *Principia*"[12] , but it had been held up to ridicule and none of us took it very seriously at the time. When I went back to rereading it, in its full rather than in the abridged form I found it a highly suggestive and innovative work. Also influential were Edgar Zilsel's "Sociological Roots of Science" (1941), G.N. Clark's *Science and Social Welfare in the Age of Newton* (1937), Victor Lenzen's essay, "Science and Social Context" (1935), and Robert K. Merton's *Science, Technology and Society in the 17th Century* (1938).[13] None of these except Hessen were explicitly Marxist (Zilsell's comes closest) however amenable they might have been to a Marxist interpretation.

Given my long-standing interest in the history of genetics, a growing interest in the history of eugenics (that movement in the early twentieth century concerned with "improvement of the human stock through planned breeding"), and a growing concern with the revival of the race and I.Q. issue in the 1970s, it seemed to me there might be a relationship between these issues. Several questions in the history of genetics had always been puzzling: (1) Why, after 35 years of neglect (1865–1900) was Mendel's work "rediscovered" and accepted so (relatively) rapidly, especially in the United States and Britain"? (2) Why was there so much interest in promoting eugenics in the first three decades of the century? (3) Why was the first eugenics organization in the United States fostered as a committee of the American Breeders' Association? (4) Was the rise of the Mendelian paradigm of heredity so divorced from agricultural interests as the standard histories of genetics by Dunn, Sturtevant, Stubbe and others would have us believe? And finally, (5) What was the relationship between genetics, eugenics and the movement for industrial and national efficiency that was so much in vogue in the later nineteenth and early twentieth century in the United States and in a number of European countries? It seemed to me that a Marxist approach might provide a very fruitful way to explore what had otherwise been presented as more or less disparate, or only loosely-connected phenomena in American economic, social and scientific history. While I have confined my own investigations of this period to the

United States, similar trends and relationships have been discovered by other historians in Britain, Germany, Scandinavia and Russia".[14]

Through the outstanding work of recent scholars (Barbara Kimmelman, Diane Paul, Richard Lewontin and Jean-Pierre Berlan, Richard Levins, Jack Kloppenburg, Paolo Palladino and Deborah Fitzgerald, among others) it has become clear that early Mendelians often had interests in agriculture, and that such interests strongly supported early work in Mendelian research programs.[15] It also became clear from my own work that many of these same agricultural interests supported the eugenics movement as part of the war against the rise of radical elements in American society (particularly the organized labor force). The question was, why would agriculturalists become so interested in Mendelian genetics in 1900 when they had shown little or no interest in it during the preceding 35 years?

What emerged is that the rise of genetics, and eugenics was part of the much larger change in economic relations associated not only with what has been called Progressivism in the United States (marked by the conscious attempt to bring the principles of rational, scientific management to bear on agriculture, society and the workplace) but specifically the attempt to industrialize agriculture in order to make it a profitable arena for capital investment. Industrialization in this sense meant not only mechanization and the shift from small to large-scale farms, but also standardizing agricultural organisms themselves, creating higher-yield more uniform plants and animals. Eugenics, as well as some agricultural research into the problems of heredity, was paid for by a very wealthy philanthropic class who wanted to increase the profitability of agriculture and to stem the tide of industrial labor organizing. These class interests pushed in many directions, but the two major ones I want to emphasize here are: the penetration of capital into agriculture on an industrial scale, and social control through rational, scientific means (especially eugenics). Eugenic methods were aimed at foreign-born ethnic group, mostly from Southern and Eastern Europe, the Mediterranean countries, and the Balkans, the very people thought to be fomenting radical discontent among organized labor. The rise of genetics itself, along with and closely connected to eugenics, was part of change in economic relations in the period 1890–1930 marked by the conscious attempt to bring concepts of scientific management to bear on agriculture, society and the workplace – all in the context of the dramatic changes associated with industrialization and urbanization in the U.S. in the second half of the nineteenth and first part of the twentieth century.

The Development of Genetics and Eugenics in the United States: A Case Study

A Marxist analysis reveals the ways in which the development of genetics and eugenics in early 20th century America was economically, technologically and politically-driven.[16] Rapid post-Civil War industrial development led to increased capital accumulation, (captured in such epithets as the "Gilded Age/the Age of the Robber Barons"). Rapid economic expansion and accumulation of wealth led to considerable social chaos and strife: periodic depressions and price fluctuations between 1873 and 1893, increasing bankruptcies in both small and medium sized businesses, accompanied by increasing monopolization among the larger businesses, and the rapid and dramatic growth of organized labor culminating in what has been called by Graham Adams and other historians, the period of the "labor wars".[17] In the "search for order" that grew out of this period, a liberal or "progressive" sector of the capitalist class, as well as a number of populist supporters, called for more control and regulation of economic and social processes by state and federal government.[18] It was the notion of rational control and "scientific management" that was the linch-pin of the progressive movement. The emphasis was on management by trained experts, whether of the economic system, the agricultural system, the trade or monetary system, food and drugs, and even what was referred to by eugenicists as the most precious resource of all: the human germ plasm.

As an area for capital expansion, agriculture had vast potential. In the 1890s it was still largely a cottage industry, with average farm size being only 137 acres, as opposed to 390 by 1969.[19] However, agriculture could be "industrialized" by increasing farm size, monocrop culture that was more amenable to large-scale mechanization, and the application of pesticides and fertilizer. The one problem was that the profit rate for agricultural products was low. Advances in scientific agriculture in the later nineteenth century had derived largely from environmental changes: increasing soil quality, fertilizers, animal nutrients and the like.[20] Although these methods increased yields, they represented continual expense, and hence cut into profit. Improving heredity seemed to answer this problem: if high yielding varieties could be produced, the initial expense of developing them would be returned year after year by higher yields – a process which was by definition self-perpetuating. In this economic climate the rediscovery of Mendel assumed an importance that had not been seen when capital expansion was confined mostly to manufacturing and industry. The profitability of new high-yield strains was not lost on Secretary of Agriculture James Wilson, when he was invited in 1910 to write the "Introduction" to the first issue of the *American Breeders' Journal*, the organ of the American Breeders' Association.[21] In saluting the Breeders' Association as a fruitful collaboration between farmers and academic scientists, Wilson pointed out the important economic gains to be realized by focusing attention on heredity:

> "Improvements by breeding are unlike those secured by adding new acres to the cultivated areas of the country, by deeper plowing, by more frequently cultivating the crop, by adding to the soil larger supplies of fertilizers, or by giving a more expensive ration to farm animals. These improvements, though they greatly increase the farmers' profits, are secured at a cost which sometimes equals the value of the added produce. But the cost of improvements through breeding usually represents only a small fraction of the added values. The increase of product secured by breeding pays the cost in a short time, and, since there is no further expense, the annual increase afterward is clear profit. The farmer will be able to retain a part of the larger production

in the form of added profit, and part will help to reduce the cost of living to those in our cities. Larger production on the farm will also give increased business for the transportation company, the manufacturer, and the merchant, and will provide the nation with a larger product with which to hold our balance of trade".[22]

The Carnegie Institution of Washington, which funded both Luther Burbank's work in Santa Rosa, California, as well as Charles Davenport's breeding and selection experiments at the Station for Experimental Evolution in Cold Spring Harbor (they later funded much of the eugenics work at the Eugenics Record Office at Cold Spring Harbor, under Davenport's direction) was also interested in promoting the study of heredity for agricultural purposes.[23] The new agriculturally-based genetics (and along with it, eugenics) shared not only the theoretical framework of Mendelian genetics, but also the progressive notion that trained experts could bring scientific principles to bear on rational management: in agriculture the management of the germ plasm of economically important crops, and socially to the management of the germ plasm of the human species. As it turned out, of course, the very feature of self-reproduction which seemed initially so attractive to capital investment in agriculture (for example, among early seed companies) later turned out to be a disadvantage. Once a true-breeding strain (in Mendelian terms a homozygote) of corn was developed, farmers needed to buy seeds only once, because they could simply retain a portion of each year's crop as the basis for their next year's production. (As breeders know, of course, pure-breeding or monozygous strains grown in the field can hybridize with strains in neighboring fields, so that maintaining a pure-bred strain over a number of generations could be a considerable problem, if it was possible at all. Thus, there were several biologically sound reasons for pushing hybrid corn, though its non-reproducibility was undoubtedly the crucial one.)

Before capital penetration could make a major profit from agriculture, some way of preventing this easy reproduction of high-yield strains had to be found. Enter the biological concept of heterosis (hybrid vigor) and the strategy of producing only high-yield hybrids. In the early years of the century a variety of corn-breeding programs were tried, using Mendelian principles. One common problem breeders faced was that in attempting to produce prize-breeding strains by inbreeding (brother-sister, offspring-parent crosses), the strains frequently lose vigor (low fecundity, loss of hardiness) in subsequent generations. Outcrossing (breeding between two strains) usually restores vigor to the lines. In Mendelian terms, pure strains are homozygous, at least for the characters important to the breeder, and hybrid strains are heterozygous.

In their 1919 *Inbreeding and Outbreeding*, E.M. East and Donald F. Jones laid out the theory of improving vigor by outcrossing and fixing desired traits through periodic inbreeding.[24] The term "hybrid vigor" was coined to emphasize what was later claimed to be a general theory of genetics, thereafter repeated in textbooks of biology, genetics and agriculture. What was less well-publicized was that often in-bred strains were just as hearty as hybrid strains once the deleterious effects of lethal genes were removed through selection. There was, however, an economic and social reason why "hybrid vigor" became the more well-known concept: it provided a theoretical basis for capital investors to enter the market. Hybrids do not breed true, hence the farmer cannot save a portion of the previous year's crop to plant the next spring. But, with hybrid strains, capital investors had a renewable market and hence a reason for expanding into agriculture. It was, in fact, for reasons related to the expansion of capital investment in agriculture that hybrid rather than pure-breeding corn became the dominant breeding strains of the mid-20th century and beyond.[25] The disastrous effect of this whole process on agriculture has been superbly illustrated in several monographs and books published over the past 15 years (listed in endnote 15). The similarity between this disguised strategy for maintaining control over germ plasm in the 1920s and 1930s and the far more blatant strategy of the Monsanto Company's proposed "terminator gene" in the late 1990s will not be lost on even the least-Marxist of agricultural historians.

But why was there also a strong interest in eugenics at the same time? This interest was clearly more than an outgrowth of agricultural metaphors of "good stock" or "good breedings". For the biologist *qua* social reformer, like C.B. Davenport, Harry Laughlin or E.G. Conklin, eugenics represented the application of a new and *predictive* science to the solution of otherwise seemingly intractable social problems. For legislators and political leaders, it provided "objective" evidence on which to base otherwise volatile political decisions concerning immigration or sterilization. The greatest threat to the maintenance of "civilization" according to geneticists, came from the rapidly-breeding, but genetically inferior working class, composed of so many "swarthy" ethnic groups, whose germ plasm would "swamp" that of the genetically superior, but slow-breeding upper class. For the wealthy owners of capital the theory of eugenics also offered psychological solace. *They* were at the top of the social heap because they were genetically well-endowed. More important, however, eugenics reinforced with "hard" science the existing racial and ethnic hierarchies of United States society. Such reinforcement – fanning the flames of racism and ethnicism – could not help but serve as one more weapon in the war of the ruling class against labor. This legacy of eugenics, whose remnants still appear in our society as Polish, Italian and Jewish jokes, and in the deeper prejudices they conceal, was to help reinforce or create class divisions during the most intense period of the labor conflict in United States history.[26]

What I am suggesting is that the economic and social context tied together many disparate developments relating to genetics, agriculture, eugenics, and the ideologies of reform and rational management in the period 1890–1930. The driving force was the development of American capitalism leading the country from an agrarian to an industrial society, with its ever-increasing need for expansion, bringing with it a need for cheaper raw materials, cheaper means of production and wider markets. The effects of this constant pressure to expand ramified to every aspect of the social process (the social superstructure, as Marx and Engels called it).

Admittedly, I am attracted to this sort of analysis because it is synthetic, in that it draws together a number of otherwise seemingly disparate developments. But it is also *explanatory*. It highlights the underlying changes within the economic system that manifest themselves in a variety of social outcomes – of which research in genetics and its human branch, eugenics, was one part. Whether the quest for such a synthetic and explanatory scenario is either possible or a worthwhile endeavor, has been a long-standing historical debate. My own view is that human history, like any other natural process, is subject to regularities and processes that can be analyzed according to underlying principles. That history follows certain "law-like" developments should in some ways come as no surprise. Human beings in all cultures develop laws (in the legal, legislative sense) that by definition regularize practices, and in response to economics and other material changes in the society, will be modified in distinct, purposeful ways. That rather specific changes are not predictable (though perhaps some larger-scale ones are) is also no surprise, but it no more undermines the idea that history can be understood as developing by regular processes than does the lack of predictability in evolution undermine our belief in or use of Darwinian theory. Marx was among the first to recognize, however, that historical "laws", in contrast to natural laws, also evolve. Historical laws are the product of cognizant human minds, and are changed for conscious reasons. They are also contingent on time and place. Thus for Marx, not only do social systems evolve, but the laws or rules governing them evolve as well, a unique recognition, and one that is only now beginning to find its counterpart in the natural sciences (in, for example, the idea that the laws governing matter and its properties in the early nano seconds of our universe were very different from those that govern the behavior and properties of matter today). Part of the difficulty bourgeois historians have had with Marx's concept of "laws of history" (many present-day Marxists do not use this nineteenth-century terminology) is that they think Marx meant them to be fixed and unalterable, like Newton's laws. Although he held a number of nineteen-century views about the natural sciences, Marx was considerably ahead of his times regarding this difference between natural and "social laws".

A Marxist approach offers a coherent picture of the world that no other historiographical paradigm has accomplished. It is theoretically powerful as a *method* of viewing and trying to understand history. It is not a "dogma" any more than Newton's principles and methods of analysis are necessarily dogma. Both are based on empirical evidence and both provide ways of understanding and analyzing complex systems. Marx's methodology must be applied, as Marx himself recognized, to specific historical developments, using concrete examples and data. I have found Marx's approach fruitful because it suggests questions to ask and avenues of enquiry to pursue. It requires more, not less of the historian. It requires the searching out of new kinds of data, and establishing connections that may be, for political reasons, ignored or even hidden in standard histories. But in my mind it makes the search more exciting, the discoveries more rewarding.

Conclusion

It would be naive to claim that my own view of history is any less socially contingent than others, and I am rather glad to own up to that. The events of the Civil Rights and anti-war movements convinced me that we as a society needed to embrace some new and radically different understanding of how our society functions if we were to ever solve the national and international problems confronting us. Moreover, my interest in the sciences suggested that history like the rest of the universe, human and non-human, might proceed according to some regular, comprehensible processes. What was especially important to me – and I emphasize this – was the necessity of integrating theory and practice in understanding history. The historical developments of the 1960s and 70s, and the more general underlying processes they revealed, meant something very different because I had experienced them directly and forcefully. To be meaningfully understood, history has to be *felt* as well as *analyzed*. That is the meaning of the unity of theory and practice. Historical work is clearly intellectual, academic and theoretical; but it is also human and emotional, and Marx has, perhaps more than anyone else, shown how these can work together – not only to understand the world, but also to change it.

Washington University

Notes

* Organized by Patrick Catt, whose dissertation at Indiana University discusses the radical science movement in the United States during the 1960s and 1970s.

1 Tom Bottomore, Laurence Harris, V.G. Kiernan, Ralph Miliband, eds., *A Dictionary of Marxist Thought* (Cambridge, Mass: Harvard University Press, 1983), p. 309.

2 Garland E. Allen, "Science, Education and Culture in Revolutionary Cuba", *American Biology Teacher*, 36 (1974), pp. 267–291.

3 Harry Magdoff, *The Age of Imperialism* (New York: Monthly Review Press, 1969); Paul M. Sweezy, *The Theory of Capitalist Development: Principles of Marxian Political Economy* (New York: Oxford University Press, 1942); Ernest Mandel, *Marxist Economic Theory*, translated by Brian Pearce (New York: Monthly Review Press: 1968).

4 Arthur R. Jensen, "How Much Can We Boost I.Q. and Scholastic Achievement?" *Harvard Education Review,* 39 (1969), pp. 1–123.

5 Mark Haller, *Eugenics: Hereditarian Attitudes in American Thought* (New Brunswick, N.J.: Rutgers University Press, 1963).

6 Garland E. Allen, "Evolution and History: History as Science and Science as History", in *History and Evolution,* Matthew H. Nitecki and Doris Nitecki, eds. (Albany: State University of New York Press, 1992), pp. 211–239.

7 See for example, Joseph Needhan, *Science in History* (Cambridge, Mass: MIT Press, 1971/1954) 4 vols; J.B.S. Haldane, "Preface" to Friedrich Engels, *The Dialectics of Nature* (New York, International Publishers, 1971/1940), pp. vii–xvi. Richard Levins and Richard Lewontin, *The Dialectical Biologist* (Cambridge, Mass: Harvard University Press, 1985).

8 Karl Marx, *Capital* (New York: International Publishers, 1967). See Volume III, Chapter 13.

9 See Levins and Lewontin, *Dialectical Biologist* (cit. note 7): esp. Chapters 1–3, 6 and 9.

10 Steven Rose, "The Rise of Neurogenetic Determinism", *Nature*, 373 (1995), pp. 380–382.

11 Marx, *Capital* I (cit. n. 8), p. 362.

12 Boris Hessen, "The Economic Roots of Newton's *Principa*", in *Science at the Crossroads* (London: n.d. [1930–31]), pp. 151–176.

[13] Edgar Zilsel, "The Sociological Roots of Science", *American Journal of Sociology*, 47 (1941–42), pp. 544–560; G.N. Clark, *Science and Social Welfare in the Age of Newton* (Oxford: Clarendon Press, 2nd ed., 1949); Robert K. Merton, "Science, Technology and Society in Seventeenth Century England", *Osiris*, 4 (1938), pp. 414–565.
[14] For Britain, see Daniel J. Kevles, *In the Name of Eugenics* (New York, Alfred A. Knopf, 1985).
[15] Barbara Kimmelman, "A Progressive Era Discipline: Genetics at American Agricultural Colleges and Experiment Stations, 1890–1920" (Philadelphia: University of Pennsylvania unpublished Ph.D. dissertation: 1987); Diane Paul and Barbara Kimmelman, "Mendel in America: Theory and Practice, 1900–1919", in *The American Development of Biology*, Ron Rainger, Keith R. Benson and Jane Maienschein, eds. (Philadelphia, University of Pennsylvania Press, 1988), pp. 281–309; Deborah Fitzgerald, *The Business of Breeding: Hybrid Corn in Illinois, 1890–1940* (Ithaca: New York, Cornell University Press, 1989); Jack R. Kloppenburg, *First the Seed: The Political Economy of Plant Biotechnology, 1492–2000* (New York, Cambridge University Press, 1988); Jean-Pierre Berlan and R.C. Lewontin, "The Political Economy of Hybrid Corn", *Monthly Review*, 38 (July–August, 1986), pp. 35–47; Paolo Palladino, "Wizards and Devotees: On the Mendelian Theory of Inheritance and the Professionalization of Agricultural Science in Great Britain and the United States, 1880–1930", *History of Science*, 32 (1994), pp. 409–444; Paolo Palladino and Michael Worboys, "Science and Imperialism", *Isis*, 84 (1993), pp. 91–102; Kathy J. Cook, "From Science to Practice or Practice to Science: Chickens and Eggs in Raymond Pearl's Agricultural Breeding Research, 1907–1916", *Isis*, 88 (1997), pp. 62–86; Kathy J. Cooke, "Twisting the Ladder of Science: Pure and Practical Goals in Twentieth-century Studies of Inheritance", *Endeavour*, 22 (1998), pp. 12–16; Garland E. Allen, "Essay Review: History of Agriculture and the Study of Heredity – A New Horizon", *Journal of the History of Biology*, 24 (Fall, 1991), pp. 529–536.
[16] A more recent version of this analysis has been published as "The Reception of Mendelism in the United States, 1900–1930. "*Comptes Rendu, Academie des Sciences, Paris, Sciences de la Vie*, 323 (2000): pp. 1081–1088.
[17] Graham Adams, *The Age of Industrial Violence* (New York: Columbia University Press, 1966); see also Robert Wiebe, *The Search for Order* (New York: Hill and Wang, 1967).
[18] An excellent discussion of the economic side of "progressivism" is found in James Weinstein, *The Corporate Ideal in the Liberal State, 1900–1918* (Boston: Beacon Press, 1968). See also Robert Wiebe, *The Search for Order* (op. cit., n. 17).
[19] *Historical Statistics of the United States, Colonial Times to 1970* (Washington, D.C.: U.S. Department of Commerce, Bureau of the Census, 1975), I, p. 457.
[20] Margaret W. Rossiter, *The Emergence of Agricultural Science. Justus Liebig and the Americans, 1840–1880* (New Haven, Connecticut: Yale University Press, 1975); James Wilson, "The New Magazine has a Place", *The American Breeders' Magazine*, 1 (1910), pp. 3–5.
[21] Wilson, "The New Magazine has a Place" (cit. n. 19), pp. 4–5.
[22] Wilson, *Ibid.*
[23] Garland E. Allen, "The Eugenics Record Office at Cold Spring Harbor, 1910–1940. An Essay in Institutional history", *Osiris* (New Series), 2 (1986), pp. 225–264.
[24] E.M. East and Donald F. Jones, *Inbreeding and Outbreeding* (Philadelphia: J.P. Lippincott, 1919).
[25] Lewontin, Richard C. and Jean-Pierre Bertan, "Technology, Research, and the Penetration of Capital: the Case of U.S. Agriculture", *Monthly Review*, 38 (1986), pp. 21–34. See also, Deborah Fitzgerald, *The Business of Breeding: Hybrid Corn in Illinois, 1890–1940* (Ithaca, N.Y.: Cornell University Press, 1990).
[26] Allen, "The Eugenics Record Office" (cit. n. 23), pp. 257–258.

GARY WERSKEY

"THE SOCIAL CONTEXT OF SCIENCE": SOC SCI 119 AS A WAY OF LIFE AND LEARNING

Abstract

This essay discusses the personal impact and contemporary relevance of Everett Mendelsohn's Harvard course on the 'Social Context of Science' – first offered in the 1960s, and continuing to attract large numbers of students today. In homage to its originality, the course is assessed as a testbed for developing teachers, initiating research, and fostering educational innovation. What was not obvious in the Sixties was how relevant the vision underpinning Mendelsohn's program would become to an increasingly cosmopolitan and knowledge-based form of capitalism in the 1990s. Business interest in such concepts and practices as learning communities, paradigms, knowledge management, scenario planning, self-organizing systems and environmental models of 'natural capitalism' are reviewed. These developments suggest the need for a new generation of historians, social analysts (and activists), management teachers and consultants to be exposed to the program and methods underpinning 'The Social Context of Science', with a mission to bring critical thinking and scholarship to bear on the needs and practices of a rapidly globalizing business community.

The Social Relations of "SOC SCI 119"

The first thing that hit you, even sitting at the back of that nineteenth century barn of a lecture theatre, was his energy. The darting eyes and quick movements of this very young Professor conveyed more intellectual and personal excitement than anyone I had yet encountered as a graduate student at Harvard. What was this man so excited about? Even by the end of his first lecture you knew not only how passionate Everett Mendelsohn was about "The Social Context of Science" (SCS) but that he wanted you to share his passion for the subject as well. Everett was in short a superb salesman as well as an innovative scholar, making a pitch for a product to which he was absolutely committed. I made an instant "buy" decision and have been living with its consequences for the last 35 years.

G.E. Allen and R.M. MacLeod (eds.), Science, History and Social Activism: A Tribute to Everett Mendelsohn, 203–214.

Although SCS has probably not been a life-changing experience for most of the hundreds of students who have enrolled in it over the decades, it has undoubtedly influenced the thoughts, careers and lives of many of us. At least this was my experience both as a student and Teaching Fellow in the second half of the Sixties. To understand immediate and longer term influences requires some appreciation of the course's social relations: its educational and academic context; the historical period in which it was developed and first taught; its significance for Mendelsohn's own career and development, and not least the intellectual interest and challenge of its syllabus.

SCS in its various guises has been offered as an advanced elective subject in Harvard College's General Education program. In the Sixties it was part of the social science stream ("Soc Sci 119"). Whatever changes have occurred in the rationale for and content of Harvard's undergraduate curriculum during the last four decades, SCS has always formed a convincing part of a "liberal" education. By spanning the arts and sciences, it has helped to fill out and enrich the social and historical understanding of undergraduates, irrespective of their majors. The course has also been made available to graduate students both in the History of Science Department and across a number of Faculties (especially popular with science educators). SCS has thus attracted students from a diverse range of ages, backgrounds and academic interests, reinforcing its effectiveness as a medium for broadening their outlooks. This has been especially evident in the lively seminar discussions organized by the course's graduate Teaching Fellows.

SCS was a particularly attractive and stimulating subject to study, research and teach in the 1960s. There was a resurgence of scholarly interest in exploring the social relations of science, both historically and in relationship to the contemporary era of "Big Science". This trend was intensified and given new focus through anti-war and other social movements that highlighted the role of scientists in supporting militarism and racism. While radical students and scientists were researching and challenging the practices of the science establishment and the "scientific-military-industrial complex", academic specialists in the history, sociology and politics of science were riding a wave of expanded support for their work. New educational and research programs flourished for a time. Harvard witnessed and supported all these developments, which inevitably found their way into SCS's syllabus and discussions.

For Mendelsohn SCS in all its changing guises has played a strategic role in the development of his multi-faceted career. As a flagship course for the History of Science Department, it has positively influenced the choice of many undergraduates' majors and minors. (The Department in fact continues to "punch above its weight" in terms of the number of students it attracts relative to its size.) SCS has also served as an important learning laboratory for Everett and his research students. Many of the grand themes in his and our work were first articulated as rough sketches in SCS lectures and section discussions. An added advantage of teaching this wide-ranging and significant subject to such a diverse and talented group of students was the much needed energy and stimulation they gave back to their teachers, not least Mendelsohn. The course was likewise a critical "practice field" for young researchers to gain some experience and training as lecturers and section leaders. (I still remember vividly the pride and excitement I felt when delivering my first SCS lecture 30 years ago, with Everett beaming encouragement from the front row.) Finally, this course was Mendelsohn's "bully pulpit", a platform where he could articulate both the intellectual and moral dimensions of his scholarly mission.

TEACHINGS AND "LEARNINGS"

So what propositions have been on offer in Mendelsohn's SCS? In the Sixties, as "Soc Sci 119", its central, now unexceptionable theme was that the rise, development and contemporary position of modern science could only be understood as an evolving product and expression of its constantly changing social context. Indeed the intimacy and interconnections between the norms, organization and output of scientific communities and other social institutions had significantly and steadily increased since the Renaissance. To illustrate his central theme's power and implications, Mendelsohn took his SCS students on a heady, semester-long journey from mediaeval to contemporary science.

We grasped the distinction between mediaeval and modern scientific work through Mendelsohn's contrast between Galenic anatomy ("any connection between body part and precept was purely coincidental") and Vesalian practice ("poised above the cadaver with his book in one hand and his knife in the other").

Drawing on Edgar Zilsel's pioneering work on Renaissance engineering,[1] Mendelsohn demonstrated the early advances that came from scholars' adopting a craftsman's experimental approach to crucial scientific and social questions, especially those bearing on ballistics and warfare.

The conflict between traditional Catholicism and the revolutionary astronomical and physical theories of Copernicus and Galileo was explored through a combination of original sources, Thomas Kuhn's only recently published work on scientific revolutions[2] and even Brecht's *Life of Galileo.*

The correspondingly easier acceptance of modern science in Protestant societies was documented through the writings of Bacon, Boyle and the Royal Society; and linked to the economic imperatives of mercantile capitalism (as documented in the 1930s by Boris Hessen and the young Robert Merton).[3]

The often challenging and contradictory relations between scientific and political transformations were a favorite theme in SCS: whether in late eighteenth-century France or twentieth-century Russia (particularly as shown by the rise of Lysenko).

Mendelsohn also made us look hard at the theories, practices and fates of scientists who attempted to exert political influence either as radical critics (the "social relations of science movement" in Britain during the 1930s) or as policy advisers (for example, Oppenheimer and others associated with the Manhattan Project).

Soc Sci 119 then concluded with a brisk overview of science's contemporary social relations: its norms and organization (viewed through the functionalist writings of Merton and Joseph Ben-David); its connections with Big Government and science policy (as seen mainly by Don Price); scientists' connections with industry; and current controversies on scientific contributions to militarism, racism and technological domination of Western culture and the Third World.[4]

From these teachings a number of "learnings", to use the jargon of organizational development consultants, emerged for SCS's different audiences. The course's underlying significance for all of us is that it challenged our own paradigms about the nature of scientific knowledge and how it was produced. SCS forced us to reconsider our common sense view of scientific work as timeless, unchanging, objective and divorced from social needs and pressures. The key to Mendelsohn's success in getting us to reframe our view of science was his unique ability to combine and apply historical and social analysis to some compelling and varied case studies. Taking a 700-year perspective made it easier to appreciate how significantly science as both a social institution and a body of knowledge had changed during its pre-modern, modern and post-modern phases. (In SCS's latest incarnation, "Science and Society in the 20th Century", the timeframe has been considerably shortened.) So did the use of sociological, psychological and anthropological perspectives and concepts to connect the behaviors of scientists with those of other social groups and communities. Finally, Everett's vivid demonstration of American science's enmeshment in the political, military and economic imperatives of the post-war world served to consolidate our conviction that scientific work could only be understood in its social context.

Otherwise the value and meaning of SCS's perspectives, methodology and case studies varied between the course's principal customer "segments".

For science majors the course represented a unique opportunity to become better scientists: more self-conscious about the nature of their work, its connections with other social institutions and the ethical choices that they might have to confront one day.

SCS offered current and aspirant secondary and tertiary teachers a new approach both to science education and to the education of scientists and engineers. This was especially important for those of us destined to become professional historians and social analysts of science, for SCS offered a curriculum and pedagogy that substantially broadened our chances of finding employment as teachers of a diverse range of subjects to a variety of audiences.

As noted earlier, SCS outlined a revolutionary and exciting research program as notable for its eclectic combination of methodologies and disciplines as it was for its easy commerce between historical and contemporary issues. For Everett's research students it was also a treasure trove of topics that inspired many worthwhile Ph.D. dissertations and promising scholarly careers.

Whatever their disciplinary or professional focus, SCS was also a boon for social activists seeking critical and analytical tools and case studies to connect scientific research to social purpose and public policy. In the Sixties these concepts were often used by radical students to "demystify" science's role in producing ideology as well as technology supportive of military, technological, economic and ideological domination.

Viewed in these ways, SCS was nothing less than a platform for a way of life and learning in a complex, apparently science-based society. That may sound overdrawn as a description of a one-semester undergraduate course, but it has certainly provided me with an enduring and enriching context for my own life's work. As the following autobiographical tribute illustrates, Everett's program has not only fulfilled all the promise I saw in it at the time but has become even more relevant to the challenges posed by an increasingly globalized and knowledge-based form of capitalism.

SOC SCI 119 As A Way of Life and Learning

There are many other former students whose careers as teachers, researchers and activists testify to Mendelsohn's and SCS's influence. Mine is summarized here simply as a microcosm or fractal of the whole.

Everett's most significant impact on me has been as a leader whose synthesis of scholarly and political commitments could serve as a model for my own career. This was not something he explicitly taught: you just picked it up by being around him. What he legitimated for me was a love of learning as both a pleasurable and exciting pursuit and an essential component of efforts to challenge social injustice. His purpose was supported through SCS's use of critical moments in social and scientific history, the course's openness to different disciplines and methodologies and its engagement with contemporary issues. Just as many of Russia's post-revolutionary writers were said to have fallen out of Gorky's greatcoat, so too did I and many other socially engaged scholars drop out of Mendelsohn's duffel coat!

SCS's influence was reflected more immediately and directly in my research interests and teaching career. Everett's SCS lecture on the British "Social Relations of Science" movement of the 1930s gave rise to 15 years of archival work and oral history on this subject.[5] Not surprisingly, this research only served to deepen my commitment to SCS's blend of scientific history and social activism. For many of the movement's central characters, including J.D. Bernal, Lancelot Hogben and Joseph Needham, were themselves some of SCS's most significant intellectual godfathers. Meanwhile, partly on the strength of this work, I was able to embark on a 20-year teaching career largely devoted to presenting SCS-based courses to a variety of science and technology students in the UK.[6] Increasingly, the focus of this teaching moved to contemporary applications of science and technology and the preparation of undergraduates for research and management careers in industry.

Running parallel to and often closely intertwined with these academic pursuits in the 1970s were a series of political interventions associated with the radical science movement. This style of scholarly engagement, best associated with the British Society for Social Responsibility in Science and the *Radical Science Journal* editorial collective, was largely carried on outside of and often in opposition to the academy.[7] (I do recall, however, a period at the University of Bath in the mid-Seventies, when I was an academic shop steward for the Association of Scientific, Technical & Managerial Staffs. ASTMS was in fact the successor to Bernal and Co's Association of Scientific Workers.) While the form my radicalism took differed from Everett's in crucial respects, there was never any loss of sympathy or respect between us at the level of scholarly or political intent. I was simply doing SCS my way!

In 1987 an opportunity arose at the University of New South Wales in Australia to synthesize my interests as a scholar-activist through a project self-consciously designed to change the social relations of engineering. Prosaically called the UNSW Co-op Program, it aimed to produce a new generation of technologically literate and socially engaged business and community leaders. Its key strategies were to: attract high-potential school drop-outs to study science and engineering, form key partnerships with public and private sector organizations to provide varied practical experience, broaden the curriculum to embrace the social and economic dimensions of technology, and expose the "Co-op Scholars" to their social responsibilities as future leaders. The program's thrust was later extended through the creation of an industry-linked Master of Business and Technology (MBT) distance education venture. When I was later able to explain these ventures to Everett, he just grinned in wry appreciation that they were both a product of SCS thinking and a candidate for yet another case study as a piece of technologically inspired social engineering![8]

The logic of this latest SCS-inspired synthesis eventually inspired me to leave the academic world altogether and work more directly with organizations to develop their future leaders and direction. I initially believed that this move might finally remove me from any direct involvement with the issues raised by Soc Sci 119 all those years ago. I could not have been more wrong. For what I have since discovered is that the social relations of science, knowledge and innovation are now emerging as the central concerns of a new form of capitalism.

Knowledge Capitalism As Science

The management literature of the 1990s testifies to an extraordinary corporate preoccupation with the creation and management of knowledge as *the* essential core capability for global commercial success in the 21st century. This focus goes well beyond long established interest in applied basic research and technical innovation as sources of competitive advantage, to embrace the notion that the knowledge corporations themselves produce about their own practices is becoming a "science" and unique source of value in its own right. The key elements of this new form of "knowledge capitalism" include:

1. Formal approaches to building knowledge businesses, developing knowledge management systems and managing knowledge workers;

2. A variety of practices generative of greater organizational learning and innovation;

3. Neo-Darwinian paradigms and metaphors of organizations as living systems both cooperating and competing in the formation of new industrial niches;

4. A focus on solutions to environmental problems ("natural capitalism") as a significant new source of business opportunity;

5. A global perspective on corporate self-regulation and social responsibility; and

6. An emphasis on understanding past and potential shifts in regional and global development ("scenario planning") as an aid to creating new industries and competitive niches.

Each of these elements deserves a brief review.

Knowledge Business as Science

A knowledge business consists of three key "structures" for the exchange and generation of knowledge: professional competence, external structures and internal structures.[9] *Professional competence* resides with the firm's knowledge workers, whose theories, information, skills, experience, values and networks enable them to generate both tangible and intangible revenues (e.g. more knowledge capital). *External structures* connect the firm with its customers and suppliers in ways that both maximize the flows of money and knowledge between them and builds the firm's brand, one of its most valuable assets. *Internal structures,* among other things, are used to generate intellectual property that can arise from knowledge workers' interactions with customers, suppliers and themselves, especially in the context of major R&D projects.

These knowledge businesses need not be large, capital-intensive or even science-based (e.g. professional services firms). What they have in common is that they exist to exchange, generate and sell knowledge, in the form of ideas, products and services that add value to their stakeholders (including their own knowledge workers' professional communities). Their market value is always far in excess of their tangible wealth, while their growth rates are substantially greater than other industry sectors. Examples of these powerhouses of knowledge capitalism are E-commerce, telecommunications, software, biotechnology, environmental solutions (see below) and management consulting businesses.

As the locus of knowledge generation and sharing gravitates to commercially driven private sector organizations, the question of what "science" is, where it happens, under whose auspices and for what ends has already become a major issue for universities and other traditional centers of basic research activity. Cash-strapped as they are in the face of increasingly costly support technologies on the one hand and declining government support on the other, it is likely that they will also be institutionally transformed to some extent in the image and interests of other knowledge businesses. This is therefore likely to represent, to say the least, a significant shift in the social context of science.

Organizational Learning and Innovation

As business leaders and strategists have grown more self-conscious about knowledge as the key source of their wealth and competitive advantage, their interest in improving their organizations' capabilities for learning and innovation has substantially increased. One sign of this development has been the "learning organization" movement associated with the work of MIT's Peter Senge.[10] A number of major corporations have been investing substantially during the past decade in projects that will significantly improve both personal and organizational learning capabilities. Some of this work has taken the form of developing work-based "learning communities" where business initiatives are treated self-consciously as experiments to test and improve the organization's basic principles, models and strategies. These insights can then be embedded in a variety of knowledge management systems, that allow the organization's/community's knowledge workers to construct better experiments and design more successful innovations in the future.

Readers versed in the SCS tradition will have already noted the parallels between these corporate efforts to construct knowledge socially and our increasing understanding of how scientific communities self-organize for similar ends. This is not surprising. For academic theorists and management consultants working in the learning organization tradition have been explicitly applying the work of leading historians and social analysts of science to assist their clients to generate knowledge and innovations more efficiently and effectively.[11] While senior executives are unlikely to have read *The Structure of Scientific Revolutions* or even seen one of the management videos featuring Thomas Kuhn, they are very familiar with the concept of "paradigms". (For example, a recent *New Yorker* cartoon shows a secretary informing her boss that "The following paradigm shifts have taken place while you were out to lunch"!) What unites the Managing Director of a knowledge business with the head of an academic research unit is that both need a vibrant and productive community of practice to attract the best knowledge workers and produce knowledge that puts their organizations at the forefront of their "industry".

The Ecology of Competition

The association between capitalism and Darwinism has been a staple of intellectual historians ever since Marx first drew the parallel in 1860. In the case of knowledge capitalism we find a strong association between neo-Darwinian metaphors and an emerging understanding of organizational development and growth. The basis for this connection has been the attraction of management theorists and consultants like Senge and Margaret Wheatley to the work of such distinguished biologists as Francisco Varela and Humberto Maturana.[12] What the former have found in the latter is confirmation that organizations' internal development and external undertakings are best understood as the products of living, self-organizing systems.[13] This paradigm of organizational dynamics in turn has been used to understand enduringly strong patterns of resistance to transformational change efforts on the part of both the "organism" as a system and its parts.[14]

Neo-Darwinism also shows itself in the work of another body of management writers interested in a new ecology of competition that is emerging as capitalism becomes increasingly globalized as well as knowledge-based.[15] Instead of the classic interplay between firms with very clear boundaries and their competitors, suppliers and customers, corporations are now entering into unforeseen alliances and joint ventures with almost anyone in their supply chain. By sharing non-critical services with competing firms and other suppliers, businesses are able to focus on their core competencies, characteristically their excellence in delivering a unique, high value-added product or service. Another arena for "co-opetition" is the investment of industry competitors in expensive start-ups and research projects where risks and knowledge can be more easily shared and better leveraged. These are where the major bets are going down in what one economist has called "the casinos of technology". Such high-risk games will have a major impact on both science and society, and they are not for the faint-hearted (or cash-poor).[16]

"Natural" Capitalism: Profiting from the Environment

Despite, or perhaps because of, decades of environmental degradation to the contrary, there is a strongly optimistic, even messianic current in recent management writings that corporations are about to reposition themselves in relation to the environment through a mixture of good science and even better technology.[17] The underlying message is that corporate profitability can be substantially enhanced both by embracing a more systematic approach to nature and resource conservation and by capturing new business opportunities arising from looming environmental catastrophes. The most credible framework for combining a conservationist ethic with a commercial ethos is the "Natural Step" process pioneered by Swedish biochemist Karl-Henrik Robert.[18] Other writers argue categorically that the most promising mega-opportunity for global business in the first quarter of the 21st century will be new products and services to clean up the environment. This expectation fits well with the advocates of "natural capitalism" and soft/green technologies who look forward to a resolution of the long-standing battles between environmentalists and developers.

Like the neo-Darwinists mentioned earlier, this species of technological utopianism will be familiar to SCS-minded readers who have been exposed to earlier examples of the triumph of optimism over experience. At least neo-Darwinism acknowledges notions of negative feedback loops, lag effects and unintended consequences when profitable technologies are introduced as fixes for the socially defined and commercially created problems of well-established natural and social ecosystems. It should be noted that one of the most influential articles in this arena gave Monsanto's biotechnology business as a prime example of the unmitigated benefits which would flow on to world food production.[19]

Corporate Social Responsibility

Movements for social responsibility in business now range from reformist think tanks like the University of Warwick's Center for Corporate Citizenship to more radical alliances like the World Business Academy, which was founded and funded by the Body Shop's Anita Roddick.[20] As with comparable earlier movements of scientists seeking to disengage their work from its "misuse" by governments and business, there are very clear limits to the debates and actions fostered by these bodies. The lowest common denominator in such activity is the need for major corporations to continue enjoying, in the language of major oil companies, a "license to practice" in their host countries. The upper limit are campaigns in support of "Fair Trade" and "green technologies" that reinforce the brands of those businesses who directly benefit from such practices. While there is nothing inherently wrong with such notions of enlightened (and consciously acknowledged) self-interest, they are unlikely to reassure others who are less sanguine about the approaching time "when corporations rule the world".[21]

Competing for the Future

Obviously knowledge capitalism is already a global reality and will become increasingly so, as cosmopolitan corporations attempt to build their businesses on the strength of world class competence, concepts and connections. To use the arresting phrase of prominent strategists Gary Hamel and C.K. Prahalad, companies are now hard at work to "get to the future first".[22] In preparing themselves for new waves of innovation and change, companies are using the techniques of "scenario planning" to identify a range of futures that could happen (including the one they most want to create).[23] Scenario building was originally developed as a planning tool at Shell in the late 1960s and progressively refined to render the company's strategies more durable in the face of an unpredictable range of events and possibilities. Scenarios can be run backwards ("backcasting") as well as forwards, to aid understanding of why and how an event or result (unforeseen or otherwise) occurred.

The interest of scenario planning for SCS devotees is that characteristically the corporations which could most benefit from this work lack both the competence and imagination to undertake it. That is why Shell and other companies have tended to commission external individuals and groups who can apply much deeper historical, economic, social, environmental and technological perspectives to issues of both regional and global development. The emerging requirement of businesses in the era of knowledge capitalism is for people who have real insight into how leading edge science and knowledge have been and could be produced.

21st Century SCS Agenda

The social context of science in the new century will be that of knowledge capitalism. What science will mean, where it is practiced, how it is applied and for what ends anywhere on earth will be influenced if not fundamentally determined by corporate imperatives to generate wealth and competitive advantage through the development of new forms of knowledge and knowledge-based products and services. Whatever one's views about the opportunities and dangers inherent in these emerging trends, the need to bring critical understanding and scholarship to bear on them is evident. (The latest version of SCS's syllabus already picks up on several of these themes.)

While the need for an undergraduate course with SCS's purpose, focus and methodology remains constant, the rise of knowledge capitalism also poses a challenge at the postgraduate level. Currently, there is a worldwide debate about the morality of offering advanced training in the humanities generally and the historical/social studies of science in particular, when the academic job market is very unlikely to absorb more than a fraction of this output.[24] Everett and I talked over this very issue the last time I saw him at Harvard in November, 1997.

My response to this challenge is to argue for the continuing education in the SCS tradition of historians and social analysts of science who can apply their competence and passion to the issues posed by the advent of knowledge capitalism. We need their presence in the academy, the corporation, management consultancies, think tanks, NGOs and, whenever appropriate, on the streets. One of the contexts where these scholars could make a great contribution is in the education of future business and community leaders. Admittedly this would require Everett and his colleagues to do something we would never have contemplated in the Sixties: walk across the Charles River and start talking to the Business School!

Educational Partnerships Australia

Notes

1 Edgar Zilsel, "The Genesis of the Concept of Scientific Progress," in *Roots of Scientific Thought*, P. Wiener and A. Noland, eds. (New York: Basic Books, 1957), pp. 251–75; and Edgar Zilsel, "The Sociological Roots of Science," *American Journal of Sociology*, 47 (1942), pp. 544–62.

2 Thomas S. Kuhn, *The Structure of Scientific Revolutions* (Chicago: University of Chicago Press, 1962).

3 B.Hessen, "The Social and Economic Roots of Newton's *Principia*," in *Science at the Cross Roads*, N.I. Bukharin *et al*, eds. (London: Frank Cass, 2nd edn. 1971), pp. 147–212.

4 See, for example, Robert K. Merton, *Social Theory and Social Structure* (Glencoe: Free Press, 1949); Joseph Ben-David and A. Zloczower, "Universities and Academic Systems in Modern Societies," *European Journal of Sociology*, 3 (1962), pp. 45–84; and Don K. Price, *The Scientific Estate* (Cambridge, MA: Harvard University Press, 1965).

5 Gary Werskey, *The Visible College: A Collective Biography of British Scientists and Socialists of the 1930s* (London: Free Association Books, 2nd edn. 1988).

[6] Gary Werskey, "The 'Non-Technical' Education of Engineers," *Metascience*, No. 4 (1986), pp. 46–59; and Gary Werskey, "Engineering People: How Japanese Electronics Firms Train Their Engineers," *Work and People*, 12 (1986), pp. 14–20.

[7] *Radical Science Essays*, Les Levidow, ed. (London: Free Association Books, 1986); and *The Radicalisation of Science*, Hilary Rose and Steven Rose, eds. (London: Macmillan, 1976).

[8] Gary Werskey, "Engineering Changes in Professional Education," in Helen Edwards and Simon Barraclough, eds., *Research and Development in Higher Education*, 2 (Sydney: Higher Education Research and Development Society of Australia, 1989), pp. 130–134.

[9] Karl Erik Sveiby, *The New Organizational Wealth: Managing and Measuring Knowledge-based Assets* (San Francisco: Berrett-Koehler, 1997).

[10] Peter M. Senge, *The Fifth Discipline: The Art & Science of the Learning Organization* (New York: Currency Doubleday, 1990). See also Peter M. Senge *et al*, *The Fifth Discipline Fieldbook: Strategies and Tools for Building a Learning Organization* (New York: Currency Doubleday: 1994).

[11] Ikujiro Nonaka and Hirotaka Takeuchi, *The Knowledge-Creating Company* (New York: Oxford University Press, 1995).

[12] Margaret Wheatley, *Leadership and the New Science* (San Francisco: Berrett-Koehler, 1993); Fritjof Capra, *The Web of Life* (New York: Anchor Books, 1996); Humberto R. Maturana and Francisco J. Varela, *The Tree of Knowledge: The Biological Roots of Human Understanding* (Boston: Shambhala, 1992).

[13] Arie de Geus, *The Living Company* (London: Nicholas Brealey, 1997).

[14] Peter M. Senge *et al*, *The Dance of Change: The Challenges of Sustaining Momentum in Learning Organizations* (London: Nicholas Brealey, 1999).

[15] Ron Ashkenas *et al*, *The Boundaryless Organization: Breaking the Chains of Organizational Structure* (San Francisco: Jossey-Bass, 1995); James F. Moore, *The Death of Competition: Leadership and Strategy in the Age of Business Ecosystems* (New York: John Wiley & Sons, 1996); Michael E. Porter, "Clusters and the New Economics of Competition," *Harvard Business Review*, 76 (1998), pp. 77–90.

[16] W. Brian Arthur, "Increasing Returns and the New World of Business," *Harvard Business Review*, 74 (1996), pp. 100–109.

[17] Amory B. Lovins, L. Hunter Lovins and Paul Hawken, "A Road Map for Natural Capitalism," *Harvard Business Review*, 77 (1999), pp. 145–158. See also Paul Hawken, *The Ecology of Commerce: A Declaration of Sustainability* (New York: HarperCollins, 1993).

[18] Karl-Henrik Robert, *The Natural Step: A Framework for Achieving Sustainability in Our Organizations* (Cambridge, MA: Pegasus Communications, 1997).

[19] Stuart L. Hart, "Beyond Greening: Strategies for a Sustainable World," *Harvard Business Review*, 75 (1997), pp. 66–76; and Joan Magretta, "Growth through Global Sustainability: An Interview with Monsanto's CEO, Robert Shapiro," *Harvard Business Review*, 75 (January 1997), pp. 78–88.

[20] John Elkington, *Cannibals with Forks: The Triple Bottom Line* (London: Capstone, 1997).

[21] David Korten, *When Corporations Rule the World* (San Francisco: Kummarian/Berrett Koehler, 1995).

[22] Gary Hamel and C.K. Prahalad, *Competing for the Future* (Boston: Harvard Business Review, 1994).

[23] Kees van der Heidjen, *Scenarios: The Art of Strategic Conversation* (New York: John Wiley & Sons, 1996).

[24] See, for example, Robert Weisbuch, "Unleash the Humanities," *WWNF Newsletter* (Spring 1998), p. 3.

III

SCIENCE, SOCIETY AND THE SOCIAL RESPONSIBILITY OF SCIENTISTS

MARK V. BARROW, JR.

NATURALISTS AS CONSERVATIONISTS: AMERICAN SCIENTISTS, SOCIAL RESPONSIBILITY, AND POLITICAL ACTIVISM BEFORE THE BOMB

Abstract

American naturalists have long served key roles in conservation causes. This paper presents case studies of three prominent scientists whose lives and careers demonstrate this pervasive, yet historically neglected phenomenon: the ichthyologist David Starr Jordan, the ornithologist Frank Michler Chapman, and the ecologist Victor Shelford. Their stories reveal a deep, abiding commitment to conservation within the American natural history community, a commitment that is not only important in its own right but also challenges the prevailing view that an interest in social responsibility and political activism in science was a product of the atomic age.

Introduction

Contrary to a persistent myth within the science studies community, scientists did not suddenly discover the notion of social responsibility with the creation of the atom bomb. There is no doubt that the development and use of this incredibly destructive weapon led to much hand-wringing, especially among some of the scientists who played a central role in its birth. There is also no doubt that the nuclear annihilation of Hiroshima and Nagasaki sparked an emotional debate that forced into the limelight the related issues of political activism and social responsibility in science.[1] The prominence of this discourse and the status of the individuals involved have led many scholars to conclude falsely that before the Manhattan project most, if not all, scientists remained aloof from society, largely oblivious to the implications of their research. Gerald Holton, a professor of physics and the history of science at Harvard and one of Everett Mendelsohn's longtime colleagues, presented a typical formulation of this prevailing view two decades ago when he argued that: "On the time scale of history, social responsibility and other social concerns as a

G.E. Allen and R.M. MacLeod (eds.), Science, History and Social Activism: A Tribute to Everett Mendelsohn, 217–233.

topic of active introspection by even a small percentage of practicing scientists is a recent notion, largely a post-Hiroshima conception".[2]

There are numerous problems with such a claim. First, it seems to equate all of science with a particular subset of it – pure or basic research – the production of fundamental knowledge of the natural world for its own sake, with little or no concern about potential applications.[3] Certainly many researchers were influenced by this lofty ideal. Yet, the thousands of scientists who routinely served on private and governmental advisory boards, who worked for state and federal agencies, or who undertook research in industrial, medical, or agricultural settings prior to World War II would be quite surprised to learn that they were indifferent to the applications of the knowledge they were producing. Applied science is and always has been intimately wrapped up with economic, social, and political concerns, and much of the science done in the last several centuries has been applied science.

A second potential challenge to the prevailing view comes from mainstream historians, who have long been providing hints that something might be wrong with the tradition of locating the emergence of social responsibility in science as late as World War II. Until the late 1950s and early 1960s most American historians viewed the Progressive movement – the prolonged period of reform agitation that stretched from the end of the nineteenth century until the First World War – in terms of a triumph of "the common people" over various narrow (and often corrupt) "special interests". With the appearance of Samuel P. Hays's *Conservation and the Gospel of Efficiency* (1959), however, they began to recognize that a major characteristic of progressivism was the ascendancy of "experts" of various sorts, including scientists, who sought to impose order and efficiency on an increasingly chaotic modern industrial society, while simultaneously advancing their own professional interests.[4] Hays' influential study focused on scientists within the federal government who lobbied for various conservation initiatives, including the creation of new federal agencies and the passage of new protective legislation; subsequent historians have examined the role of psychologists, geneticists, chemists, and other scientists who were central to Progressive reform.[5]

Sociologists studying professions have provided a third approach that might have led to more questioning of the notion that scientists first began to grapple with the issue of social responsibility in the shadow of the bomb. For decades now a key theme in the vast body of sociological literature devoted to professionalization – studies in which occupations like medicine, law, and the ministry predominate – is the claim that one of the distinguishing characteristics of modern professions is an orientation toward public service.[6] Now, either the professionalization of science is entirely anomalous (an unlikely prospect), or we need to start paying more attention to the ways in which scientists have interacted with the public in the years before the bomb.[7] I would argue that the latter proposition makes the most sense.

That is not to suggest that scholars have entirely ignored the issues of social responsibility and political activism in science prior to the bomb. In *The Visible College* (1978), a book that began as a dissertation under the direction of

Everett Mendelsohn, Gary Werskey convincingly documents the extensive political activities of several left-leaning British scientists during the 1930s, including J.B.S. Haldane, J.D. Bernal, Joseph Needham, Herman Levy, and Lancelot Hogben.[8] Similarly, Peter J. Kuznick's *Beyond the Laboratory* (1987) presents an analogous case of American scientists who became politicized during the Great Depression, a period of profound social and intellectual ferment on both sides of the Atlantic. Working under the auspices of newly established organizations, like the American Committee for Democracy and International Freedom and the American Association of Scientific Workers, prominent scientists in the United States positioned themselves on the "forefront of both the antifascist and social reform movements" of the 1930s.[9] And William Akin's study of the technocratic movement has demonstrated how engineers and scientists sought to re-orient society along more rational, scientific, and efficient lines in the years leading up to World War II.[10]

These are important studies that present difficult challenges to those who would link the emergence of social consciousness and political activism in science to the development of nuclear technologies, but they fail to go far enough in recognizing how many scientists behaved as socially engaged political actors in the years before the Second World War. More particularly, they ignore an important, pervasive tradition of naturalists who had long been active in conservation causes.[11] To illustrate this phenomenon here, I rely on case studies examining the lives and careers of three prominent American naturalists who were also well-known conservationists: the ichthyologist David Starr Jordan, the ornithologist Frank Michler Chapman, and the ecologist Victor Ernest Shelford. All three were active in the period between the end of the nineteenth century and the beginning of the Second World War, and all three made fundamental scientific contributions that were widely acclaimed by their peers. At the same time all three were centrally involved in conservation campaigns that required them to go before the public as experts in their given fields. While Jordan, Chapman, and Shelford might be exceptional in the amount of time and effort they devoted to activities in the public sphere, they are hardly unique in their desire to apply their expert knowledge to civic affairs. Many of their colleagues were also firmly dedicated to conservation causes. Whether motivated by a desire to impose order on an unregulated and chaotic society, a longing to advance their emerging profession, a sense of civic duty, or an emotional attachment to nature, naturalists provided crucial leadership in an ongoing campaign to protect American flora and fauna from the juggernaut of unrestrained economic, technological, and territorial growth.

David Starr Jordan and the Conservation of Human and Aquatic Resources

David Starr Jordan (1851–1931) spent his childhood in rural upstate New York, an environment that provided him with ample opportunity to explore his youthful interest in fishing, collecting plants, and gazing at the stars.[12] His parents were moderately prosperous farmers and successful teachers who encouraged their children's curiosity about the natural world while instilling them with a strong sense of morality and civic duty. After receiving a special exemption to study at the Gainesville Female Seminary, Jordan entered the newly established Cornell University in 1869 with visions of becoming either a botanist or a sheep herder.

Jordan seemed to thrive at Cornell and later claimed that his experiences there "exerted a controlling influence" over his entire subsequent career.[13] Besides having contact with inspiring professors – like the geologist C.F. Hartt and the zoologist Burt Wilder – Jordan relished the opportunity to teach botany to his fellow undergraduates. During this period he also became enthralled with the writings of the American transcendentalist Henry David Thoreau, which were finally beginning to gain a wide audience, and the educational theories of Cornell's reform-minded president, Andrew D. White.[14] Impressed with his accomplishment, Jordan's instructors granted him an M.S. in 1872, just a little over three years after he first set foot on the Cornell campus.

The year after graduation Jordan attended an experimental summer school that the famed naturalist Louis Agassiz organized on Penikese Island, off the coast of Woods Hole, in Buzzard's Bay, Massachusetts. This experience also proved critical in Jordan's development as a naturalist. Agassiz, a charismatic scientist and visionary institution-builder, inspired Jordan to take up the study of fish, an undertaking he quickly and enthusiastically embraced. Jordan soon established a reputation as the "greatest living authority on ichthyology", and by the end of his life, he had authored or co-authored more than 600 articles and books on fish and named more than 2,500 new species, a truly phenomenal accomplishment.[15] In addition to extensive contributions to descriptive taxonomy, Jordan came up with theoretical insights into the role of isolating mechanisms in evolution, including Jordan's law – the notion that the species most closely related to each other tend to be found just beyond the barriers that separate their populations.[16]

Unfortunately for Jordan, good teaching positions in natural history were scarce at the end of the nineteenth century, even for someone with his ambition and credentials. Following graduation from Cornell, he held jobs at a series of high schools and small colleges until 1879, when he received an invitation to become professor of natural history at Indiana University. Six years later he reluctantly agreed to serve that institution as America's youngest college president. Jordan overhauled the curriculum, introduced "majors" and electives, and enacted other important reforms during his six years as president at Indiana.

These initiatives eventually gained the attention of Leland and Jane Lathrop Stanford, who were in the process of creating a university as a memorial to their deceased son. The Stanfords were particularly keen to have their new university provide its students with "training for usefulness in life", an educational philosophy with which Jordan was in full agreement.[17] In 1891 he became founding president of Stanford University. He attracted an impressive faculty, designed the curriculum, recruited students, and managed to shape the institution into a thriving concern, despite troublesome financial and legal challenges that engulfed its early years. In a circular outlining its guiding principles, Jordan announced that "work in applied science was to be carried on side by side with the pure sciences and humanities and be equally fostered".[18] Nine years later, during a lecture to a Japanese audience, he argued that "the final end of education is not learning or official position, but service to humanity".[19] Jordan remained as president at Stanford until 1913 and as chancellor until 1916.

In a break with precedent, Jordan remained actively engaged in research and teaching even after he became a university president. One of his favorite courses was on "Bionomics", which explored the relationship between the biological and human sciences. The popular course, which he first taught at Indiana and continued offering regularly for the next thirty years, began with the "laws of organic life" and concluded with eugenics and the relationship between biology and ethics.[20]

Jordan's interest in the application of scientific knowledge to society was not confined to the college lecture hall; he also became an outspoken proponent of eugenics. In addition to regularly promoting the idea in public lectures and publications, in 1906 he became chair of the first formal eugenics organization in the United States, the American Breeders Association Committee on Eugenics, which included several other prominent American scientists and inventors.[21] By his own account, four years later Jordan was also instrumental in obtaining funding for the Eugenics Record Office, which under the leadership of the Harvard-trained biologist Charles B. Davenport, became the nerve center for the American eugenics movement.[22] Supporters of that movement, which included most practicing geneticists at the time, became instrumental in the passage of a long series of state laws authorizing the sterilization of patients in publicly owned mental facilities and federal legislation that restricted immigration.[23] These policies were exactly the kind of reform many turn-of-the-century progressives advocated. Through the systematic application of the science of genetics, eugenicists hoped to impose order and control on the chaotic process of human breeding.

Jordan also became a leading American pacifist. Although he had long been interested in educational reform, not until after the Spanish-American War – a conflict that he felt could have been avoided – did his "mind began to turn more directly to matters of government – national, international, and municipal".[24] His frequent pleas for world peace and international arbitration were informed not only by a distaste for the human suffering wrought by war but also by its eugenic implications. Throughout history, Jordan argued, warfare had resulted in a kind of reverse selection, a racial degeneration in which the strongest, bravest, and most ambitious young men tended to be killed off, leaving the less fit behind to pass on their negative traits to future generations.[25]

Jordan found other opportunities to apply his scientific knowledge through conservation efforts on behalf of aquatic animals.[26] By the end of the nineteenth century overharvesting, habitat destruction, and pollution were wreaking havoc with fish populations in the waters between the United States and

Canada. The political boundaries that cut through these populations hindered early attempts at reversing the decline but did nothing to prevent the regular movement of fish and fishermen across them. In 1896 an initial Joint Commission Report copiously documented the deterioration of the boundary water fisheries but offered only a handful of specific recommendations for restoring their productivity. Not until twelve years later did U.S. and Canadian officials finally negotiate a treaty creating a two-member International Fisheries Committee (IFC) with the authority to draft protective regulations.

Jordan accepted an invitation to serve as the American representative on the IFC "as a matter of pleasure and honor, and also as a matter of duty".[27] In 1909, after traveling through the areas to be affected by the regulation, reviewing the concerns of commercial fishers, and meeting with state and federal officials, Jordan and his Canadian counterpart issued a set of sixty-six regulations covering all of the waters from Passamaquoddy Bay to Puget Sound. Behind the scenes Jordan struggled aggressively in an effort to get his regulations enacted into law, but he and other supporters were unable to rally American conservationists around their cause.[28] In the end he was dismissed from the IFC when his outspoken views clashed with those of high-ranking State Department officials, and the enabling legislation for the Inland Fisheries Treaty failed to gain passage in the U.S. House of Representatives.[29]

Jordan had better success with his efforts to protect the northern fur seal, an economically valuable marine mammal that was born on American- and Russian-owned islands in the Bering Sea, but spent most of its life in international waters. In the 1880s Canadians began taking seals on the high seas. This practice was within their legal rights, but it led to exorbitant waste and a precipitous decline in fur seal populations. With the possibility of extinction looming on the horizon, in 1896 the U.S. and Great Britain dispatched scientists to the eastern Bering Sea to study the problem. Heading the American delegation was David Starr Jordan, chosen more for his solid scientific reputation than for his knowledge of seals. In a four-volume report Jordan and his colleagues defended the practice of hunting surplus male seals on land while condemning pelagic sealing as unsustainable because it depleted females more quickly than they were replaced. Canadian officials initially refused to accept this conclusion. Facing a continuing impasse, increased hunting pressure from Japanese sealers, irrefutable evidence of the ongoing decline of the fur seal population, and growing public opposition to the practice of pelagic sealing, in 1909 the United States government appointed Jordan to chair a new Fur Seal Advisory Board that included several other well-known American naturalists.[30] Jordan, who had been promoting the idea of international arbitration through his work for the peace movement, argued that the U.S. should invite Canadian, Russian, and Japanese officials together to reconcile their differences. Jordan's proposal soon came to fruition, and by 1911 representatives of these four governments hammered out a landmark agreement that helped reverse the long-term decline in fur seal populations.

A man of strong conviction and boundless energy, David Starr Jordan belies the image of the scientist working in cloistered isolation from society, indifferent to the results of his study. As his reputation for scientific research and educational reform grew, Jordan increasingly worked privately and appeared before the public in the name of eugenics and aquatic animal conservation. While not always successful, Jordan nonetheless remained committed to the idea that expert knowledge was critical to solving the challenges facing modern society, a commitment he shared with other progressive reformers.

Frank Michler Chapman: Ornithologist and Apostle for the Birds

"That science which is sufficient to itself has no excuse for existence. If our studies of birds have no bearing on the progress and welfare of mankind they are futile". So argued Frank Michler Chapman (1864-1945) in an address before his colleagues assembled at the Fourth International Ornithological Congress in 1905.[31] Although Chapman had less formal education than Jordan, he clearly felt a similar sense of obligation to use his expert knowledge in the service of humanity, particularly in the cause of wildlife conservation. As the author of popular field guides, a leader in the Audubon movement, and founding editor of its official journal, Chapman played a leading role in awakening American sympathy for birds, "the most eloquent expressions of Nature's beauty, joy, freedom". At the same time he managed to make fundamental contributions to the science of ornithology, both through his lengthy service to the American Museum of Natural History and his pioneering studies on the biogeography of South American birds.

Chapman was born in Englewood Township, New Jersey, at a time when it served as a sparsely populated bedroom community for New York City.[32] His father was a prosperous attorney with an office on Wall Street and a gentlemanly farmer who possessed a "strong sense of civic duty"; his mother was an avid gardener and gifted musician.[33] As a youth Chapman enjoyed exploring, hunting, and collecting birds in the countryside surrounding his family's forty-acre estate, but he had little sense of what he wanted to do with his life and no desire to attend college. Following graduation from Englewood Academy in 1880, the sixteen-year-old Chapman began working as a collection agent at the American Exchange National Bank of New York, an institution for which his father had once served as lawyer.

In the spring of 1884 Chapman volunteered to become a migration observer for the American Ornithologists' Union, a decision that marked a turning point in his life. Although he had a lengthy commute and a six-day work week at the bank, he still managed to put in two and a half hours in the field each day. Chapman was ecstatic when A.K. Fisher, the ornithologist in charge of the project, declared his migration report to be the finest in the eastern United States. He had now found a calling, and much to the consternation of his "mystified colleagues", he soon gave up his bank position in the hope of pursuing a career as a professional ornithologist.[34] After volunteering at the American Museum of Natural History, in 1888 he was hired as an assistant to

J.A. Allen, who became an important father-figure and tutored him on the finer points of technical ornithology.[35] Chapman remained at the institution for more than fifty years, serving as chair of the bird department from the time it was created in 1908 until his retirement in 1942. The last two decades of Chapman's reign have since become known as "the Golden Years" of the bird department. He increased the staff by recruiting many world-class ornithologists to curate the collections, he organized or sponsored dozens of expeditions, he revitalized the exhibits through the introduction of habitat groups, he cultivated relationships with many generous patrons, and he built up the study collection into the second largest in the world.[36] According to Ernst Mayr, who joined Chapman's staff in 1931: "There were (in the 1930s and 1940s) an intellectual excitement and a level of professional competence and ornithological universality at the American Museum that had nowhere existed previously and perhaps can never again be duplicated".[37]

Chapman also managed to make important scientific contributions of his own. Mayr argues that Chapman had a much broader vision of ornithology than many museum curators at the time, most of whom seemed content simply to create new genera and species.[38] He developed a keen interest in the living bird, its habitats, evolutionary origins, and distribution, and more than once privately expressed frustration at feeling pressured to collect when he was in the field. His monographs on the biogeography and ecology of the birds of Colombia (1917), Peru (1921), Ecuador (1926), and Venezuela (1931) were widely acclaimed at the time they were published and are now considered classics.[39] Later in life he initiated an entirely new line of work: detailed life histories of tropical island species.[40] His colleagues recognized his ongoing scientific contributions by electing him to membership in the National Academy of Sciences (1921) and by granting him many awards.[41]

Chapman's museum and scientific work would have been more than enough to keep most other people busy, but he also managed to pursue the equivalent of a second career as one of the nation's best-known bird popularizers and conservationists. He was in great demand as a lecturer, where he often displayed his pioneering bird photographs and motion pictures, while his series of innovative field guides were bestsellers. Through these and other means he introduced hundreds of thousands of Americans to the joys of birdwatching, an accomplishment that was central to his deep commitment to the cause of bird protection.[42] Chapman was an early and active member of the AOU bird protection committee (organized in 1884), a founder of the New York Audubon Society (organized in 1897), and longtime member and chair of the board of directors of the National Association of Audubon Societies (organized in 1905). Even more time-consuming was his work for *Bird-Lore,* the official journal of the Audubon movement, which he founded in 1899 and continued to edit until 1934. In all these capacities, he railed against the commercial exploitation of birds, lobbied for protective legislation, and worked for the creation of private, state, and federal avian sanctuaries. While many of Chapman's colleagues were also active in the cause of bird protection, no one did more than he did to "spread the gospel of birdlore".[43]

VICTOR ERNEST SHELFORD, ANIMAL ECOLOGY, AND THE PRESERVATION OF WILDERNESS

Victor Ernest Shelford (1877–1968) was also something of a late bloomer. But what he lacked in initial focus, he soon made up for in enthusiasm and achievement. By the end of his long, productive career he was lionized as the "father of animal ecology in the United States" and a tireless, effective advocate for wilderness preservation.[44] Working first under the auspices of the Ecological Society of America and later of the Nature Conservancy (both of which he had helped to found), Shelford served as an outspoken crusader for identifying and protecting undisturbed natural landscapes.

Like Chapman and Jordan, Shelford was born to rural circumstances and enjoyed the outdoors from a young age, although he never developed a taste for either farming or hunting.[45] Because he began teaching while still a student, he failed to receive his high school diploma until age twenty-two. He then spent two years at West Virginia University before transferring to the University of Chicago, where he earned an S.B. in 1903 and a Ph.D. in zoology in 1907.

When Shelford arrived at Chicago the university was only a decade old, but its biology program had already earned a reputation as one of the strongest in the United States. He soon fell under the spell of several gifted instructors on the faculty. The geneticist Charles B. Davenport encouraged Shelford to take up the study of animal ecology and suggested his dissertation project on tiger beetles; the developmental biologist Charles M. Child took him on long rambles through the countryside, provided him with his "first serious inspiration in natural history", and taught him the experimental techniques of physiology that became so central for Shelford's ecological research; perhaps most importantly, the ecologist Henry C. Cowles introduced him to physiological ecology and enchanted him with regular field trips to the dunes on the shores of Lake Michigan, an environment that provided a unique opportunity to study the process of ecological succession.[46] The dunes not only supplied Shelford with data for his dissertation and other early published papers, but also gave him a window onto the effects of industrialization on the landscape; his concern that this unique environment was falling victim to development was a key factor in his decision to become active in the wilderness preservation movement.

After earning his Ph.D., Shelford stayed on at Chicago until 1914, when he began his long association with the University of Illinois. There he combined experimental work in the physiology of animals with an attempt to understand their distribution from the perspective of natural communities, an approach that pervades his first two books, *Animal Communities in Temperate North America* (1913) and *Laboratory and Field Ecology* (1929).[47] Soon after his arrival at Illinois, he began a long and often trying collaboration with America's preeminent plant ecologist, Frederic E. Clements. Their attempt to merge plant and animal ecology, previously distinct enterprises with separate concepts and terminologies, eventually came to fruition in their co-authored book *Bio-Ecology* (1939). Besides his many widely acclaimed publications,

Shelford exerted a tremendous influence on the science of ecology through the more than two dozen Ph.D. students who worked under his direction, including many who went on to distinguished careers of their own.[48] Although forced to retire in 1946, Shelford remained active in research and conservation work until the end of his life. He published his final book, *The Ecology of North America* (1963), when he was eighty-six years old. Five years later the Ecological Society of America granted Shelford an "Eminent Ecologist" Award, the organization's highest honor.

Despite a heavy schedule of teaching, writing, and fieldwork at Illinois, Shelford always found time to devote to conservation. He was a founding member and the first president of the Ecological Society of America, organized in 1915. Two years later he convinced the ESA to establish a Committee on the Preservation of Natural Conditions, which he then chaired for much of the next three decades.[49] By 1921 more than fifty ecologists were contributing to the committee's increasingly ambitious efforts, which included actively lobbying federal and state agencies, politicians, and the public about the need for preservation. According to Shelford, undisturbed natural areas were "living museums" that were critical for reconstructing how healthy biological communities functioned and for discovering how to mitigate the damage humans had done to the environment. He argued that such areas should be set aside for scientific study in every national park and every national forest and that protective buffer zones, where development was severely restricted, should be established around them.[50] In making his pitch for the preservation of intact natural areas for scientific research, Shelford was an early proponent of abandoning the individual species approach that had characterized the wildlife conservation movement until that point. By 1926 Shelford and his colleagues had identified 600 areas in North America that were either "preserved" or "worth preserving" and incorporated their extensive inventory into a book, *Naturalist's Guide to the Americas*.[51]

Initially Shelford worked with the blessing of the ESA, though the organization provided only minimal financial support for his preservation work. But as his committee became increasingly large and active, criticism began to mount from those who felt that open advocacy of political positions was inappropriate for a scientific organization. Serious concern was voiced in the early 1930s, when the committee began attacking the U.S. Biological Survey's well-funded predator control program.[52] In 1943, after several ESA officers complained that conservation activities were dominating the society, Shelford decided to survey the membership to gage the level of support. Eight-four percent of the 400 ESA members who returned Shelford's survey voiced sympathy for the ESA's preservation work. One year later, in an editorial in *Science* that was clearly aimed at his ESA colleagues, Shelford castigated scientific societies that refused to become involved in applications of the knowledge their members produced:

> To minimize misconstructions and misrepresentations, public application of scientific principles and the needs of future research should be urged by

> specialists themselves. Human society, which supports research, will hold scientific men and the societies which they constitute responsible for the failure to urge the application of their knowledge directly and simply whenever it is in the interest of society to do so. No scientific society devoted to research should fail in fulfilling this obligation.[53]

Despite the earlier mandate, in a subsequent referendum held in 1945, ESA members voted to emasculate the preservation committee by barring it from taking part in any "political, pressure group, or propaganda activity".[54] Ironically, just as scientists working in the shadow of the bomb began embracing the issue of social responsibility, the ESA abandoned its previous commitment to wilderness preservation.[55]

Undaunted, Shelford and several of his supporters decided to form a new scientific organization to continue the preservation work of the former ESA committee. In 1946 eighty-three ecologists, including Shelford and many of his former students, served as charter members of the Ecologists' Union, whose goal was "preservation of natural biotic communities, and encouragement of scientific research in preserved areas".[56] The new organization continued to work on many of the issues the ESA had addressed and vigorously lobbied Congress and governmental agencies to make its position known. Within three years the Ecologists' Union boasted 294 members and Shelford was named a patron. In 1950 George Fell, a former student of Shelford, and Richard Pough, of the American Museum of Natural History, convinced the leaders of the organization to open membership to non-scientists and to begin raising money for a revolving fund to purchase natural areas, thereby sparing them from the threat of development. Thus, the Nature Conservancy, one of the largest and most successful environmental groups in the United States, was born.[57]

Conclusion

If Jordan, Chapman, and Shelford were unusual, it was only in the level of their commitment to conservation and the visibility of the campaigns with which they were involved. Before the Second World War countless American naturalists were active in efforts to establish state and national parks, wilderness areas, and wildlife preserves; to gain protective legislation for plants, animals, and landscapes; and to mobilize the public in support of their conservation initiatives. To achieve their goals they assumed leadership positions in (and often helped to found) local, state, and national conservation organizations; they served on state and federal conservation advisory committees; they created conservation committees within their own professional societies; they delivered speeches and published popular natural history accounts to cultivate public sympathy for conservation; and they appeared before legislative committees, wrote editorials, and even lobbied in person on behalf of protective legislation. Swept up in the reform ethos of the progressive era, anxious to bolster the standing of their nascent professions, informed by an emotional attachment to nature and deep sense of civic duty, naturalists deployed their expertise on behalf of more measured, efficient, and sustainable uses of the natural world. While the profound commitment of American naturalists to conservation has not gone entirely unnoticed, especially among environmental historians, it clearly deserves more scholarly attention than it has received.

Acknowledgments

I would like to thank Mark Largent, Kathleen Jones, Gar Allen, and Roy MacLeod for reading drafts of this essay and providing numerous helpful suggestions. Most of all, I want to thank Everett Mendelsohn for his warm encouragement, sage guidance, and unflagging support. I will always be extremely grateful for everything he has done.

Virginia Tech

Notes

1 On atomic scientists' early political activism in the United States, see Alice K. Smith's classic study, *A Peril and a Hope: The Scientists' Movement in America, 1945–1947* (Chicago: University of Chicago Press, 1965). Smith published a précis of her argument in "Scientists and the Public Interest, 1945–1946", *Newsletter on Science, Technology and Human Values*, 24 (June 1978), pp. 24–31. See also, R.R. Wilson, "Hiroshima: The Scientists' Social and Political Reaction", *Proceedings of the American Philosophical Society*, 140 (1996), pp. 350–357; Elizabeth Hodges, "Precedents for Social Responsibility among Scientists: The American Association of Scientific Workers and the Federation of American Scientists, 1938–1948" (Ph.D dissertation, University of California-Santa Barbara, 1983); and Lawrence Badash, *Scientists and the Development of Nuclear Weapons: From Fission to the Limited Test Ban Treaty, 1939–1963* (Atlantic Highlands, NJ: Humanities Press, 1995). Studies that place the scientists' reaction to the bomb into larger contexts include: Allan M. Winkler, *Life Under a Cloud: American Anxiety about the Atom* (New York: Oxford University Press, 1993); Paul Boyer, *By the Bomb's Early Light: American Thought and Culture at the Dawn of the Atomic Age* (New York: Pantheon Books, 1985); and Lawrence S. Wittner, *The Struggle Against the Bomb*, 2 vols. (Stanford: Stanford University Press, 1993–1997).

2 Quoted in Peter J. Kuznick, *Beyond the Laboratory: Scientists as Political Activists in 1930s America* (Chicago: University of Chicago Press, 1987), p. 3, which provides other examples of this argument. See also Edward Schils, "Scientists in the Public Arena", *American Scholar*, 56 (1987), pp. 185–202; Paul T. Durbin, *Social Responsibility in Science, Technology, and Medicine* (Bethlehem, Penn.: Lehigh University Press, 1992), p. 24; and Rae Goodell, *The Visible Scientists* (Boston: Little, Brown, 1977), pp. 40–42.

3 On the difficulties of the pure and applied distinction in science, especially as they relate to the American context, see Nathan Reingold, "The Scientist as Troubled American", *American Industrial Hygiene Association Journal*, 40 (1979), pp. 1107–1113; idem, "American Indifference to Basic Research: A Reappraisal", in *Science American Style* (New Brunswick, NJ: Rutgers University Press, 1991), p. 54–75; and Gerard Radnitzky, "Science, Technology, and Political Responsibility", *Minerva*, 21 (1983), pp. 234–264. See also, Alan Beyerchen, "On the Stimulation of Excellence in Wilhelmian Science", in *Another Germany: A Reconsideration of the Imperial Era*, Jack R. Dukes and Joachim Remak, eds. (Boulder, Colorado: Westview Press, 1988), pp. 139–168.

4 Samuel P. Hays, *Conservation and the Gospel of Efficiency* (Cambridge, Mass.: Harvard University Press, 1959). Robert Wiebe's, *The Search for Order, 1877–1920* (Westport, Conn.: Greenwood Press, 1967). Robert Wiebe's influential book did much to further the trend of stressing the role of experts in progressive reform. See also, Arthur S. Link and Richard L. McCormick, *Progressivism* (Arlington Heights, IL: Harlan Davidson, 1983).

5 See, for example, David J. Rothman, *Conscience and Convenience: The Asylum and its Alternatives in Progressive America* (Boston: Little, Brown, 1980); Kenneth Ludmerer, *Genetics*

and American Society: A Historial Appraisal (Baltimore: Johns Hopkins University Press, 1972); Mark H. Haller, *Eugenics: Hereditarian Attitudes in American Thought* (New Brunswick, NJ: Rutgers University Press, 1963); Daniel J. Kevles, *In the Name of Eugenics: Genetics and the Uses of Human Heredity* (New York: Knopf, 1985); James Harvey Young, *Pure Food: Securing the Federal Food and Drugs Act of 1906* (Princeton: Princeton University Press, 1989); and Mark Largent," 'These are Times of Scientific Ideals': Vernon Lyman Kellogg and Scientific Activism, 1980–1930" (Ph.D dissertation, University of Minnesota, 2000).

[6] See, for example, several convenient review articles: J.B. Morrell, "Professionalization", in *Companion to the History of Modern Science*, R.C. Olby et al., eds. (London: Routledge, 1990), pp. 980–989; Patricia Roos, "Professions", in *Encyclopedia of Sociology*, Edgar F. Borgatta and Marie L. Borgatta, eds. (New York: Macmillan, 1992), 3, pp. 1552–1557; Ivan Waddington, "Professions", in *The Social Science Encyclopedia*, 2nd edn., Adam Kuper and Jessica Kuper, eds. (London: Routledge and Kegan, 1996), pp. 677–678. See also, *The Sociology of Professions: Lawyers, Doctors, and Others*, Robert Dingwall and Philip Lewis, eds. (London: Macmillan, 1983); and Thomas L. Haskell, *The Authority of Experts: Studies in History and Theory* (Bloomington, Indiana University Press, 1984).

[7] In an article that continues to be widely cited, Everett Mendelsohn offered an early historical treatment of the professionalization of science: "The Emergence of Science as Profession in Nineteenth-Century Europe", in *The Management of Scientists*, Karl B. Hill, ed. (Boston: Beacon Press, 1964), pp. 3–48.

[8] Gary Werskey, *The Visible College: A Collective Biography of British Scientists and Socialists of the 1930s*, reprint ed. (London: Free Association Books, 1988 [1978]). See also, Maurice Goldsmith, *Three Scientists Face Social Responsibility: Joseph Needham, J.D. Bernal, F. Joliot-Curie* (New Delhi: Centre for the Study of Science, Technology and Development, 1976).

[9] Peter J. Kuznick, *Beyond the Laboratory* (cit. n. 2), p. 2.

[10] William E. Akin, *Technocracy and the American Dream: The Technocrat Movement, 1900–1941* (Berkeley: University of California Press, 1977). Other publications that emphasize scientists' concern with social and political issues before the Second World War include: Gregg Mitman, *The State of Nature: Ecology, Community, and American Social Thought, 1900–1950* (Chicago: University of Chicago Press, 1992); Daniel Kevles, "Notes on the Politics of American Science: Commentary on Papers by Alice Kimball Smith and Dorothy Nelkin", *Newsletter on Science, Technology and Human Values* 24 (June 1978), pp. 40–44; John Ziman, "Social Responsibility in Victorian Science", in *Science, Technology, and Society in the Time of Alfred Nobel*, Carl Gustaf Bernhard, Elisabeth Crawford, and Per Sörbom, eds. (Oxford: Pergamon Press, 1982); Hodes, "Precedents for Social Responsibility among Scientists" (cit. n. 1); and C.H. Waddington, *The Scientific Attitude* (Harmondsworth, UK: Penguin Books, 1941). Despite these and other studies, a tendency to view the emergence of social responsibility in science as a post-war phenomenon continues.

[11] I agree with those who stress the bomb's importance in providing the issue of social responsibility in science with an increased sense of urgency and a heightened visibility. Yet, while there are very few explicit discussions of the topic before World War II, there are numerous pre-war examples of scientists behaving as responsible actors committed to applying their expertise to the problems of society. In this paper I stress one such category of scientists who were clearly socially aware and politically active before the Manhattan Project, even if they did not invoke the idea of "social responsibility" to describe their sense of obligation to act in the public sphere.

[12] Useful biographical sketches of Jordan include: Elizabeth N. Shor, "David Starr Jordan", *Dictionary of Scientific Biography*, 7 (1973), pp. 169–170; B.W. Evermann, "David Starr Jordan, the Man", *Copeia* (December 1930), pp. 93–105; idem, "David Starr Jordan", *Dictionary of American Biography*, 10 (1933), pp. 211–214; and Keith R. Benson, "David Starr Jordan (1851–1931) Papers at Stanford University Archives", *Mendel Newsletter*, 26 (1986), pp. 1–5. One of the most comprehensive sources of information on the busy life and career of Jordan is his two-volume autobiography, *Days of Man: Being Memories of a Naturalist, Teacher, and Minor Prophet of Democracy* (Yonkers-on-Hudson, NY: World Book Company, 1922). See also, Edward McNall Burns, *David Starr Jordan: Prophet of Freedon* (Stanford: Stanford University Press, 1953).

[13] Jordan, *Days of a Man* (cit. n. 12), p. 95.

[14] On White's innovations in higher education, see Laurence Veysey, *The Emergence of the American University* (Chicago: University of Chicago Press, 1965),

[15] Evermann, "David Starr Jordan" (cit. n. 12), p. 212. A bibliography of his publications may be found in Alice N. Hays, *David Starr Jordan: A Bibliography of his Writings, 1871–1931* (Stanford: Stanford University Press, 1952). On Jordan's influence on ichthyology, see Carl L. Hubbs, "History of Ichthyology in the United States after 1850", *Copeia* (March 1964), pp. 42–60, who

argues (on p. 50) that: "[t]hroughout the final quarter of the nineteenth century David Starr Jordan – a man of massive frame and even vaster intellect – was the dominant figure in American ichthyology".

[16] Although Agassiz was personally opposed to evolution, his comparative approach to zoological research was quite amenable to evolutionary explanations, and virtually all of his students quickly accepted the idea. See Mary P. Winsor, *Reading the Shape of Nature: Comparative Zoology at the Agassiz Museum* (Chicago: University of Chicago Press, 1991). On Jordan's law, see David Starr Jordan, "The Law of Germinate Species", *American Naturalist*, 42 (1908), pp. 73–80. On Jordan's contributions to evolutionary theory, see Davis Magnus, "In Defense of Natural History: David Starr Jordan and the Role of Isolation in Evolution" (Ph.D dissertation, Stanford University, 1993), which is the only recent study of Jordan by someone trained in the history and philosophy of science.

[17] Jordan, *Days of a Man* (cit. n. 12), I, p. 485.

[18] Jordan, *Days of a Man* (cit. n. 12), I, p. 357–358.

[19] Jordan, *Days of a Man* (cit. n. 12), II, p. 79.

[20] Jordan, *Days of a Man* (cit. no. 12), I, p. 298. On the development and various uses of the term "bionomics", see Mark Largent, "Bionomics: Vernon Kellogg and the Defense of Darwinsim", *Journal of the History of Biology*, 32 (1999), pp. 465–488.

[21] Barbara Kimmelman, "The American Breeders' Association: Genetics and Eugenics in an Agricultural Context, 1903–1913", *Social Studies of Science*, 13 (1983), pp. 163–204.

[22] Jordan, *Days of a Man* (cit. n. 12), II, p. 298. Garland E. Allen, "The Eugenics Record Office at Cold Spring Harbor, 1910–1940: An Essay in Institutional History", *Osiris*, 2 (1986), pp. 225–264, which stresses Davenport's role in founding the ERO.

[23] A good introduction to the vast literature on eugenics is Diane Paul, *Controlling Human Heredity, 1865 to the Present* (Atlantic Highlands, NJ: Humanities Press, 1995). See also the studies listed in note 5. For starting points on the connection between eugenics and conservation (a subject that clearly deserves more study), see: Gray Brechin, "Conserving the Race: Natural Aristocracies, Eugenics, and the U.S. Conservation Movement", *Antipode*, 28 (1996), pp. 229–245; Helen L. Horowitz, "Animal and Man in the New York Zoological Park", *New York History*, 65 (1975), pp. 426–455; and Donna Harraway, *Primate Visions: Gender, Race, and Nature in the World of Modern Science* (New York: Routledge, 1989), pp. 54–58. On the relationship between eugenics and progressivism, see: Donald K. Pickens, *Eugenics and the Progressives* (Nashville, TN: Vanderbilt University Press, 1968); and Garland E. Allen, "Eugenics and American Social History, 1880–1950", *Genome*, 31 (1989), pp. 885–889.

[24] Jordan, *Days of a Man* (cit. n. 12), I. p. 618.

[25] See, for example, David Starr Jordan, *The Blood of a Nation: A Study of the Decay of Races through the Survival of the Unfit* (Boston: Beacon Press, 1902); *The Human Harvest: A Study of the Decay of Races through the Survival of the Unfit* (Boston: Beacon Press, 1907); and *War and the Breed: The Relation of War to the Downfall of Nations* (Boston: Beacon Press, 1915).

[26] His first major effort along these lines involved an extensive report on fish populations along the Pacific coast that the Census Bureau and the U.S. Fish Commission requested in 1880. Working with his graduate student Charles Henry Gilbert, Jordan cataloged 400 species (including 80 that were new to science), disentangled the confusing salmon taxonomy, and provided the first economic study of the salmon industry. Jordan, *Days of a Man* (cit. n. 12), I, pp. 226–227. The account of Jordan's conservation activities that follows relies heavily on Kurkpatrick Dorsey, *The Dawn of Conservation Diplomacy: U.S.-Canadian Wildlife Protection Treaties in the Progressive Era* (Seattle: University of Washington Press, 1998). See also, Michael Smith, *Pacific Scientists and the Environment, 1850–1915* (New Haven: Yale University Press, 1987).

[27] Quoted in Dorsey, *Dawn of Conservation Diplomacy* (cit. no. 26), p. 64.

[28] He was unwilling, however, to fulfill a request by the Secretary of State to appear in the Senate to lobby on their behalf, an action he thought improper, especially in the wake of Gifford Pinchot's recent firing for lobbying the Senate on behalf of forestry conservation. Jordan, *Days of a Man* (cit. n. 12), II, p. 271.

[29] See the analysis in Dorsey, *Dawn of Conservation Diplomacy* (cit. n. 26), pp. 51–104.

[30] Including C. Hart Merriam, Charles H. Townsend, Leonard Stejneger, and Frederic Lucas.

[31] Frank M. Chapman, "What Constitutes a Museum Collection of Birds?", *Proceedings of the Fourth International Ornithological Congress, London, June 1905* (London: Dulau and Co., 1907), p. 146.

[32] While there is no modern biography of Chapman, there are numerous published accounts of his life and work, including Robert H. Welker, "Frank Michler Chapman", *Dictionary of American Biography* supplement 3 (1941–1945), pp. 61–62; Ernst Mayr, "Frank Michler Chapman",

Dictionary of Scientific Biography, 17, supplement 2 (1990), pp. 152–153; William King Gregory, "Biographical Memoir of Frank Michler Chapman, 1864–1945", *Biographical Memoirs of the National Academy of Sciences*, 25 (1949), pp. 111–143 (which includes a bibliography of his writings); and Robert Cushman Murphy, "Frank Michler Chapman, 1864–1945", *Auk*, 67 (1950), which includes references to several other biographical sketches. These accounts all rely heavily on Chapman, *Autobiography of a Bird-Lover* (New York: Appleton-Century, 1933).

[33] Chapman, *Autobiography* (cit. n. 32), p. 3.

[34] Chapman, *Autobiography* (cit. n. 32), p. 42

[35] Chapman, *Autobiography* (cit. n. 32), p. 62. Chapman's father had died in 1876.

[36] On the history of the bird department, see Frank M. Chapman, "The Department of Birds, American Museum: Its History and Aims", *Natural History*, 22 (1922), pp. 307–318; Wesley Lanyon, "Ornithology at the American Museum of Natural History", in *Contributions to the History of North American Ornithology*, William E. Davis and Jerome A. Jackson, eds. (Cambridge, Mass.: Nuttall Ornithological Club, 1995), pp. 113–144.

[37] Ernst Mayr, "Epilogue: Materials for a History of American Ornithology", in Erwin Stresemann, *Ornithology from Aristotle to the Present* (Cambridge, Mass.: Harvard University Press, 1975), p. 372.

[38] Ernst Mayr, "Frank Michler Chapman" (cit. n. 34), pp. 152–153.

[39] "Distribution of Bird-Life in Colombia", *Bulletin of the American Museum of Natural History*, 36 (1917); "Distribution of Bird-Life in the Urubamba Valley of Peru", U.S. National Museum, Bulletin no. 117 (Washington: Government Printing Office, 1921); Distribution of Bird-Life in Ecuador, *Bulletin of the American Museum of Natural History*, 55 (1926); and "The Upper Zonal Bird-Life of Mts. Roraima and Duida", *Bulletin of the American Museum of Natural History*, 63 (1931), pp. 1–135.

[40] These he published in various technical articles and in two popular books: *My Tropical Air Castle* (New York: D. Appleton, 1929); and *Life in an Air Castle* (New York: Appleton-Century, 1938).

[41] Including the following: the medal of the Linnaean Society of New York (1912); the Elliot Medal of the National Academy of Sciences (1917); the Roosevelt Medal (1928); the John Burroughs Medal (1929); and the Brewster Medal of the American Ornithologists' Union (1933).

[42] On Chapman's bird guides and their importance to the development of birdwatching and the Audubon movement, see Mark Barrow, *A Passion for Birds: American Ornithology after Audubon* (Princeton: Princeton University Press, 1998), pp. 156–158. On the Audubon movement, see Robin Doughty, *Feather Fashions and Bird Preservation: A Study in Nature Protection* (Berkeley: University of California Press, 1975); and Frank Graham, Jr., *The Audubon Ark: A History of the National Audubon Society* (New York: Alfred A. Knopf, 1990).

[43] Chapman, *Autobiography* (cit. n. 32), p. 84.

[44] S. Charles Kendeigh, "Victor Ernest Shelford, Eminent Ecologist, 1968", *Bulletin of the Ecological Society of America*, 49 (1968), pp. 97–100. On Shelford's place in the history of ecology, see also, W.C. Allee, Orlando Park, Alfred E. Emerson, Thomas Park, and Karl P. Schmidt, *Principles of Animal Ecology* (Philadelphia: W.B. Saunders, 1949), pp. 43–72.

[45] For biographical information on Shelford, see the laudatory biography: Robert A. Croker, *Pioneer Ecologist: The Life and Work of Victor Ernest Shelford, 1877–1968* (Washington, D.C.: Smithsonian Institution Press, 1991); Sara Fairbank Tjossem, "Presevation of Nature and Academic Respectability: Tensions in the Ecological Society of America, 1915–1979" (Ph.D disseration, Cornell University, 1994); William C. Kimler, "Victor Ernest Shelford", *Dictionary of Scientific Biography*, 18 (supplement 2) (1990), pp. 811–813; John D. Buffington, "Obituary: Victor Ernest Shelford, 1877–1968", *Annals of the Entomological Society of America*, 63 (1970), p. 347; Sheldon Miller, "Victor E. Shelford Papers", *Mendel Newsletter*, 13 (1977), pp. 4–6; and Kendeigh, "Victor Ernest Shelford" (cit. n. 44).

[46] Quoted in Croker, *Pioneer Ecologist* (cit. n. 45), p. 12. On Shelford's mentors and biology at Chicago during this period, see Eugene Cittadino, "A 'Marvelous Cosmopolitan Preserve': The Dunes, Chicago, and the Dynamic Ecology of Henry Cowles", *Perspectives in Science*, 1 (1993); pp. 520–559; Jane Maienschein, "Whitman at Chicago: Establishing a Chicago Style of Biology?" in *The American Development of Biology*, Ronald Rainger, Keith R. Benson, and Jane Maienschein, eds. (Philadelphia: University of Pennsylvania Press, 1988), pp. 151–182; Sharon M. Kingsland, "Toward a Natural History of the Human Pysche: Charles Manning Child, Charles Judson Herrick, and the Dynamic View of the Individual at the University of Chicago", in *The Expansion of American Biology*, Keith Benson, Jane Maienschein, and Ronald Rainger, eds. (New Brunswick, NJ: Rutgers University Press, 1991), pp. 195–230.

47 Over the years Shelford conducted his ecological research in a bewildering variety of locations, including a specially constructed vivarium at the University of Illinois, the Puget Sound Biological Station, and on legendary cross-country field trips with his students. See Keith R. Benson, "Experimental Ecology on the Pacific Coast: Victor Shelford and His Search for Appropriate Methods", *History and Philosophy of the Life Sciences*, 14 (1992), pp. 73–91.

48 A complete list of his graduate students, including twenty-five Ph.D students and twenty-four master's students, is found in Croker, *Pioneer Ecologist* (cit. n. 45), pp. 161–164.

49 Shelford's conservation work for the ESA is detailed in Croker, *Pioneer Ecologist* (cit. n. 45), pp. 120–146; and Tjossem, "Preservation of Nature and Academic Respectability" (cit. n. 45), pp. 33–64, both of which I have relied on in the account that follows. See also, Geoffrey McQuilkin, "Saving the Living Laboratory: The Natural Area Preservation Efforts of the Ecological Society of America" (Senior honors thesis, Harvard University, 1991).

50 Ecological Society of America, Committee on the Presevation of Natural Conditions, *Preservation of Natural Conditions* (Springfield, Illinois: Barnes, 1921); Victor Shelford, "The Nature Sanctuary Idea", *Audubon Magazine*, 43 (1941), pp. 503–510; idem, "The Preservation of Natural Biotic Communities", *Ecology*, 14 (1933), pp. 240–245; idem, "National Parks", *Science*, 53 (1921), p. 431; idem, "The Preservation of Natural Conditions", *Science*, 51 (1920), pp. 316–317.

51 *Naturalist's Guide to the Americas*, Victor E. Shelford, ed. (Baltimore: Williams and Wilkins, 1926).

52 On the Biological Survey's predator control program, see Thomas Dunlap, *Saving America's Wildlife* (Princeton: Princeton University Press, 1988). In 1931 ESA officials divided Shelford's committee in two: the first – the Committee on the Study of Natural Conditions, charged with fact-finding, and the other – the Committee on the Preservation of Natural Conditions, charged with writing letters and resolutions to government bureaus, attending hearings, and other active work. Shelford chaired both committees until the mid-1930s.

53 Victor E. Shelford, "The Conflict between Science and Biological Industry", *Science*, 100 (1944), pp. 450–451; see also his earlier letter, "Twenty-five Year Effort at Saving Nature for Scientific Purposes", *Science*, 98 (1943), pp. 280–281.

54 Quoted in Croker, *Pioneer Ecologist* (cit. n. 45), p. 143.

55 The committee was formally abolished at the next meeting of the ESA in 1946.

56 The story of the founding of the Ecologists' Union and its transformation into the Nature Conservancy is told briefly in Croker, *Pioneer Ecologist* (cit. n. 45), pp. 144–146; and Dorothy Behlen, "Thirtieth Anniversary Issue: A History", *Nature Conservancy News*, 31 (July/August), pp. 4–18.

57 For sympathetic histories of the Nature Conservancy, see Noel Grove, *The Nature Conservancy: Preserving Eden* (New York: Abrams, 1992); and William D. Blair, Jr., " A Look Back", *Nature Conservancy*, 41 (November/December 1991), pp. 10–21. As of December 1998, the Nature Conservancy boasted 900,000 members and claimed to have provided protection for 10.5 million acres in the United States and "assistance" in the protection of 60 million acres outside the United States. Figures are from *Home Page of the Nature Conservancy*, available from: <http://www.tnc.org>, 22 February 1999. For a more cynical view of the organization's activities, see Tim Lake, "The Nature Conservancy or the Nature Cemetery: Buying and Selling Perpetual Care as Environmental Resistance", in *Ecocritique: Contesting the Politics of Nature, Economy, and Culture* (Minneapolis: University of Minnesota Press, 1997), pp. 56–74.

58 Some of the studies that examine naturalists' involvement in conservation include: Barrow, *Passion for Birds* (cit. n. 44), which also briefly mentions the issue of social responsibility and political activism in science; Dunlap, *Saving America's Wildlife* (cit. n. 52); Dorsey, *The Dawn of Conservation Diplomacy* (cit. n. 26); Donald Worster, *Nature's Economy: A History of Ecological Ideas*, 2nd edn. (Cambridge: Cambridge University Press, 1994); Janet Foster, *Working for Wildlife: The Beginning of Preservation in Canada*, 2nd edn. (Toronto: University of Toronto Press, 1998); Alfred Runte, *Yosemite: The Embattled Wilderness* (Lincoln: University of Nebraska Press, 1990); and Richard W. Sellars, *Preserving Nature in the Natural Parks: A History* (New Haven: Yale University Press, 1997). One of the reasons I feel fortunate in having Everett Mendelsohn as my graduate advisor is because he strongly encouraged my interest in linking the history of science and environmental history. For many years he has regularly taught a graduate seminar on the history of American environmentalism, and he has routinely included environmental topics in his popular undergraduate course, "Science and Modern Society". Besides myself, several other recent graduate students have written dissertations with environmental themes under his direction (all at Harvard University): Peder Anker, "Ecology of Nations: British Imperial Sciences of Nature, 1895–1945", 1999); Mary A. Cooper, "Inventing the Indigenous: Local Knowledge and Natural History in Early Modern German Territories" (1998); Catherine M. Roach, "Mother Nature/Human Nature: An

Environmental Anthropology Drawing on Christian Thought, Psychoanalysis, and Feminism" (1997); David I. Spanagel, "Chronicles of a Land Etched by God, Water, Fire, Time, and Ice" (1996); Mark Madison, "Green Fields: The Agrarian Conservation Movement in America, 1890–1990" (1995). Also relevant is the dissertation that Charlotte M. Porter completed under him just over two decades ago: "The Excursive Naturalists or the Development of American Taxonomy at the Philadelphia Academy of Natural Sciences, 1812–1842" (1976).

ROBERT N. PROCTOR

HUMAN EXPERIMENTAL ABUSE, IN AND OUT OF CONTEXT[1]

> "There is no objection to human experimentation when there is no invasion of human rights."
> *Albert Leffingwell, anti-vivisectionist, 1916*

Abstract

The twentieth century is littered with examples of human experimental abuse. There are the well known Nazi experiments of the 1940s, of course, and the Japanese biowarfare experiments in Manchuria, but there are also similar abuses in the United States, Australia, and other parts of the world – many of which proved fatal. How do we account for such abuses? What kinds of social arrangements have encouraged experimental mistreatment, and how can such abuses be prevented? I explore such questions here, along with the question of why radiation has figured so prominently in experimental abuses, especially in the United States and the former Soviet Union. Some points are made about the dehumanizing of experimental subjects and the symbolic history of experimentation; a caution is also raised against the dangers of exculpatory relativism, the tendency to justify experimental abuses by situating (exotocizing) them in some distant or alien context, when rules and expectations were different.

Introduction

Human experimental abuse does not arise in a vacuum. There are causes of such abuses and, like cancers, more or less effective sources of promotion and suppression. Promoting factors could include: professional competition, especially the pressure to publish; military urgencies, used to justify the testing of dangerous practices, perhaps with the rationale that the greater good of society is at stake; governmental secrecy, intended to deceive foreign powers or potential critics or the experimental victims themselves; commercial incentives, including pressures to introduce new drugs and other medical technologies; the climate of "perpetual emergency" in medicine fostered by secular fears of dying and the valuation of health and longevity over other values; and the more

G.E. Allen and R.M. MacLeod (eds.), Science, History and Social Activism: A Tribute to Everett Mendelsohn, 235–254.

general disregard for individual lives and liberties endemic in the human condition, based on narrow-minded ethnocentrism, broad-minded brutality, or any of a thousand other rationales.

Suppressing or inhibiting factors, if I may continue the cancer analogy, might include the absence of the elements just mentioned, of course, along with lack of opportunity, subject reluctance or resistance, traditions of activist journalism, legal defense,[2] and patient rights protection, bioethical codes and sanctions, and enforcement of human rights more generally. Human rights activists have done important work in exposing abuses, making some of the more egregious violations unpalatable for political leaders. Elections and appointments can make a difference: in the United States, for example, President Clinton's appointment of Hazel O'Leary as Secretary of Energy did a great deal to open the lid on American nuclear secrecy.[3] One can even speculate that, had Clinton not been elected, the Presidential Advisory Committee on Human Radiation Experiments might never have been formed. The formation of that committee and the publication of its 1996 *Final Report* were important in publicizing instances of abuse, and in strengthening safeguards against inappropriate experimentation.[4] There have also been times, however, where violence has been crucial to the elimination of abuse: Japanese leaders of the infamous Unit 731 in Manchuria admitted after the war that Allied bombing was responsible for ending some of the world's most cruel human experiments – more on that in a moment. We can think about this in light of the failure of U.S. military authorities to order the destruction of the rail lines leading to Auschwitz.[5]

Not everyone, though, has been equally likely to become the object of a dangerous scientific experiment. Experimenters are selective, and prejudiced. We should also recall that the conditions that give rise to experimental abuse are not necessarily those that give rise to human experimentation generally. There are some interesting ambiguities in whether we classify a given form of practice as an "experiment"; the symbolic history of human experimentation is also a topic I shall touch upon.

Early Experimental Medicine

Human experimentation as we most often think of it dates largely from the second half of the nineteenth century.[6] Isolated cases occur prior to this time,[7] but laboratory science was not a routine part of medical education until the age of Pasteur, Koch and Lister, and there was little pressure to investigate new medical practices. The end of the nineteenth century saw a burst of energy in this area, as evidenced by the establishment of the *Journal of Experimental Medicine* (in 1896), new mandates for laboratory training in medical schools, research institutes like the Rockefeller Institute for Medical Research (opened in 1904), and new techniques of medical intervention – notably the gastric tube, X-ray diagnostics, therapies involving radioactive isotopes, new anesthetics such as ether, and new vaccines and diagnostic tests following the triumph of the germ theory of disease. The germ theory was particularly important, reinforcing as it did the "generic theory" of treatment, the idea that people were pretty much all the same, and that disease was caused not by an imbalance of humors but rather by a specific pathogenic agent to which one had been exposed. And since there were many different kinds of pathogens, and many possible kinds of antibiotics, there was suddenly a great need to test these various antibiotics in human populations.

Since experimental medicine is largely an offspring of the latter part of the nineteenth century, it is not surprising that that is when we find the first great wave of human experimental abuses. The earliest known efforts to transplant cancerous tissues into healthy human bodies date from the 1890s,[8] for example, as do the most notorious experiments to determine whether certain extracts of diseased bodies might serve as vaccines. The first strong reactions against human experimentation also emerge at this time. As Susan Lederer has shown in her book on this topic, much of the opposition to human experimentation actually stemmed from people primarily concerned about animal welfare. The fear was that anyone cruel enough to experiment on dogs and cats would not hesitate to vivisect humans.[9] African-American folktales of "night doctors" begin to appear about this time, a reaction, apparently, to the not-altogether-unfounded fear of being captured and used in medical experiments. Opposition also begins to emerge in Europe. In Germany, the sensational case of Albert Neisser, a Breslau professor of dermatology who in 1892 injected healthy children with syphilis to test the effectiveness of a vaccine, generated broad public outrage, prompting the Prussian Ministry of Religious, Educational and Medical Affairs in 1900 to issue the world's first explicit regulations of human experimentation.[10]

Experimental abuse was also fostered by increasing hospitalization. The number of hospital beds in the U.S. increased by about a factor of eight from 1873 to 1909, a consequence of urbanization, charity, and the increasing powers of state and local governments for the care for its ailing citizens.[11] The half a million patients in American hospitals by the First World War provided a new source of "material" for human experimentation – a moral situation complicated by the fact that clinical research hospitals usually offered higher standards of care. (We find this also in other situations of human experimental abuse, the plutonium injection experiments, for example.)[12] The growth of other forms of institutional confinement–notably prisons, orphanages, and psychiatric asylums–provided further opportunities for experimental abuses, as did the medical novelties generated by the military technologies of the First World War.

Commercial pressures should also be mentioned in this context. Financial inducements have long been a way of gaining the cooperation of experimental subjects. Susan Lederer points out that the economic recession of the late 1920s seems to have increased the pool of people willing to subject themselves to science for pay: the *New York Times* in 1927, for example, ran an ad by a man offering himself for medical experimentation for fifty dollars per week. Employment bureaus were also known to send out unemployed men to laboratories to become paid subjects of research.[13] Allen Hornblum has shown how in the 1960s, prisoners in Philadelphia's largest county jail could earn extra cash by allowing themselves to be painted with diverse chemical and cosmetic agents. Opportunities for earning money were scarce in the prison, and prisoners actually protested when, in the 1970s, the "skin for cash" exchange was eliminated.[14]

Financial inducements were not of course necessary when people could be bought and sold against their will. In the antebellum South, slaves were sometimes bought for experiments, and as late as 1916 an American physician proposed that Chinese prisoners condemned to death be purchased for use in human experiments.[15] Criticism of the practice of paying people to submit to injurious experiments was common in the early years of the twentieth century, though when inmates of a Mississippi prison in 1915 were fed a restricted dietary regime to induce pellagra (release from prison being their reward), the more common criticism was that hardened criminals were being let off the hook too easily.[16]

The impact of commercial relations goes beyond the offering of financial inducements, however, given that pharmaceutical companies rely heavily on clinical trials to bring out new products. Commercial drug testing has a long history, and in our own century, the pressure to introduce new and ever more effective drugs has often led to abuses. Prior to World War I, for example, the Rockefeller Institute for Medical Research trademarked a skin test it had developed for syphilis; two pharmaceutical companies were licensed to produce the extract, though a Rockefeller scientist came under sharp criticism for having tested the product on 46 children and 100 adults with ailments unrelated to syphilis.[17] The prospect of infecting innocent children with syphilis, as one might imagine, aroused a great deal of protest.

Commercial pressures expanded rapidly with the growth of the pharmaceutical industry after the Second World War. Testing (in the U.S.) was promoted by new Food and Drug Administration requirements that elaborate tests be completed prior to drug certification;[18] testing was also increased as stricter epidemiologic standards required ever larger and more highly controlled populations. Drug testing became big business, with hundreds of millions of dollars potentially hanging on the approval of a single drug. Five thousand different chemotherapeutic agents were tested in the 1950s against cancer alone. Services were established to organize the performance of large-scale clinical trials, and journals such as *Science* even today publish the advertisements of companies offering their services in this area. Bioethicists often worry about the *quality* of human experimentation, but one can also worry that the sheer *quantity* of experiments now being performed brings with it new opportunities for abuse.

Europeans have recently cautioned against the "experimentalization of society" – which presents an interesting contrast with the U.S., where the whole idea of "the experimental" still has an overall positive ring, as in "experimental music", "experimental education", and so forth. Experimental therapies seem to have the implicit imprimatur of being "on the frontier" or "extra-special", which may help explain why it has taken so long for safeguards to be built into the protection of experimental subjects.

The Dehumanization of Experimental Subjects

Human experimental abuse is sometimes regarded as a purely ethical problem: a problem of the virtuous or odious actions of individual agents, whose solution requires only the establishment of proper codes of experimental conduct. Power relations are crucial, however, given that the history of experimental abuse is by and large a history of disproportionate harms to the socially vulnerable or marginal, mirroring larger patterns of neglect and mistreatment outside the experimental orbit. Self-experimentation has an old and venerable history, but it is difficult to find a case where a cruel or degrading experiment has been performed on an unwitting physician or politician (while in power): most of the victims have been prisoners, orphans, prostitutes, low-ranking soldiers, disempowered ethnic or racial minorities, or various and sundry "enemies of the state". Scientists for many years were open about their views on such matters: Claude Bernard's famous 1865 injunction never to perform an experiment "which might be harmful ..., even though the result might be wholly advantageous to science", for example, allowed two important exceptions: people condemned to death, and people expected soon to die from disease.[19] Defense attorneys at Nuremberg would later cite such exemptions to rationalize the use of concentration camp prisoners as experimental subjects, citing also the numerous cases where American prisoners had died from injuries suffered at the hands of medical experimenters.

Experiments on prisoners and the terminally ill are notorious, but others low on the status hierarchy have suffered similar fates. Eighteenth century inoculations were often tested on slave populations controlled (e.g. in Jamaica) by European planters.[20] Albert Neisser, the German co-discoverer of the leprosy bacterium, tested his first vaccines against syphilis on three young girls and an equal number of female prostitutes. The famous yellow fever experiments organized by Walter Reed in Cuba were performed on Spanish emigrants, enticed by the offer of 100 dollars in gold to allow themselves to be bitten by the mosquitos suspected of carrying the disease. In Nazi Germany, the mentally and physically handicapped became the objects of numerous medical experiments prior to becoming victims of the "euthanasia" operation. In the Tuskegee study, conducted by the U.S. Public Health Service from 1932 to 1972, where men with advanced-stage syphilis were allowed to go untreated, the victims were all poor black males in the rural south.[21] At the Willowbrook State School on Staten Island, live hepatitis vaccines were tested on mentally retarded children with no prospects of benefiting from the tests.[22] Dr. Carl Janson of the Charity Hospital in Stockholm in 1891 decided to use orphans (foundlings) in his experiments with black smallpox pus because, as he put it, "the most fit subjects, calves, were obtainable only at considerable cost".[23] The British government chose Australian soldiers to use in its mustard gas experiments in Queensland during the Second World War; the men were technically volunteers, but few must have realized that the experiments could prove fatal.[24]

In these and many other cases, there was a consequential power *asymmetry* between the experimenting scientist and the experimental subject. This is sometimes overlooked in discussions of human experimental ethics, though it shouldn't take much thought to realize that consent among equals is a very different thing from consent extracted from people without a lot of options. Formal rules governing the treatment of human subjects don't carry much weight if they are not viewed as fully human. In many cases, experimental abuse is accompanied by an implicit (and sometimes explicit) dehumanizing, a presumption that the suffering of the subjects does not matter, or even that the victims in question deserve such treatment. You can see this in the language of human experimental abuse, where subjects are described as *Menschenmaterial*, "logs" and the like. The medical experiments organized by Lt. General Shiro Ishii in Manchuria carried this to an extreme: the Chinese prisoners sacrificed in the course of Japan's biowarfare research were derisively characterized as "logs" (*maruta*) – a reference not just to the disguise of the facility as a sawmill, but also to the disregard for the value of the lives of the victims.[25]

Nazi and Imperial Japanese Medical Experiments

It is important to understand the context within which human experiments have been performed. I say this because there is a tendency to regard human experimental abuse as a matter of personal perversion, individual excess, or a culture of national sadism. The Nazi experiments are a case in point. Much of our revulsion for human experimental abuse stems from the outwash of the Nazi period, a benchmark for human experimental cruelty and the backdrop against which a great deal of modern bioethics has developed. In the majority of cases, Nazi medical experiments were conducted with specific military medical objectives in mind. In the cold water freezing experiments organized by SS physicians at Dachau, for example, Russian POWs were immersed in icy water to determine how long a pilot downed in the North Sea or English channel could survive in water of a given temperature.[26] Was there any chance of him remaining alive after, say, 3 or 4 or 7 hours in water of, say, 5 degrees Celsius? Was it worth sending out a submarine to attempt a rescue under such conditions? What kinds of protective clothing could be developed to slow the freezing process?[27] Luftwaffe physicians knew that it cost more to train a pilot even than to build his plane, so there was a great deal of interest in such questions.

Similar reasoning can be seen in most of the other medical experiments conducted during the Nazi period. Prisoners were forced to drink salt water at Dachau in 1944, to find out how long a man at sea might survive without fresh water. The low-pressure experiments conducted by Drs. Holzlöhner and Rascher, during which some 80 prisoners died, stemmed from the Luftwaffe's interest in practical problems connected with high-altitude flight. Sterilization experiments – involving X-rays, natural herbs, and irritants such as formaldehyde or CO_2 gas – were conducted as part of the Nazi program to eliminate racial undesirables from the breeding stock of Germany and its occupied territories.[28] Inoculation experiments were part of the effort to develop vaccines against the infectious diseases Germans were encountering in the newly-conquered territories of Africa and Eastern Europe. Other experiments involved forcing prisoners to eat food substitutes derived from chemical or manufacturing waste, the point being to explore whether previously unused organic products (blood from slaughterhouses, for example) could be turned into food resources.

The point in most of these experiments was a specific military-medical interrogation. The questions scholars were trying to ask were generally speaking practical; this was not an "unfettered quest for knowledge" or "science without values". Nazi physicians were also selective in their choice of human subjects: experiments were conducted not on "healthy" Germans but on Russian prisoners of war and Jews and other "lives not worth living" in German hospitals and concentration camps.[29] These people were deliberately chosen because they were regarded as cheaper than animals; there was also the added inducement that by killing a Russian or a Jew in the course of an experiment, you saved on whatever food would have been used to keep that person alive. *Vernichtung durch Versuchen.*[30]

A similar logic can be seen in the deadly Japanese experiments conducted in Manchuria from 1932 to 1945. Though less well known than the Nazi experiments, they are equally chilling. The Japanese army in 1932 gave Shiro Ishii 200,000 yen and a command of 300 men to construct a biological warfare facility near Harbin, where Ishii and his staff exposed thousands of captured Chinese to anthrax, plague, glanders (a bacterial skin disease of horses, mules and goats, to which humans are also vulnerable), and dozens of other pathogens.[31] Many of the victims were political prisoners – especially communists, but also Kuomintang sympathizers, common criminals, and people who otherwise ran afoul of the law.

The point in most of these experiments was to develop the knowledge base for biological warfare. The Japanese command expected an outbreak of war with Russia, and bioweapons were envisioned as a counter-weight to the numerically superior Russian army (and whatever remnants of the Chinese army might be encountered). Cold was a major problem, so Ishii's men induced frostbite in a group of Chinese prisoners to test remedies for defrosting. Prisoners were forced to stand outside in freezing weather with arms exposed, until their arms had frozen to the point that they sounded like wood when struck with a stick. Experiments showed that soaking in water of 37 degrees was the most effective way of rewarming a frozen arm or leg.[32]

Practical military objectives were also behind Ishii's construction of more elaborate facilities in the mid-1930s, including at least four other stations for biowarfare production and several remote field test sites. Japanese military authorities were pleased with his work, and by 1936 he controlled an annual budget in excess of ten million yen. (Ishii boasted that year that he had "no worry whatsoever" about the availability of research funds.) We cannot characterize the experiments as "scientifically valueless": American military authorities apparently thought enough of the work to order detailed summaries of the major findings; Sheldon Harris in his recent history of this era records Ishii's boast, after the war, that his work in Manchuria had generated over 200 patents.[33]

A commonly-heard estimate for the number of fatalities from Japanese experimentation in Manchuria is 3,000, though Harris points out that this is surely too low. It does not include the deaths from experiments prior to 1941 (recall that Ishii started human experimental BW work in 1932); nor does it include the deaths from four other BW facilities supervised by Ishii (apart from the central Ping Fan facility). Considering these brings the total into the neighborhood of 12,000 immediate experimental victims, though there are others one could plausibly count. Deaths from plague-infested animals released from field tests Harris estimates at 30,000, and fatalities from plague germs released in combat could have been as high as 200,000. Harris concludes that Japanese military scientists turned Manchuria "into one giant biological laboratory".[34]

It is perhaps worth recalling in this context that abstract ethical codes may have little force, if the people performing such experiments do not regard their subjects as fully human. Germany, after all, had the world's strongest ethical guidelines regulating human experimentation: the 1931 guidelines issued by the Reich Health Council, requiring voluntary and informed consent along with many other protections, were never formally dismissed.[35] What was new in the years of the most heinous experimental crimes was an officially-sanctioned attitude that broad classes of people were not to be regarded as fully human.

American and Soviet Radiation Experiments

Disregard for the health and dignity of the individual is a common theme in the history of experimental abuse. It needn't be of a racial, sexual, or nationalist character, however, as is clear from the experiments reviewed by President Clinton's Advisory Committee on Human Radiation Experiments,[36] or the atomic tests undertaken by other nuclear powers in this same period. France conducted nuclear tests in the Algerian desert, the Soviet Union conducted tests in Kazakhstan and the Arctic, the United States conducted experiments in Nevada and in the South Pacific. In each case, on-the-ground military maneuvers were conducted to simulate conditions on a nuclear battlefield – including exposure to radiation. Realism required that soldiers and even civilians be exposed.

The goal in many such tests was to see how soldiers would perform under the novel conditions of nuclear war. In the Soviet Union in 1954, for example, a 20-kiloton atomic bomb was exploded near Totskoye in the Southern Urals less than two miles from 45,000 Red Army troops, to see how soldiers would be able to carry out their duties in the immediate aftermath of an atomic explosion. Films of the test, which have been preserved, indicate that soldiers received more than ten times the radiation allowed for American troops at that time for a whole year – though it must also be said that American tests are known where exposures greatly exceeded official limits. Civilians were deliberately exposed in the Soviet test; many of those who fell ill or died from radiation injuries had remained in the vicinity of the blast or collected firewood from the downed trees.[37]

The United States conducted at least 235 atmospheric tests between 1944 and 1962, exposing some 200,000 troops and an unclear number of civilians to varying levels of radiation. (Much depends on whom and how you count, since everyone on the earth has received some slight radiation from such tests – which is one reason it doesn't make much sense just to say that someone has been "exposed": the key is what kind of radiation, by what route and, of course, how much). Military exercises often were performed in connection with these tests: during Operation Upshot-Knothole, for example, 12 American military "volunteers" were ordered to observe the flash of an atomic test, along with 700 rabbits, to determine whether men could continue to fight after being blinded – temporarily – by a nuclear explosion.[38] The men used in these and other tests were officially classed as "volunteers", but it is unlikely that many were fully informed of the long-term health risks. The officers in charge were no doubt equally ill-informed in many cases. Participation in such tests – like training with live ammunition or actual combat duty – was simply one of the duties presumed by military service in this era.

Contamination even of civilians, though, was often deliberate. There is evidence that the 15-megaton H-bomb "Bravo Shot" exploded near the Marshall Islands in the Pacific in 1954, for example, was deliberately undertaken with the knowledge that indigenous peoples living in the area would be exposed to fallout, and that "health effects" from the blast would be studied.[39] The same is true of many of the Soviet tests carried out at Semipalatinsk in Kazakhstan, though these are much less well known.[40]

The logic and scope of post-WWII radiation experimentation has to be understood in the context of Cold War fears of global nuclear war. Radiation health effects were often regarded as state secrets, and medical records of many of the victims were kept under wraps. The health records of more than a hundred thousand American radiation workers were likewise held in confidence – until political pressures forced disclosures in the 1990s. The fact that the Atomic Energy Commission (AEC) was conducting human experiments was itself supposed to be secret: as a 1947 AEC document put it, the purpose of secrecy was to avoid both "adverse effects on public opinion" and potentially damaging lawsuits from victims.[41]

One of the oddest twists of this story is that even when U.S. military authorities did adopt ethical guidelines governing the use of experimental subjects, the guidelines themselves were classified. Secretary of Defense Charles E. Wilson's now-notorious memo of February 1953, for example, outlining how the Nuremberg Code should be adopted by U.S. military scientists, was declared top secret.[42] The question of how to interpret this Borghesian, Kafkaesque fact is likely to perplex bioethicists for quite some time.

Fleshing out this context also requires that we appreciate the larger cultural attachment to all things radiant in the early atomic age. The 1950s was an era of atomic euphoria: bartenders sold atomic cocktails, hairdressers coiffed atomic hairdos, and nuclear power was equated with sexual danger ("blond bombshells" and, of course, the "bikini" itself – explosive, dangerous – named for the central Pacific atoll where H-bombs were tested).[43] The Atomic Energy Commission, responsible for both building the atomic arsenal and safeguarding the health of atomic workers (the "fox guarding the henhouse"), promoted every imaginable use for atomic power, from nuclear-powered subs and rockets to atomic cars and airplanes.[44] Glenn T. Seaborg, the discoverer of plutonium and director of the AEC, at one point envisioned plutonium-heated bathing suits. There were plans to use atom bombs to move mountains and to dig canals ("Project Ploughshares"); AEC architects even proposed redesigning American cities in long linear strips to make them less vulnerable to atomic attack. Country singers crooned atomic kitsch, and nuclear war planners worried about things like "VD Control in Atom-bombed Areas".[45]

Radiation was supposed to be not just a source of ultimate weapons and unlimited power, but the key to new kinds of instruments, new kinds of probes, new means of reckoning time. The AEC supported the use of radioactive tracers in medicine, and radioisotopes for dating archaeological sites and palaeontologic strata. If there was a technical problem you wanted to solve in the 1950s, there is a very good chance you could obtain AEC funding to attack it. By the end of the 1950s, the AEC was spending $3.5 million per year on human genetics, a reflection of its concern for the long-run, potentially species-altering effects of atomic radiation on the human genome.[46] There was a great deal of optimism connected with the new-found power of the atom: among medical professionals, for example, radiation was more commonly regarded as a *cure for* cancer than as a cause of it. In the disciplines of radiology and (later) health physics, strong professional interests encouraged an appreciation of the curative powers of radiation while downplaying potential hazards. Scientific conservatism fostered an additional presumption of the absence of harm in the absence of evidence of harm.[47]

It is hardly surprising that many of the experimental abuses of this era were tied to radiation. Radiation was a technology in search of applications. Appreciating this helps us to understand the nature of the experiments conducted. Many were pursued not with the narrow goal of fighting or surviving a nuclear war, but rather to help understand how the average worker, housewife, and consumer might better adapt to the brave new world of atomic energy. Studies were launched to explore how radioactive isotopes in different chemical forms move through the body: to assist in the monitoring of occupational exposures, for example, or to determine which among the various chemical forms of a given radionuclide might be more or less safe to transport, store, or even to eat or breathe. These were all questions that "required" testing, and were often difficult to answer without direct observation of exposed human subjects.

But there were also radiation experiments that had nothing to do with health effects per se. In a number of cases, radioactive tracers were used to explore basic aspects of human metabolism. In the now-notorious Fernald School experiment, for example, learning disabled children were fed radioactive oatmeal as part of a project to determine whether certain kinds of cereals will block the uptake of calcium and iron. Conducted from 1946-1956 with the support of MIT and the Quaker Oats Company, the experiment involved 120 pupils at the Fernald School in Waltham, Massachusetts, who were told they were participating in a "science club".[48] Radioisotopes were used – in very low doses – to trace the movement of chemical nutrients through the body.[49] The same was true of many other radiation experiments, though the boundary between "tracer" experiments and experiments designed to measure radiation health effects was not always well defined.

In the Vanderbilt Experiment of 1945–47, for example, 830 pregnant women were fed radioactive iron (Fe-55 and Fe-57), to determine whether iron supplements would be of value to gestating mothers. Metabolic pathways were studied, but volumetric measurements were also made to document the storage of the tracer isotopes. In 1997, when Vanderbilt University was sued for compensation for injuries and deaths suffered by the experimental victims (at least four children died by the university's own admission), university lawyers tried to argue that since this was merely a "tracer" experiment, no harm was envisioned. Vanderbilt epidemiologists had already admitted an excess of cancer deaths among the offspring of the experiment, so the case boiled down to whether the experimenters could have and should have known of the potential risks. The case was eventually settled in favor of the plaintiffs when it became clear that the experimenters had acted recklessly, endangering the lives of the women and their unborn children. (The mothers of the dead children were awarded $10.8 million and an apology. I was the primary historian serving as expert witness for the plaintiffs; Susan Lindee was the chief historical witness for the defense. I was able to show that the idea of a "zero-threshold" dose-response relation for radiation-induced cancer should have been known to the experimenters, since one of the authors of that principle – Max Delbruck – had recently emigrated from Germany to the U.S., becoming head of the very department in which Paul Hahn, the chief radioiron experimenter, had been working.)[50]

It is probably not fair to regard many of the excesses of this era as "experiments" – the X-raying of feet to fit shoes or of scalps to treat ringworm, for example, or the massive accidental and routine releases of radioactive iodine from the Hanford Reservation in Washington in the late 1940s[51] – some are no doubt better classed as environmental or medical malfeasance, involving varying degrees of negligence, ignorance, or criminality. Oftentimes, though, such exposures were used as opportunities to study health effects that might have been difficult for practical or ethical reasons to generate deliberately. We shall examine these in a moment.

The radioeuphoria of the era helps us understand why so many experimenters were cavalier in their handling and mishandling of radionuclides.[52] Radiation experimental abuse must be regarded not just as an abstract lapse of personal moral or professional vigilance, but also as a consequence of broader Cold War fears of nuclear combat, military secrecy, AEC largesse, and the radioeuphoria associated with the newly exposed powers of the atom.

Experiments of Opportunity

Human experimental abuse is most commonly regarded as a violation of the doctor-patient relationship. But there are also important abuses that occur where physicians are not involved, where the harms or wrongs inflicted have nothing to do with doctors or medical practice. I've already mentioned the example of atomic tests, but there are others. Among these are experiments "after the fact", so to speak, in the wake of a nuclear reactor disaster, for example, where the exposure is not deliberate. In such cases, the accident or "event" is recognized as an opportunity to gain valuable knowledge.[53] These are sometimes considered human experiments, because there are hypotheses developed for testing, controlled observations, and much of the other paraphernalia we associate with experimental practice. There is also the possibility for abuse, since subjects may be harmed, or may not be aware of their participation, or may be unaware of the nature of the risk or measures that could reduce that risk. The ethical breach is particularly grave if a decision is made to "study" an endangered population while deliberately withholding treatment that could have helped alleviate the injuries in question.

One such experiment was the Public Health Service's study of uranium miner health on the Colorado Plateau in the 1950s and 1960s. Uranium mining had been shown to cause lung cancer prior even to the discovery of radioactivity; two Austrian physicians had shown in 1879 that the *Schneeberger Krankheit* endemic among miners in the Bohemian *Erzgebirge* in southern Saxony was lung cancer, and that about 3/4 of all miners in the region were dying from the disease. Suspicions that the root cause was the "radium emanation" (radon) in the mine arose shortly after the First World War, and by the 1930s there was a broad medical literature on radiation-induced cancer as a hazard of European uranium mining.[54]

American epidemiologists were aware of this literature when U.S. military authorities ordered the escalation of uranium production, following the bombing of Hiroshima and the gearing up for large-scale nuclear weapons production and expectations of a peacetime nuclear economy. AEC and Public Health Service scientists repeatedly asked for epidemiological studies of the hazards of uranium mining in the late 1940s, but were turned down. Wilhelm Hueper, the father of American occupational carcinogenesis, was actually barred from undertaking a study of uranium miner health and safety in the early 1950s.[55] Radiation was a very sensitive topic in these Cold War years.

When the Public Health Service finally did begin a study of the miners, the hypothesis being tested – that miners would be dying from lung cancer – was never publicized. The miners themselves were never told about the European evidence, and steps were actually taken to bar them from learning about the danger. None of this was unique to America. Nuclear secrecy blocked publicity of the uranium mining hazard in both the east and the west: in Czechoslovakia, for example, where many more miners were employed – including tens of thousands of political prisoners – the findings of uranium mine epidemiology were declared top secret, apparently out of a fear that foreign powers might learn, from the numbers of cancers being produced, how pure the uranium was and how much was being produced.[56]

Were the uranium mining epidemiological studies "human experiments"? They were sometimes defined as such at the time. Since the 1940s, epidemiologists had prided themselves on being "experimental": similar methods had been used to nail down the cigarette-cancer link (beginning in the 1930s), and hopes were high that the "science of epidemics" could one day rank with the older medical sciences in terms of experimental rigor. In the uranium case, "results" in the form of excess lung cancer deaths began to appear in the late 1950s, and by the late 1960s the hypothesis of a major danger had been amply confirmed. This was hardly surprising to any of the principals, since European physicians had already established the nature of the hazard and its probable cause, decades earlier, with both human and animal data. The fact that the miners were not told, and that protective measures were so feeble, justifies our calling this an abusive human experiment (one could of course use stronger language). The end result was the loss of about 4,000–5,000 American lives to lung cancer – along with three or four times this many in Czechoslovakia and East Germany. These were all deaths that could have been prevented by a less cavalier attitude toward the health and safety of those who were putting their lives on the line.

Experimental Cruelty and Its Symbolisms

Why does human experimental abuse evoke such horror? Jan Phillip Reemtsma has distinguished between cruelty "cold" and "hot", and the distinction helps us understand part of our revulsion for science-based ("clinical") brutalities. Popular and legal rhetoric distinguishes similarly between "cold-blooded murder" and "crimes of passion" – and experimental cruelty is invariably treated more like the former than the latter. Why, though, are such passions evoked, beyond what one might expect from the degree of physical harm done to the victims?

One reason, I believe, lies in the cultural authority of medicine, authority that magnifies the sense of betrayal (or violation) felt when that power – uniquely bound with intimacy – is abused. Physicians have unique rights to touch and manipulate our bodies, and when that relationship is callously or cruelly wielded, there is a quantity of wrong we feel that can be disproportionate to the physical harms produced. Lawyers and bioethicists routinely distinguish between "wrongs" and "harms" in this sense – human experimental subjects may be wronged, even if they are not physically harmed. And when wrongs and harms are combined, the separate effects are magnified.

We also have to realize, however, that patients are often co-conspirators in this process. People presumably would not be so willing to serve as subjects of experiments were they not already convinced that "doctor knows best" or that "the cause of the nation comes first". Financial incentives or false hints of personal health rewards can help subjects become victims, but so can exhortations to "do one's duty". In the 1930s, participating in an experiment –

especially a dangerous one – was often seen as tantamount to heroism. Prisoners could have their sentences commuted, and doctors who suffered or died in the course of self-experimentation were proclaimed martyrs.[57] Curtailing experimental abuse can be difficult in such a climate.

Doctors are supposed to have sworn an oath to "do no harm". Non-therapeutic human experimentation, even if not injurious, entails a novel, or at least non-routine, element of doctor-patient relations insofar as subjects are asked to provide, from their bodies, something that may be of gain to a physician's resumé or the good of society or the Future of Science. This instrumentalization of the human body – sick or well – is surely one source of our revulsion for abusive experimentation, and it helps us see why human experimentation opens novel opportunities for abuse. Human experimental abuse (as opposed to simple error or negligence) is probably only possible when the health and well-being of an individual is subordinated to some orthogonal goal – like the advance of medical knowledge, or the augmentation of a nation's military strength, or the lengthening of a scholar's academic resumé.[58]

There are surely further elements to this revulsion, however (if we can indeed speak of revulsion, since surveys often indicate that most Americans are quite willing to undergo experimentation if they believe it might help them.) Even in the extreme case of Nazi concentration camps, where murder was an everyday occurrence, there still seems to be a special revulsion reserved for killings that took place in the course of experiments (even though these were a tiny fraction – less than a tenth of one percent–of all of the murders committed by the Nazis). Perhaps, again, this is because of their chillingly modern form, combined with the central participation of medical doctors. Science is supposed to be the path of the future, the path *to* the future; perhaps we are disturbed to think that the future will be paved with such cruel and unusual misdeeds. A barnyard murder or crime of passion we condemn as cruel, but there seems to be something uncanny about clinical or industrial killing that makes it more terrifying. Barnyard hatchet murders are not planned by bureaucrats, recorded by secretaries, approved on medical forms, and administered by nurses and MDs wearing white coats to symbolize purity. In human experimental cruelty, there is arguably not even the same kind of closure as with other forms of cruelty: experiments, after all, are supposed to be repeatable.

I raise this rather speculative question because I think it is important to reflect upon what might be called the "symbolic history" of human experimentation: what has experimentation meant in different political or cultural contexts? How have different cultures understood the dangers of experimental abuse? What kinds of contexts tend to generate experimental abuse, and what kinds do not? We need a comparative anthropology of experimentation; we need to better understand the meaning of experimental abuse if we are going to be able to better organize its prevention.

Exculpation by Contextualization

Finally, I want to say something about the limits of informed consent and the dangers of trying, too hard, to understand the "context" within which experimental abuses have occurred.

Informed consent is arguably the single most important element of the ethical treatment of human subjects. We do have to recognize, however, as Yale University psychiatrist Jay Katz and others have pointed out, that informed consent is often difficult to define and easy to sidestep.[59] How much "information" does a potential subject have to know to be informed? What kinds of persons are incapable of being informed, and what kinds are incapable of granting consent? A great deal of attention has been given to the need for informed consent by subjects; I would like to draw attention to the equally important duty for experimenters to be informed.

This is important to appreciate, given the diversity of scientific opinion concerning something as simple as the health effects of low-level radiation. There are hundreds of radiation experts in the United States, for example, who still today will argue that low-level radiation in certain forms is actually good for you (the so-called "hormesis thesis").[60] When scientists such as these perform experiments, what can one expect from patients in the way of "informed consent"?

What do we do when the scientists themselves are unaware of potential hazards – through negligence or professional myopia? A prominent U.S. physician and director of the University of Chicago's Center for Clinical Ethics recently claimed that the linkage between cancer and radiation "wasn't really clear until the mid-1960s", and cited this as "one potential defense of the (radiation) experiments".[61] Medical historians have shown that radiation was already recognized as a cause of cancer in the first decade of the twentieth century, and that by the end of the 1940s there were hundreds of publications attesting to this fact. The amnesia expressed in such denials helps to exculpate the perpetrators of this era, but it also gives us yet another clue as to how experimental abuses arise in the first place. Physicians and scientists can be uninformed in matters concerning the history of medicine, so much so that they may actually be ignorant of the dangers to which they are exposing their subjects. This is yet another reason to stress not just the informed consent of the subject, but the training of potential experimenters in the history and sociology of experimental abuse. As of 1994, the U.S. Department of Energy (DOE) was still conducting more than 200 experiments involving low-dose radioactive tracers in humans.[62] It would be interesting to interview the investigators and the subjects involved in this research, asking them what they believe about the hazards of radiation. So far as I am aware, such a project has never been conducted, although it might reveal some interesting information about the limits of the concept of "informed consent". The same thing could be done for other types of experiments – in gene therapy, for example–where opinions about possible harms remain sharply divided.

Let me conclude by drawing attention to two further obstacles in coming to grips with human experimental abuse.

The first has to do with what we could call the "myth of the unfettered quest for knowledge". The argument here is that experimental abuse occurs when the thirst for "positive knowledge" overwhelms moral sensibilities; that the cruelty of the Nazi experimenters, for example, was made possible by an ethos of normlessness, resulting in a science unconstrained by values. There is surely some truth in this apparent oxymoron: in the Japanese case, for example, Dr. Ishii encouraged his colleagues in his infamous Unit 731 by reminding them of the "dual thrill" of scientific research on human subjects: there was the pleasure of successfully building "a powerful military weapon against the enemy", of course, but there was also the thrill of probing for "the truth in natural science ... and the unknown world".[63] Pure science and military urgencies combined in General Ishii's experimental ethics; there is no evidence of a concern for the rights of his subjects.

Human experimentation, however, is never totally unconstrained. Nazi physicians did not perform experiments on "healthy" Germans, but rather only on persons whom they considered "lives not worth living". Ishii and his entourage experimented on Chinese peasants and other foreigners, never on other Japanese. American atomic tests were almost always performed on enlisted men, rarely on officers. Abusive human experiments are generally performed on people whom the experimenting authorities feel cannot be aware of the danger, or cannot put themselves in a position to do anything about it. Experimental abuse is fostered by an atmosphere of medical arrogance, a disregard for individual human worth that allows the suffering of individuals to be discounted or devalued. Scientific research is always constrained by norms: the important ethical question is: What kinds of norms?

A second obstacle to understanding – and proper action – has to do with what one might call "exculpation by contextualization". There are many different versions, but let me single out one, a kind of exculpatory relativism, according to which we can only blame institutions, or epochs of history, but not individuals. As most commonly heard, the argument is that the actions of individuals have to be understood "in the context of the times". That of course is true, as any historian is aware, but the question is then: what are the appropriate contexts? Everett Mendelsohn, whom we honor in this volume, has always taught that scholars – including historians! – have obligations to the present as well as to the past. One way of failing at this obligation would be to exonerate by over-contextualizing: if only institutions are culpable, then the Nuremberg Tribunal should have held the SS responsible, but not Holzlöhner or Mengele or any other individual agent. The Japanese army command would deserve blame, but not General Ishii.

The question is sometimes phrased – inappropriately, in this instance – as a need for "balance". A recent news report in *Science* ("Radiation: Balancing the Record"), for example, asks whether the abusive experiments conducted by the AEC and other military agencies "have a historic context and scientific value that provides justification for these seemingly inhumane acts".[64] If historic context and scientific value are all one needs to "justify" an experiment, then perhaps we should exonerate the perpetrators of the Nazi medical crimes, General Ishii's experiments, and every other human experimental abuse, since all have a "historic context" and few are without some "scientific value".

Contexts are crucial, but contexts do not exculpate – or at least should not. We need to appreciate the origins of abuses, in order to know where blame is due and to better understand how to prevent their recurrence. The ethics of human experimenters are likely to reflect the ethics of the broader society in which those experiments are conducted, but societies are never homogeneous, and people can disagree.[65] There is always hope for dissent, resistance, improvement, redemption. It is therefore vital to retain the moral strength we need to issue judgements – including moral guidelines and/or condemnations over vast stretches of time and space – so that our understanding does not become uncritical sympathy, so that our explanations do not become exculpations.

Pennsylvania State University

Notes

[1] This paper was originally presented at the conference on "Humans in Experiments", sponsored by the Hamburger Institut für Sozialforschung, Hamburg, Germany, 28 June, 1995.

[2] Carol Pogash, "Chain Reaction: An In-House Activist Prepares Lieff, Cabraser & Heimann for Two Radiation Class Actions", *California Lawyer* (March 1995), pp. 48–52, 90.

[3] Keith Schneider, "Disclosing Radiation Tests Puts Official in Limelight", *New York Times*, 6 January, 1994.

[4] Ruth R. Faden, ed., *Final Report of the Advisory Committee on Human Radiation Experiments* (New York: Oxford University Press, 1996).

[5] See Richard H. Levy, "The Bombing of Auschwitz Revisited: A Critical Analysis", *Holocaust and Genocide Studies* (Fall 1997), pp. 129–70.

[6] The historiography of human experimentation is large and growing; a good early overview is N. Howard-Jones, "Human Experimentation in Historical and Ethical Perspectives", in *Human Experimentation and Medical Ethics*, F. Bankowski and N. Howard-Jones, ed. (Geneva: XVth CIOMS Council Conference, 1982), pp. 453–95; compare also Susan E. Lederer, *Subjected to Science: Human Experimentation in America before the Second World War* (Baltimore: Johns Hopkins University Press, 1995); the essays in *The Nazi Doctors and the Nuremberg Code: Human Rights in Human Experiments*, George J. Annas and Michael A. Grodin, ed. (New York: Oxford University Press, 1992); Barbara Elkeles, *Der moralische Diskurs über das medizinische Menschenexperiment im 19. Jahrhundert* (Stuttgart: Gustav Fischer, 1996); and Paul Weindling, "The Origins of Informed Consent: The International Scientific Commission on Medical War Crimes, and the Nuremberg Code", *Bulletin of the History of Medicine*, 75 (2001), pp. 37–71.

[7] Londa Schiebinger, "Sex and Race in Eighteenth-Century Human Experimentation" (unpub. ms.).

[8] In 1891, for example, Victor Cornil reported on an experiment designed to test the contagion theory of cancer by transplanting tissues from a woman's cancerous breast into her healthy breast. The experiment was a "success", insofar as tumors were thereby induced in the healthy breast. See M.V. Cornil, "Sur les greffes et inoculations de cancer", *Bulletin de l'Académie de Médecine*, 25 (1891), pp. 906–909.

[9] Lederer, *Subjected to Science* (cit. n. 6), p. 79.

[10] Michael A. Grodin, "Historical Origins of the Nuremberg Code", in *Nazi Doctors*, Annas and Grodin, eds. (cit. n. 6), pp. 127–28.

[11] Charles E. Rosenberg, *The Care of Strangers: The Rise of America's Hospital System* (New York: Basic Books, 1987).

[12] Eileen Welsome, "The Plutonium Experiment", *Albuquerque Tribune*, Nov. 15–17, 1993; Geoffrey Sea, "The Radiation Story No One Would Touch", *Columbia Journalism Review* (March/April 1994), pp. 37–40.
[13] Lederer, *Subjected to Science* (cit. n. 6), p. 120.
[14] Allen M. Hornblum, *Acres of Skin: Human Experiments at Holmesburg Prison* (New York: Routledge, 1998). Jessica Mitford's 1973 book helped eliminate the use of prisons as experimental laboratories; see her *Kind and Usual Punishment* (New York: Knopf, 1973), esp. the chapter titled "Cheaper than Chimpanzees".
[15] Southern medical journals in the pre-Civil War era used to post want-ads for slaves to be used in testing new medical preparations; see Todd Savitt, "The Use of Blacks for Medical Experimentation in the Old South", *Journal of Southern History*, 48 (1982), pp. 331–48.
[16] Lederer, *Subjected to Science* (cit. n. 6), pp. 110–11.
[17] Lederer, *Subjected to Science* (cit. n. 6), p. 83.
[18] David J. Rothman, *Strangers at the Bedside* (New York: Basic Books, 1991), p. 53.
[19] Claude Bernard, *An Introduction to the Study of Experimental Medicine* (New York: Dover, 1957), p. 101.
[20] Schiebinger, "Sex and Race" (cit. n. 7).
[21] James H. Jones, *Bad Blood: The Tuskegee Syphilis Experiment* (New York: Free Press, 1981).
[22] David J. Rothman and Sheila M. Rothman, *The Willowbrook Wars* (New York: Harper and Row, 1984).
[23] Lederer, *Subjected to Science* (cit. n. 6), p. 51. Recognition of the dangers of experimentation with vulnerable populations ("cruelty to the helpless") led one U.S. Senator, Jacob H. Gallinger, in 1900 to propose a bill outlawing experimentation in the District of Columbia on anyone regarded as incapable of giving consent, including newborns and children under the age of twenty, pregnant women or women who had given birth within the previous year, the aged, insane, or feebleminded, and people with epilepsy. Prisoners, interestingly, were not included in his list. The bill was defeated after lobbying from the medical research community (pp. 72, 143–44).
[24] Bridgette Goodwin, *Keen as Mustard: Britain's Horrific Chemical Warfare Experiments in Australia* (St. Lucia: University of Queensland Press, 1998); also her 1998 ABC-TV documentary film by the same title.
[25] The Chinese experimental subjects were generally killed within a month of their service as human guinea pigs; see Sheldon H. Harris, *Factories of Death: Japanese Biological Warfare, 1932–1945, and the American Cover-Up* (New York: Routledge, 1994), p. 39.
[26] Michael H. Kater, *Das "Ahnenerbe" der SS 1933–1945* (Stuttgart: Deutsche Verlags-Anstalt, 1974), pp. 231–245; Wolfgang Benz, "Dr. Sigmund Rascher: Eine Karriere", in *Medizin im NS-Staat: Täter, Opfer, Handlanger*, Wolfgang Benz and Barbara Distel, eds. (Munich: Deutscher Taschenbuch Verlag, 1993), pp. 190–214; Gerhard Baader, "Das Humanexperiment in den Konzentrationslager: Konzeption und Durchführung", in Michael Wunder, ed., *Menschenversuche: Wahnsinn und Wirklichkeit* (Cologne: Kölner Volksblatt, 1988), pp. 48–69.
[27] The curious and, to my mind, simple-minded debate conducted a few years ago in the U.S. over "whether to use the Nazi data" generally missed the fact that much of the concentration camp data has already been used – in life jacket design, for example, where the modern stress on neck-warming traces back to Nazi experiments.
[28] Sterilization experiments were performed using X-rays, chemical injections, and extracts from the plant, *Caladium seguinum*. Horst Schumann's X-ray experimentation in Auschwitz was designed to see whether rapid techniques of sterilization could be developed for use in the German campaign to limit the birth of racial undesirables. One thousand men and women had their genitals irradiated for up to 15 minutes, after which they were returned to work and eventually castrated to examine the effects of the rays. Claus Clauberg rented subjects from Auschwitz for use in experiments involving injection of irritants such as formalin into the cervix to induce sterility; other experiments included: experiments to heal phosphorous burns, experiments to determine the efficacy of aconite nitrate bullets, and experiments to determine the value of antibacterial agents. See Alexander Mitscherlich and Fred Mielke, *Doctors of Infamy: The Story of the Nazi Medical Crimes* (New York: Henry Schuman, 1949).
[29] The topic has never been systematically investigated, but the number of prisoners killed in the course of Nazi medical experiments was probably relatively small – perhaps only one or two thousand. One should also not forget, however, that other medical means were used to kill POWs. At Buchenwald, for example, 8,000 Russian prisoners were killed in the course of supposed "medical exams": prisoners were asked to stand up to a device designed to measure their height, and then shot in the back of the head with a small caliber gun concealed immediately behind the device. Medicine in this context functioned as a disguise, to increase the efficiency of the killing

process. See my *Racial Hygiene: Medicine Under the Nazis* (Cambridge: Harvard University Press, 1988), p. 221.

[30] I am not aware of experiments performed on healthy German soldiers in the Nazi era; indeed it appears that prior to the outbreak of war in 1939, the experimental abuses of Nazi Germany are not noticeably different from those of other industrial nations. There may even have been more experimental abuse in the United States in the 1950s than in Germany in the 1930s.

[31] In one early experiment, three communist Chinese peasants were injected with bacteria from fleas from plague-ridden mice captured near the Soviet border. All three developed fevers and were anatomized while still unconscious to examine the course of the disease until death. In another gruesome series of experiments, prisoners were injected with potassium cyanide, jolted with electric shocks, or forced to breathe phosgene gas to see how much it took to kill a man; see Harris, *Factories of Death* (cit. n. 25), pp. 13–100.

[32] Harris, *Factories of Death* (cit. n. 25), pp. 28, 70. Fans were used to speed up the freezing process, and prisoners prior to the experiment were forced to dip their arms in water. Frostbite experiments were sometimes conducted in conjunction with weapons tests: in January of 1945, for example, ten Chinese prisoners were exposed to both freezing weather and shrapnel bombs loaded with gas gangrene, the point being to see whether the weapons under development would work in very cold conditions. The Ping Fan facility also conducted Nazi-like low pressure experiments to test the performance of pilots in conditions of high-altitude flight.

[33] *Ibid.*, p. 28. Work at the camp was disrupted by a 1934 prisoner riot and escape, during which 12 prisoners made it safely into a forest where they met up with and joined a band of partisans (p. 29). The Beiyinhe facility was abandoned and destroyed in 1937; the remaining prisoners were "sacrificed" on the spot to prevent word of the operation from spreading (p. 30).

[34] Harris, *Factories of Death* (cit. n. 25), p. 5.

[35] Hans-Martin Sass, "Reichsrundschreiben 1931: Pre-Nuremberg German Regulation Concerning New Therapy and Human Experimentation", *Journal of Medicine and Philosophy*, 8 (1983), pp. 99–111; Grodin, "Historical Origins" (cit. n. 10), pp. 128–32.

[36] Final Report , Faden, ed. (cit. n. 4).

[37] Marlise Simons, "Soviet Atom Test Used Thousands As Guinea Pigs, Archives Show", *New York Times*, 7 November 1993. The Soviet Union conducted 137 atmospheric tests over roughly this same period.

[38] Philip J. Hilts, "Panel Finds Wide Debate in '40s On the Ethics of Radiation Tests", *New York Times*, 12 October 1994. Pilots were asked to fly through the radioactive exhaust of experimental nuclear rockets; see "U.S. Staged Nuclear Rocket Accident in Nevada in 1965", *Facts on File World News Digest,* Dec. 31, 1994, p. 986F3. Nonhuman experiments were also conducted: thousands of pigs, monkeys, and beagles, for example, were sacrificed by the Atomic Energy Commission to determine radiation health effects. The "megamouse" experiment, operated over nearly half a century at various labs of the AEC and DOE to determine the long-term effects of low-levels of radiation, consumed several million mice. Many of these experiments were cruel: in the experiments conduct by the Defense Nuclear Agency, for example, monkeys trained in simple flight manipulations were strapped into seats and heavily irradiated to determine how much radiation a pilot could withstand while continuing to perform his mission.

[39] Thomas H. Saffer and Orville Kelly, *Countdown Zero* (New York: Putnam's Sons, 1982); H. Josef Hebert, "Contamination Called Deliberate", *CDT* (AP), Feb. 25, 1994.

[40] See Geoffrey Sea's forthcoming book, *Proving Grounds: The Desecration and Redemption of the Steppe.*

[41] Jason Vest, "U.S. Aides Proposed Using A-Test Troops for Research", *Washington Post*, 6 July 1994.

[42] Hilts, "Panel Finds Wide Debate" (cit. n. 38).

[43] The classic source here, of course, is Kevin Rafferty's 1982 film, "Atomic Café"; compare also Elaine May, *Homeward Bound: American Families in the Cold War Era* (New York: Basic Books, 1988).

[44] Gil Whittemore, "In the Shadows of Hiroshima and Nuremberg: Debates Over the Ethics of Human Experimentation in the Development of Nuclear Powered Bomber, 1946–51", *Sociology of Science Yearbook,* 12 (1988), pp. 431–62.

[45] C.W. Clarke, "VD Control in Atom-bombed Areas", *Journal of Social Hygiene*, 37 (January, 1951), pp. 3–7.

[46] M. Susan Lindee, *Suffering Made Real: American Science and the Survivors at Hiroshima* (Chicago: University of Chicago Press, 1994).

[47] See my *Cancer Wars: How Politics Shapes What We Know and Don't Know About Cancer* (New York: Basic Books, 1995), pp. 261–65.

48 Faden, ed., *Final Report* (cit. n. 4). In 1997, MIT and the Quaker Oats Company agreed to pay 100 of the Fernald School survivors $1.85 million in damages; see "MIT, Quaker Oats to Settle Experiment Suit", *US News*, Dec. 31, 1997, <www.cnn.com/US/9712/31/radioactive.oatmeal/>.

49 One of the earliest U.S. human radiation experiments by the U.S. military involved the Navy's injection of radioactive antimony into hospital patients in 1945. The metal, used to killed parasites, was injected in a radioactive form to allow it to be tracked through the body; see "Robert Burns, "Navy Test Injected Radiation into Humans", AP January 13, 1994. It is not clear whether the "volunteers" for the experiment were told about the radioactivity and its possible consequences.

50 See my "Expert Witnesses Take the Stand", *Nature,* 407 (2000), pp. 15–17; also my "Rebuttal Expert Witness Report", Testimony for the Plaintiffs in "Emma Craft, *et al.*, Plaintiffs, vs. Vanderbilt University, *et al.,* Defendant", Case # 3-94-0090, Jan. 29, 1998, 71 pages.

51 Richard Stone, "Scientists Study 'Cold War' Fallout", *Science*, 262 (1993), p. 1968.

52 Spencer R. Weart, *Nuclear Fear: A History of Images* (Cambridge: Harvard University Press, 1988). Radiation injuries (e.g. cancer) often appear only years after exposure, making it difficult to associate cause and effects. There were also reasons to believe, early on, that radiation might somehow explain the healing powers of mineral spas. Mineral baths had been treasured for millennia, and when high levels of "radium emanation" were discovered in such baths in the first decade of the twentieth century (e.g. in the *Erzgebirge* of southern Saxony), the suspicion arose that the radioactivity in those springs might somehow be responsible for their healing powers. It should also be recognized, however, that cautions against the excessive inhalation of the vapors from such spas appeared soon after radioactivity was discovered; see my *Cancer Wars* (cit. n. 47), pp. 174–80.

53 Experiments of opportunity were occasionally conducted at Ishii's Ping Fan Biological Warfare facility. In May and June of 1943, for example, a Chinese prisoner knocked out a guard, stole his keys, and freed more than 100 prisoners. The guards flooded the grounds with what they thought was a relatively harmless tear gas, chloropicrin, but found instead that the substance was actually fatal when breathed in such quantities. Though all of the prisoners died, the experimenters consoled themselves with the knowledge that valuable information would be gained from the dissection of the suffocated Chinese. See Harris, *Factories of Death* (cit. n. 25), p. 71; compare also Harry M. Marks, *The Progress of Experiment: Science and Therapeutic Reform in the United States, 1900–1990* (Cambridge: Cambridge University Press, 1997), pp. 172–81 for a more general discussion of "experiments of nature".

54 Wilhelm C. Hueper, *Occupational Tumors and Allied Diseases* (Springfield, Illinois: Thomas, 1942), pp. 435–68.

55 See my *Cancer Wars* (cit. n. 47), pp. 36–48.

56 See my *Cancer Wars* (n. 48), pp. 194–95.

57 Lederer, *Subjected to Science* (cit. n. 6), pp. 126–38.

58 Other opportunities for abuse remain open, but the one presently under consideration does not.

59 Jay Katz, "The Consent Principle of the Nuremberg Code: Its Significance Then and Now", in *The Nazi Doctors and the Nuremberg Code: Human Rights in Human Experiments*, George J. Annas and Michael A. Grodin, ed. (New York: Oxford University Press, 1992), pp. 227–39. Recognition of the dangers of medical coercion date back at least to the 1880s, when the gynecologist James W. Etheridge asked that caesarean sections be performed only with the consent of the patient "obtained without direct or indirect coercion". Etheridge also forecast a time when consulting a woman about such questions would no longer be necessary: that time would come, he maintained, when the procedure was judged "routine", and not extraordinary. See Lederer, *Subjected to Science* (cit. n. 6), p. 13.

60 See my *Cancer Wars*, pp. 161–65, 172.

61 For Dr. Mark Siegler's remarks, see Gina Kolata, "Memory of Era May Temper Judgment of Radiation Tests", *New York Times*, 1 January 1994.

62 Michael Ross, "O'Leary: New Radiation Test Subjects Informed" (LA Times), CDT, January 26, 1994.

63 Harris, *Factories of Death* (cit. n. 25), p. 44.

64 Charles C. Mann, "Radiation: Balancing the Record", *Science*, 263 (1994), p. 47.

65 David J. Rothman, "Government Guinea Pigs", *New York Times*, 9 January 1994.

MARK B. ADAMS

NETWORKS IN ACTION: THE KHRUSHCHEV ERA, THE COLD WAR, AND THE TRANSFORMATION OF SOVIET SCIENCE

Abstract

Based on more than a quarter century of evolving work, this "thought piece" is a re-evaluation of Soviet science during the Khrushchev era (1954–1964). It focuses on a series of remarkable events during the so-called "Thaw" (1955–1958) that give evidence of a fundamental transformation then occurring in Soviet science under the influence of the Cold War. These developments suggest the critical role played by personal networks in the evolution of Russian science. The piece concludes with some methodological implications for studying Soviet science, the Cold War, and the history of science writ large.

Introduction

I began my study of Soviet science as a student of Everett Mendelsohn, during the Khrushchev era, even before that era had ended. At that time, in 1963, the "big news" in science was the deciphering of the genetic code: the quest for the so-called "secret of life" kept the lights in the bio lab buildings on my campus burning through the night – so of course I studied genetics. It was also the age of "Sputnik": the Russians had beat us into space, and we all knew it. For my "history and science" major, then, it was perhaps natural that my "history" would be Russian and my "science" biology. When it came time to do my undergraduate thesis, however, I learned to my astonishment that, at least officially, Lysenko and his crackpot pseudoscientific biology still reigned supreme in the land of Sputnik.

How could it be, I wondered, that a scientific powerhouse like the Soviet Union – the country that had launched the first satellite – had somehow missed out on the central scientific breakthrough of the age? Just then, the "little October revolution" of 1964 unfolded and Khrushchev was ousted from power. I followed with rapt attention as the objects of my rather arcane research

G.E. Allen and R.M. MacLeod (eds.), Science, History and Social Activism: A Tribute to Everett Mendelsohn, 255–276.

became big news, Lysenko's depredations were publicly bruited, and Soviet genetics reemerged. It was intoxicating to see my private passion make the headlines of the *New York Times*. I was hooked, and have been ever since.

More than a third of a century has passed since then, and much has changed. The history of science has emerged as a substantial international scholarly discipline, and within it, so too has the field of the history of Russian science. Molecular genetics has begun to show its remarkable potential, and, in the process, has become a big business. The Khrushchev era is no longer front page news (the only recent reminder of him was when his son became an American citizen). Indeed, even "the Soviet experiment" itself seems to have come to its uncertain end, and the Cold War that influenced so much over the past half century is now history.

With the Cold War over, we can now look back at the "Soviet experiment" and the Khrushchev era with a new perspective. Those forty intervening years (and especially the past ten) have provided much to sharpen our vision and unclutter our view: the waning of ideological sensitivities, the new access to archival sources, intensified interactions with Russian colleagues, and the efflorescence of work in the social history of science – all have provided new ways to understand that history and its broader significance. Together at long last, Russian, European, and American scholars are beginning to unravel the story of Soviet science and the "Soviet experiment" in its full dimensions. Our divisions are no longer ideological, but rather interpretive and methodological, and we struggle to develop a common language and framework and to figure out the central storylines.

In our story of Soviet science, as it eventually unfolds, I suspect that the Khrushchev era will loom rather large, much larger than it has thus far. Lasting barely a decade (1953–1964), and perched between the eras of Stalin and Brezhnev, it is often treated as punctuation – as the denouement of a story about Stalinism, or the preface to a story about something that was to follow (e.g. the dissident movement). Yet that decade was at the very cusp of Soviet history: 1956 was the precise midpoint of the seventy years of Soviet rule. Immediately before the Khrushchev era had been a quarter-century of Stalinism, with its turmoil, turbulence, and terror; immediately after would come twenty years of Brezhnevism, with its bland bureaucratic modernity. Just before it, Trofim Lysenko's "Michurinist biology" was enthroned; just after it, Lysenkoism was abandoned. This was the first full decade of the "atomic age" and the Cold War, of de-Stalinization and the Hungarian Revolt, of Sputnik and its aftermath, of the "U2" and the "Cuban missile crisis", when the Cold War stabilized and "normalized" into "missile gaps" and the "space race". It was, in short, a key decade of transition, when the Cold War settled in and Stalinism somehow became transformed into Brezhnevism.

In the course of my own studies of Soviet science over the past forty years, I found it illuminating to move from a comparative disciplinary perspective to the analysis of the institutional matrix of Soviet science. Then, in the late 1980s, as the Cold War was flagging and the Soviet Union was beginning to unravel, I returned to my earlier study of the Khrushchev era, and this time was able to obtain much more information about events, not only from personal archives and documents, but also from candid interviews with firsthand participants.

Just within the last few years, I have come to see those events in a somewhat new way, one which, I believe, has broad implications for the way we study science and its history. In this short essay, I want to sketch that view by discussing four matters that I now see as closely interrelated: the Khrushchev era, the Cold War, the transformation of Soviet science, and the key role played by a somewhat neglected factor – *networks* – which, I believe, has helped to shape modern science.

"The Thaw"

What happened in Soviet science during the Khrushchev era, and what can it tell us about Soviet science as a whole, and about the history of science more generally? To begin exploring these questions, let us focus for a moment on a cluster of events in Soviet science. They all took place during the time of what has been aptly called the "thaw" (*ottepel*) – a three-year period between late 1955 and late 1958, beginning just as Khrushchev was coming to power, and remembered for his "de-Stalinization" campaign, the launching of Sputnik, and a brief time of cultural and political liberalization, which largely came to an end in December 1958. It was during this time, I will argue, that things were set into motion that would effect the key *transition* from what we know as "Stalinism" into what became "Brezhnevism".

Before discussing these key transitional events, let me set the scene by briefly rehearsing the "legacy" of Stalinism – the "status quo ante". In early 1953, the Michurinist campaign, supplemented by the "results of the Pavlov session" and numerous other offshoots, was still in full swing. Lysenko's "Michurinism" ruled Soviet biology, and his followers had taken over all key administrative posts. The geneticists had been stripped of their institutions and, in most cases, their jobs. The elections in the Academy of Sciences in 1953 had flooded its Biology Division with corresponding members of Michurinist persuasion (or, at least, coloration). Lysenkoism was as thoroughly in place, administratively and ideologically, as anything could be. Then quite suddenly, in the spring of 1953, two portentous events happened that would open an age of new possibilities: Stalin died; and Watson and Crick published their double-helix model for the structure of DNA. In the decade that followed, there were at least six political zigzags: it looked, for a brief time, that Lysenko would lose his legitimacy in the mid-1950s. When things settled out, however, he had managed to cultivate a personal relationship with Nikita Khrushchev, and, on the official level, his "Michurinist genetics" remained the politically correct, state-sanctioned variant until after Khrushchev's ouster in October 1964.

When Lysenkoism was officially debunked in 1966 and Soviet genetics reemerged as legitimate science, a curious fact became immediately apparent: although called other names, and protected in odd places, Soviet genetics was in fact reestablished not in 1966, but a decade earlier, in 1956! Now, this fact has been noted before – by me among others – but I'm not sure we have fully come to terms with its implications.

What happened around 1956 that got Soviet genetics going again? The answer, in its most concrete form, is that not one, but *many* things happened, almost simultaneously. There were at least twenty "initiatives" taken in those months, and each is a story unto itself, complete with its own setting, characters, plot, drama, and narrative line. I have managed to reconstruct almost all these stories in some detail, but I am not going to tell any of them here. Instead, I want to simply tick them off, and then ask what they have in common, and what patterns they form.

Here, then, are a few of the things that happened between roughly December 1955 and December 1958:

Under botanist V. N. Sukachev, the Botanical Society launched a journal campaign criticizing and scientifically disproving various of Lysenko's theories.

A "private" letter signed by 300 biologists and many physical scientists was delivered to the Central Committee, calling for an end to Lysenko's stranglehold and the reestablishment of genetics. (It became known as the "Letter of 300".)

Lysenkoist A.I. Oparin was replaced as head of the Academy's Biology Division by biochemist Vladimir Engelhardt, who used his office to support genetics and molecular biology.

At his Institute of Physical Problems, Petr Kapitsa sponsored a major "event" on molecular genetics, where subsequent Nobel physicist Igor Tamm and geneticist Nikolai Timoféeff-Ressovsky spoke to an overflow audience.

Moscow mathematician A. A. Liapunov, a Soviet pioneer of "cybernetics", created a "home seminar" on genetics.

Timoféeff-Ressovsky began holding summer workshops on experimental biology and genetics at Miasovo in the Urals.

Under its rector, mathematician Alexander D. Aleksandrov, Leningrad University appointed Mikhail Lobashev to head its Genetics Department, and genetics began to be taught there once again.

The Academy's Institute of Biophysics got a new director, biophysicist Gleb M. Frank, and was reorganized to include a new Laboratory of Radiation Genetics, headed by the one remaining geneticist in Academy ranks – Nikolai Dubinin.

Nikolai Semenov (1956 Nobelist in chemistry) opened a new Laboratory of Mutagenesis at his Institute of Chemical Physics.

The Institute of Atomic Energy, directed by the "father of the Soviet A-bomb", Igor Kurchatov, opened a Radiobiology Section which soon included some five or six separate genetics laboratories.

Engelhardt created a new Institute of Radiation and Physicochemical Biology (later renamed the Institute of Molecular Biology).

The Academy of Sciences launched a major discussion and conference on the philosophy of science that gave scientists primacy in interpreting science according to "dialectical materialism" and officially endorsed the role of physical scientists in biology.

Under the leadership of mathematician Mikhail A. Lavrent'ev, the Academy of Sciences created a wholly new "Siberian" Division of the Academy and began construction of a science city, *Akademgorodok*, which included a new Institute of Cytology and Genetics, directed by Dubinin.

Although far from exhaustive, this list gives a flavor of the variety of initiatives that took place during the months of the "thaw".

We may note three features of these initiatives at the outset: they were undertaken by a whole range of the most distinguished Soviet scientists, spanning mathematics, physics, chemistry, and biology; they were breaking out all over, within a wide range of institutions, divisions, and locations; and almost all survived the end of the "thaw" intact. When we think about it, this is rather remarkable: just a few short months after the peak of "High Stalinism", with its show trials, strident ideological campaigns, and arrests, the government's official line was being tested and challenged on a wide range of fronts. Furthermore, these initiatives, initially so fragile and vulnerable, were able to build such momentum and become so successfully consolidated and established, in such a short time, that even when the "thaw" ended in late 1958, they were able to continue and expand.

In the early 1960s, when I began to study these events, it took some digging to decipher them and to figure out how and why they belonged together. That is hardly surprising: even in the 1960s, when some of the basic facts became known, the Government and Party never did manage to admit that they had made a mistake about genetics, so, aside from details of Khrushchev's misbehavior (which conveniently helped to legitimate his ouster), official published accounts about the resurrection of genetics were generally triumphal in tone and uninformative about process. In addition, of course, the Party and Government still ruled, so, not surprisingly, contemporary scientists were anything but forthcoming about how they were able to manipulate the system to achieve their own objectives, especially since these objectives were, at least officially, at variance with state policy at the time. And, of course, tactics that worked wonderfully to hide, protect, or legitimate genetics in Khrushchev's Russia also served to mask the trail from innocent historians: one had to know, for example, that "genetics" sometimes meant "genetics", and sometimes did not; that "biochemistry" or "biophysics" *could* mean genetics; and that "radiobiology" and "physicochemical biology" often *did* – as did (to cite a particularly ornate example) "the biochemistry of physiologically active high-molecular-weight polymer compounds"!

In recent years our knowledge of many of these cases has become much more detailed. In the journal *Priroda* especially, documentation concerning one or another of these events has tended to appear every few issues, often accompanied by published excerpts from the archives and substantial biographical and historical reminiscences. These are marvelously rich and informative. In historical terms, however, they have had the effect of turning one triumphal drama – the struggle for the rebirth of Soviet genetics – into fifteen or twenty distinct tales, each with its own chronology and plot, its moments of humor and drama, and its turning points, and each built around its own hero.

That is, no doubt, exactly the way these events look when viewed closely. But what do we see when we look at all of them, together, and from a distance?

The Primacy of Networks

Suppose we eschew the exotic, "exceptionalist" emphasis that has so often colored Western discussions of Soviet science, and try to look at it more or less the way we might look at any set of events in the history of science in any country.

Viewed from the perspective of the modern social history of science, the individual events or initiatives I have enumerated take on a somewhat different character, one that is rather more familiar. What was involved in the struggle over genetics was a *disciplinary* crisis involving competing groups and the theories, practices, traditions, and paradigms they represented. In 1948, one group was given complete control of the discipline by the central authority and patron of science. Part of the resultant crisis – and one of the reasons why Lysenko's new hegemony was so successful – was Lysenko's mandated takeover of all the *institutions* where the discipline was practiced. Under the new conditions of the mid-1950s, with Khrushchev's government and Party willing to endorse Michurinism, it became difficult for Soviet geneticists to retake the institutions they had lost. So, most of the initiatives during the "thaw" understandably involved creating new institutional niches, and in some cases, whole new institutes – away from, outside of, or beyond their rivals' reach.

Thus far, our story has been about *disciplines* and *institutions*. This is natural enough: both are central to the social structure of science, and both leave public traces. Disciplines have their associated societies, journals, texts, and international reference groups and congresses. Institutions have their budgets, buildings, organizational charts, publications, and patrons. Each has, by its very nature, essentially public dimensions, each creates its own public record and leaves its paper trail. And the interrelation between the two is not simple: a single discipline can be an aggregation formed around an object of study, theory, and/or method, and it encompasses members based in many different institutions of different sorts; likewise, a single institution can encompass members of many different disciplines. Each social form has its own special characteristics, politics, trajectory, and dynamics.

But this social picture, structured around institutions and disciplines, is clearly incomplete. Consider: on their own, the geneticists did not have the clout to retake their discipline *or* its institutions, nor to create new institutional niches for themselves. Where do disciplines and institutions come from, after all, and what makes them change and evolve? In particular, why were so many seemingly separate and apparently unrelated initiatives, by so many, directed at the same end, happening at the same time? Put somewhat differently, we may ask: with the relevant *disciplines* and *institutions* seized, what other factor could come into play to reverse this apparent stranglehold?

The answer, I think involves a third dimension – one which has been, until quite recently, the most difficult to document: *networks*. Here, I am not referring to anything arcane or technical – not to the "networks" of the sociologist, dynamist, or social studies theoretician, much less the computer specialist – but to the looser, more evocative meaning the word has come to have in everyday language, one familiar to every kind of historian: personal networks. A personal network is a much looser, less coherent "structure" than either an institution or a discipline. Developed out of extended family, old school ties, mutual experience, hobbies, private passions, and shared interests, such networks involve ramifying contacts that are multiple and complex – as are all the free associations that underlie civic society. They can also form "nodes" or "ganglia" where various networks interface and new connections are made – sometimes in the form of informal circles, private societies, clubs, salons, soirées, and the like, sometimes in more organized forms, ranging from things we might call "movements" to interest groups, political organizations, and even "mafias". And some of these, in turn, might eventually gain further structure as would-be disciplines or proto-institutions.

For all their looseness – indeed, perhaps because of it – such networks can be remarkably influential, resilient, and enduring. Note, however, that, as a form of social organization, networks are *by their very nature* more private than public. They are informal, voluntaristic, "personal" associations whose nature and ramifications can shift and change quickly over time. They leave historical traces, to be sure – but such traces are likely to be scattered in personal photographs, reminiscences, diaries, and private letters. Further complicating the matter is the fact that the *legitimacy* of disciplinary and institutional actions often may require *hiding* the networking behind them. (In justifying a new appointment to his government patron, for example, an institute director is more likely to stress needs and qualifications than to admit, "He's the son-in-law of a colleague who gave my daughter-in-law a job".) So powerful are networks, indeed, that both disciplines and institutions have often formally implemented procedures to minimize or regulate their influence – for example, blind peer review, or anti-nepotism rules. Nor should we imagine that such networks are simply "scientific" or "professional" or "disciplinary" – they can take many different forms, as can all voluntary personal associations.

Networks cut across disciplines and institutions with promiscuous impunity

– and a shifting, evolving, interconnected web of such networks is almost impossible for any agency to control. In Russia, in particular, one suspects that the political authorities may have tried. From their own "underground" years, the Bolsheviks were no doubt well aware of the importance of "private" networks; by the late 1920s, their new state had effectively seized control of all disciplines and institutions, both by government seizure of all funding and appointments, but also by seeding all organizations with Party members who reported directly up its separate chain of command. When even these measures did not produce the desired results, the authorities may have begun to suspect that something like interest groups and networks were subverting their well-laid plans. Indeed, judging from what we know about the kinds of questions asked by the interrogators of the victims of the various Stalinist purges (who are you related to, who do you know, when did you see them, what did you talk about, who do *they* know, and so forth) one might suspect that it is precisely these kinds of personal networks that the secret police agents were trying to uncover. In a system where official sources are not trusted, and public speech has to be politically correct, we might well expect that networks would become all the more important.

We have long known that networks have been remarkably consequential in Russian history, although we have not put it in exactly those terms. In political history, we know of "the Decembrists", who seem, in retrospect, to have been much closer to a socially active network of likeminded acquaintances than a Party or movement. Much has been spoken and written about the "Russkii kruzhok" or "Russian circle", which appears so often in Russian cultural history. In Russian musical history, for example, there is the famous Balakirev Circle – Balakirev, Borodin, Kui, Mussorgskii, and Rimsky-Korsakov – also referred to as the *Mogúchaia kúchka* (the "Mighty Little Group"). In Russian science, such informal groupings have sometimes been included under the loose, catch-all term of "school", even when there were no formal teaching relations involved.[7] Other groupings we are learning about brought people of more diverse interests together.[8] Soviet conditions soon discouraged private organizations as being potentially subversive; one of the reasons for the arrest of Sergei Chetverikov in 1929 and the dispersal of his group in population genetics, we may recall, was the so-called "Drosophila Screeching Society" he led – a discussion club that met regularly in private homes and apartments, with a closed membership and new nominees subject to blackballing.[9] If Stalinism discouraged these kinds of gatherings, however, it soon animated and reinforced other forms of networking, including the famous "rumor networks" and "grapevines" that operated throughout the Soviet period; it was these networks that managed to inform Western sources (not always accurately) about arrests or executions during the darkest days.[10]

During my year-long research trip to Moscow in 1976–1977, in the midst of the Brezhnev era, I witnessed many examples of networks in action. One of the notable events of my stay was the fire in the Hotel "Rossiya". At a secret meeting in the Kremlin, I learned, the hotel director had been rebuked for sloppiness and arrested. The news was not public: I heard it from a friend during a specially arranged walk in Gorky Park on a cold winter's day. As it happened, for a brief time in the mid 1930s, her father had been an up-and-coming Party figure, so she had been able to attend a special Party primary school. There she had become good friends with a classmate, and had subsequently kept up with her chum over the years. During the succeeding four decades, as it happened, he had managed to survive and rise through the Party ranks, until he had found himself at that "secret" Kremlin meeting that decided the hotel director's fate. The news of what happened there must have rippled through Moscow networks remarkably fast: as best I can tell, despite the complicated arrangements, I learned what had happened only four hours after the secret meeting had ended!

A second example involves an impeccable scholar and scientist of my acquaintance, a man in his sixties, whom I saw on an almost daily basis. My evening visits to his Moscow flat were usually interrupted by frequent phone calls from his sundry acquaintances, old friends, and contemporaries, usually with the purpose of exchanging the latest news. There was almost no one I had heard of or wanted to talk to whose name and home phone number could not be found in his thick, dog-eared directory. One evening, out of the blue, he asked me: "Do you believe in flying saucers?" It seems he had heard from many of his sources that, the previous Tuesday, flying saucers had landed in one of Moscow's major squares, stopped all clocks and machines, and made it so that none of the people there could talk about it afterwards. "Of course, I didn't believe the rumors", he said, "until today's paper" – at which point he showed me a sober announcement of one sentence on the front page of a leading Moscow daily. It read: "Rumors that flying saucers recently landed in Moscow are completely unfounded". "Do you see?" he pressed. "If this weren't true, why would they take the trouble to deny it?"

Of course, this episode illustrates the profound and longstanding Russian distrust of political authority, but, on reflection, it tells us rather more. Consider: a "hard-headed" scientist had come to credit not merely the possibility of flying saucers in the abstract, but to take seriously the possibility *that they had actually landed in Moscow*, because of information he had heard through his network. Surely, all of Moscow must have been alive with these rumors for the government to feel obliged to deny them in print so promptly ... which, in turn, had the result of lending them credence. This example reminds us that, far from undermining personal networks, totalitarian systems can actually strengthen and reinforce them as the only reliable loci of trustworthy information. Indeed, one suspects that no matter what the government had said or done in this particular case, the network account would have gained legitimacy.

I do not believe that my experiences with Russian networks are all that unusual: everyone who has spent serious time in Russia has almost certainly experienced something similar. Indeed, much of what Western scholars know firsthand about Russia has derived from their good luck in tapping into or being "adopted" by personal networks, by being introduced and passed along to "friends of friends". Knowing, as we all do, how powerful, wide ranging, and vigorous personal networks in Russia can be, it is curious we have not paid more attention to their influence on maintaining and shaping Soviet science.

In many of my earlier studies, I now realize, there were strong traces of networks in action. For example, my studies of the founding of the Kol'tsov institute revealed the rich Moscow network of scientists, educators, philanthropists, civic activists, and entrepreneurs to whom Kol'tsov repeatedly turned for funds, staffing, support, and protection.[11] Other, comparable networks in St. Petersburg, many involving the Bekhterev Institute, would make possible Iurii Filipchenko's institutional success in creating genetics and eugenics in that city.[12] In my work on the history of Russian eugenics, I have been able to show that the same networks which created Russian eugenics in the 1920s managed to survive its official "demise" in 1930 intact, and went on to create "medical genetics" four years later as a politically acceptable substitute.[13] In a study of the Soviet nature-nurture debate, I was able to trace a continuous tradition of interconnected networks that created and sustained Russian genetics, originating around 1900 and still active in the 1980s and 1990s.[14]

I had long suspected that networks played a key part in the rebirth of genetics during the Khrushchev years, and had spotted telling patterns.[15] Only in the last few years, however, has it been possible to chronicle and detail some of the roles networks played in the events in the 1950s. To cite but one example in the "Liapunov home seminar" one can trace the interlocking family ties of the Liapunov family, show how it became involved with the genetics matter, how and when it interconnected with other networks, and even track the young students who visited Liapunov's home in the early 1950s as, some years later, they moved into jobs in virtually every newly created center of genetics in Moscow.[16] A similar account can be given about each of the events I cited above. Suffice it to say that our disciplinary and institutional chronicle is now greatly enhanced by what is becoming clear about networks.

Transforming Soviet Science

By focusing on the role of networks, a picture emerges of what was happening in Russia in the 1950s that informs our view of its role at the cusp of the Soviet experiment.

Before I sketch that picture, however, let us recall, for purposes of contrast, how the story of Soviet science is so often told. The traditional story has been one of repression, of brilliant scientists fighting for the truth who fell victim to the system – Vavilov, Vernadsky, Sakharov, and the many others who are treated variously as heroes, saints, or martyrs. According to this traditional view, their academy was taken over and Bolshevized, forced to a new ideological line, purged and harassed. That new "USSR" Academy of Sciences – a vast, centralized, bureaucratized, Bolshevized, politicized, hierarchical structure which Vucinich has justly termed an "empire of knowledge"[17] – has been seen as the instrument devised by the Party-State apparatus in order to control science, to create a truly *Soviet* science subservient to its will. Hence the

1948 Lysenko meetings and the extraordinary Michurinist campaign that followed.

The new picture is in sharp contrast. The first thing we may note is that for all its sweeping, drum-beating character, the Michurinist campaign seems to have left the Soviet scientific community in general, and its geneticists in particular, both thoroughly unconverted and remarkably uncowed. Not only did some find haven in "secret" research, or elsewhere in the Academy's vast system: at the earliest possible opportunity, even while Stalin lived and was preoccupied with his "Doctors' Plot",[18] they were maneuvering to retake their discipline and oust Lysenko. One reason, I think, is because the networks that had been there before – some of them dating from pre-revolutionary days – remained as strong and vital as ever. By government edict, the Lysenkoists could seize the discipline and the institutions where it was practiced, but their reach did not extend to the networks that had built them in the first place. By tracing how the "letter of 300" made the rounds among a whole series of networks (at least six), couriered by their shared members, we can better understand how three hundred biologists and many physical scientists could manage to sign what amounted to a petition – which, not so long before, might well have bespoken "conspiracy" – and to do so within two years of Stalin's death.

In all of the other "initiatives" during the thaw, there were also networks in action. Particularly consequential were networks that linked geneticists with physicists, chemists, and mathematicians. There were, of course, good intellectual reasons why such scientists should suddenly take an active interest in genetics, given the flurry of activity centering on DNA: a molecule, base pairs, bonding, a helix, a code, information – now, this was their meat. It is no wonder, for example, that information theorists (such as S.L. Sobolev or A.A. Liapunov) or synthetic organic chemists (such as I. Knuniants, the developer of "plastics" in Russia, or Academy President Alexander Nesmeianov) would be interested.[19]

But reinforcing and informing this natural scientific curiosity were a series of personal networks of various kinds.

Complex family relations formed some of these links. Of course, we all know of the case of the Vavilov brothers: one, the leading Soviet geneticist, who died in prison in the early 1940s, while the other – a physicist – became President of the Academy of Sciences a short time later.[20] This is only the tip of the iceberg. To cite just a few examples: The founder of *Akademgorodok*, Lavrent'ev, was related by marriage to a someone who had worked in T.H. Morgan's lab. Liapunov's wife was the daughter of Academician Nametkin, a chemist, and the family used his dacha (located not far down the path from Kapitsa's). The newly appointed director of the Institute of Biophysics I mentioned above, Gleb M. Frank, had worked along side Semenov and others in the Ioffe Institute in the 1920s; his brother was the physicist (and future Nobel Prize winner) Ilya M. Frank. Given the inbreeding of the Russian intelligentsia, we should not be surprised to learn that family ties often played an important role.

This networking also involved old school ties. World-renowned mathematician A. Kolmogorov and geneticist Nikolai Dubinin had been classmates. Many of Russia's physicists had known their contemporary, Georgii Antonovich Gamov, before he had emigrated and become the renowned American astrophysicist George Gamow; in the early 1950s, he was publishing important articles on the DNA coding problem.[21] A particularly important role was played by physicist Igor Tamm (another subsequent Nobelist), who headed up the "theory" section of the so-called "Installation" [*ob"ekt*], the top-secret monastery where the Soviet bomb was being designed; when he became aware of the work of Western physicists on DNA and genetics (apparently through a chance encounter with field biologists on a mountain-climbing holiday), he naturally consulted an old buddy from his school days, biologist B.M. Zavadovsky, to fill him in.[22] Tamm soon became an outspoken campaigner for genetics, spreading the word among his physics colleagues at the Installation and elsewhere; his efforts would lead to the "Kapitsa-fest" mentioned above (where Tamm and Timofeef-Ressovsky spoke to an overflow audience) and, ultimately, to the creation of those five laboratories of genetics at the Installation's base in Moscow, Kurchatov's "sacrosanct" Institute of Atomic Energy.

As the above examples suggest, these networks were not only interdisciplinary, but also international. Timofeeff-Ressovsky's work in Germany on radiation genetics during the 1930s had been known to Soviet physicists well before he had been spirited home from Berlin in 1945 by the KGB: while he was still imprisoned, his work had been prominently mentioned by the great physicist Erwin Schrödinger in his postwar book, *What is Life?*, which was quickly translated into Russian on the urging of Soviet physicists. It was through their influence that Timofeeff was rescued from the Gulag (where he had almost gone blind from malnutrition) and eventually relocated into radiation research at a small station at Miasovo. (While on a rock collecting expedition in the Urals, Liapunov stumbled upon Timofeeff, whom he thought had died; when it became generally known that he was still alive, young students and old friends flocked to his new base – hence the Miasovo summer seminars.) Indeed, leading Soviet physicists became so involved with genetics that, in the mid-1950s, one of them – L.A. Artsimovich – smuggled drosophila cultures back to the Soviet Union from an international astrophysics congress.[23] Meanwhile, in the West, such prominent Russian émigrés as Theodosius Dobzhansky, I.M. Lerner, George Kistiakowki, and George Gamow were making contacts through their own extensive networks in American and world science with their former Russian friends and colleagues.

Like the geneticists, then, many Soviet physical scientists were part of networks that were both interdisciplinary and international, and they had their professional interests peaked by the DNA news that was just then transfixing so many of their Western colleagues. But, *unlike* the geneticists, they were in a position to do something about it.

The final years of "high Stalinism" witnessed the rise of a new and powerful group of physical scientists who were associated with "the bomb".[24] Science had become vital to Soviet power, and those who could provide the strategic products vital to the KGB, military, and Politburo – the A-bomb, the H-bomb, the rockets – became increasingly vital components of the Soviet Union's political-military-industrial complex. As they succeeded with their magic, they gained prestige, power, a measure of ideological immunity, and privileged access to the highest Party-state leaders. We now know that in 1950, physics was scheduled to undergo its equivalent of a Lysenko meeting – complete with scripted oratory and apologetic recantations. At the very last moment, the meeting was canceled, by order of Beria: the bomb-makers were apparently too important to the State to be trifled with.[25]

Rewarded with quick and disproportionate entry into Academy membership in the elections of 1953, and with real clout in higher government, military, and security echelons, these men had largely taken over the Presidium of the Academy of Sciences by 1956. From there they began to do for genetics what geneticists could not do for themselves. On some issues, the group was divided – the question of pure versus applied research, for example – but on the question of genetics, DNA, and Lysenko, they acted almost as one. We should not imagine these men as oppressed victims or closet dissidents. As a rule, they were Soviet patriots, and constituted a very privileged elite indeed, very near the pinnacle of Soviet power. They knew (or had learned) how to deal with the government and Party – indeed, some of them had joined during the war – and sincerely sought to provide it with all the weapons and such that it wished, and more.[26]

It was in the Khrushchev era, then, that a new dialectic of power, always implicit, came to the fore. If the Party-state had created the Big Academy as the instrument to control science, it had also thereby managed to create an instrument that had become the monopolistic supplier of what was now a top-priority product. And that "instrument" was now headed by people with very definite ideas about what they wanted for Soviet science: they wanted respect, they wanted ideologists and bureaucrats off their backs, and they wanted intellectual control over their own enterprise. If testimonials about the superiority of Marxist or dialectical materialist approaches to science were called for, fine – they would be happy to produce what was required themselves, thank you very much. They were willing to stay out of politics, and they wanted politicians to stay out of science – so long, of course, as they continued to fund it lavishly. And of course, they wanted Lysenko and his Michurinist ideologues neutralized or out, and genetics in.

And so the game unfolded. Lysenko had Khrushchev's ear on agriculture and biology, true; but Kurchatov, Lavrent'ev, Kapitsa, and others had his ear on other things, such as rockets, sputniks, and bombs. Their strategies: Open labs in areas under their own control. Maneuver to isolate Lysenko in the biology division, where they would, with Engelhardt's help, chip away at his stronghold. Open a new Siberian division, part of which would involve a genetics institute of their own design, and create an administrative structure for it that closed Lysenkoists out. Orchestrate a reform in dialectical materialist philosophy that made the scientists themselves its chief spokesmen on matters of science (a goal that they accomplished in 1958, thereby legitimating their meddling in biology). In carrying out this policy, some of the group (for instance, Semenov) "adopted" particular geneticists as their tutors, advisors, or assistants.

Of course, Lysenko and his allies knew perfectly well what was happening (after all, they had their own networks!), and used every lever they had with Khrushchev to unravel it. With that help, they were able to keep their strongholds, and to mobilize Khrushchev to oust Dubinin from directorship of the Siberian institute, and even attack him publicly and at some length at a December 1958 Central Committee meeting. Remarkably, however, Khrushchev's tirade at that meeting proved inconsequential: Dubinin stayed on for a time; the Siberian institute's genetics did not change, only the public descriptions of it; and Dubinin continued heading his Moscow laboratory as before. This was a harbinger. Despite the end of the "thaw" and Khrushchev's increasing alliance with Lysenko, every single center of reborn genetics that was created during the mid-1950s continued to grow and flourish in the years that followed – irrespective of the Lysenkoists' persistent attempts to dislodge them. We may wonder: why were Lysenko's own "Michurinist" networks, so extensively involving agriculture, the farming bureaucracy, and Party ideologists, unable to overcome these initiatives to oust him? In retrospect, it seems clear that one of the main factors was the Cold War: for the Politburo and the military and KGB leadership, under such conditions, agriculture and ideological "purity" were lower priorities than bombs, satellites, and missiles – and it was these scientists, not Lysenko and his people, who were providing them.

Learning from their experience, key members of the Soviet scientific leadership continued to press their campaign ever more strongly. Piggy-backing on Khrushchev's fondness for Akademgorodok, they tried to win his support for the construction of another science city closer by, at Pushchino – this one to be devoted entirely to "physicochemical biology" – but Khrushchev demurred, saying it was just too expensive. (They would eventually get it – a few months after Khrushchev's ouster.) Having learned from their 1957 Siberian success, in 1963 they engineered a major reorganization of the Academy, transforming its 8 divisions into 15, grouped into 3 larger "sectors", each run by a vice-president. Its chief architect was Nikolai Semenov, who, as a result of the reorganization, now became head of the sector that administered all of chemistry *and* biology. Lysenko ended up even further isolated, in a much reduced division of "general biology". Along side it was now a new, powerful, hybrid division with the awkward name "Biophysics, Biochemistry, and the Chemistry of Physiologically Active Compounds" – for which read, "molecular biology".

Meanwhile, with Nirenberg's announcement in 1961 that the genetic code had been broken – an announcement at the International Congress of Biochemistry which, not coincidentally, was being held in Moscow (again, international networks at work!) – the campaign against Lysenko gained momentum. Now new rhetoric, articulated by leading Soviet physicists, came to the fore: that the imminent "control of life" would be even more important than the atomic bomb! As a counterstroke, Lysenko – now isolated in a smaller division where he was even more firmly in control – sought to reinforce it, and gained Khrushchev's leave for four new Academy slots in his division for his own handpicked nominees. In 1964, when the Academy's General Assembly refused to elect two of them – a move unprecedented, to the best of my knowledge, since the advent of Stalinism – Khrushchev was heard to be considering the abolition of the Academy and its replacement by a state committee.[27] That was just one of the indictments against him during his ouster a few months later, in October 1964.

Thanks to their steadfast efforts in the Khrushchev era, by the mid-1960s the Soviet Academy's new leaders (and the networks they represented) had achieved their basic social agenda for Soviet science. So long as they stayed clear of politics, science was theirs: they could manage their ever-growing scientific empire, exercise freedom of scientific thought, increasingly travel abroad, and generally enjoy the secure perks of a highly privileged Soviet caste (complete with dachas, special stores and clubs, chauffeured limousines, luxury apartments, and salaries higher than most members of the Central Committee enjoyed). In an era of Big Science, theirs was the biggest, and the continuing competition with the West in military technology and the "space race" kept their prestige and priority high, and their coffers full.

Throughout the Brezhnev era, under the conditions of the waxing and waning Cold War, Soviet science continued to grow, cultivating itself as an increasingly chubby symbiont on the body-politic. The major upsets in that symbiosis occurred when scientists, more often than politicians, breeched the unwritten contract – as with the occasional forays to "liberate" the social sciences, or when certain scientists used their freedom and privileged status to contest state policy, and eventually the legitimacy of the entire system itself. We should not judge the scientific leadership too harshly for being unsympathetic to these efforts. After all, having spent their careers shutting the door between science and politics, it is hardly surprising that such men as Engelhardt and Semenov did not welcome the efforts of the Sakharovs to open that door all over again.

In short, the Khrushchev era did indeed see the remaking of Stalinist science into its Brezhnev variant. This happened because of a new generational scientific elite, representing the views and interests of a diverse set of highly resilient and well developed interdisciplinary and international networks. Empowered by the importance of science for the Cold War, these leaders became proactive and strategic in renegotiating the symbiotic relationship between science and the State. What they gained was substantial control over the conduct of science. Thanks to their efforts during the Khrushchev years, Soviet science under Brezhnev was bigger, fatter, freer, less ideologically vexed, and much more secure than it had been under Stalin. (That does not mean it was *better*: Soviet physicists won a hefty slew of Nobel prizes under Brezhnev, to be sure – but, almost without exception, the prizes were for work they had done under Stalin.)

Dare we admit what, in retrospect, has become all too obvious? – In the Soviet Union as in the West, the Cold War was probably the best thing that ever happened to science and technology. Without it, the Soviet scientific elite could hardly have carried such weight with the political leadership, nor negotiated for themselves so much autonomy. For a half-century thereafter, each community, the American and the Soviet, could use the threat of the other to keep the immense and growing resources of their own countries flowing to science and technology. With so many worthy and unworthy claimants to scarce resources in both countries, there is no chance that such vast sums could have found their way to science and technology in *either* country in the absence of the Cold War. Nothing has ever before given science such a lien on the wealth of nations, or made scientific and technical development such a high state priority, in so much of the world, for so long.

Final Thoughts

In closing, let me draw a few conclusions from this, and put them as provocatively as I can. I have four: one about Cold War history, a second about Soviet science, a third about its relation to the broader history of science, and the fourth about networks.

First, it seems to me that Cold War science, technology, and medicine are ripe for international and comparative study. Ironically, the very conditions that seem to have revitalized the study of the history of American science and led to such an efflorescence of excellent work in that field have also left it strangely incomplete. In many ways, the social, ideological, financial, and even substantive development of American science and technology during this period has been shaped by its place in the great bi-polar dialectic of the Cold War. Yet, to read some of the Americanist literature, one might think all that happened was entirely the product of local home-grown politics, with the "Soviet Union" being some sort of convenient Platonic construct. Even in studying things American, there is much to gained by being less provincial.

We can learn much by comparative study, working internationally to unearth the relevant archives and put them into proper perspective. Studies are well underway, as one can see from the remarkable CNN "Cold War" television series recently aired. Even so, there would seem to be several vital pieces of this history that are still missing. For example, if my analysis is right, the rebirth of Soviet genetics and the rise of molecular biology may well have received support from the Soviet military-industrial-research-security complex, not only indirectly (via their physicist supporters), but more directly, because of its implications for biological warfare. I am not aware of any substantial documentation, publications, or archives chronicling Soviet biological warfare research during any period, let alone the 1950s; yet, from what I know, I feel certain that something important must be there. Until we get a better handle on Soviet secret and sharashka research of all sorts, and its counterparts in the United States, our understanding will be fundamentally incomplete. It would also help if we could know what each superpower was learning about the other's research through their intelligence "networks".

Second, in analyzing Soviet, Eastern, and totalitarian science generally, it is time for us to stop imagining that everything was visited upon science from above. In fact, successful national science has always been symbiotic with the state, and has always found ways to serve its interests. It *must*, in order to acquire the resources it needs for its sustenance and growth – and with the postwar emergence of "Big Science" these needs have become vast. It is true that the Soviet Party-state developed techniques for controlling every aspect of scientific life more complete than even the most severe critics of totalitarianism might have feared; but it is also true that the scientific community proved more resourceful at manipulating that system to serve its own agendas than even the most optimistic advocate of academic freedom might have hoped.[28] As the higher government archives become increasingly available, it will become more and more difficult to balance the view from above with the view from below. Yet we must, for however much every boss would like to hope that "What you SAY is what you get", we all know it just isn't so.

Third, I would make a plea for a reexamination of Soviet science generally. For too long, Western studies have emphasized the unique and exotic features of Soviet science – Lysenkoism, Marxist ideology, the Terror, the Gulag, the Soviet health care system – the things that have made Soviet science appear unique, strange, alien, pathological, or particular. Boggled by lists of Russian names few American historians have ever heard of (and fewer still can pronounce), and bewildered by the shifting meanings and arcane lexicons of Soviet "Newspeak", Western historians of science have tended to feel that the history of Russian science is somehow marginal to anything they need to know about or be interested in, a "specialists' specialty" with little to tell us about the history of science writ large, in America or elsewhere. Few would have seen the world this way before the First or Second World War: to a certain extent, I would suggest, this attitude is itself an artifact of the Cold War. The explosive rise of our own discipline in the West is a largely Cold War phenomenon, primed by the post-Sputnik funding for science; although the Cold War is over, we may still be suffering from the effects it has had on the way we have seen and interpreted things. Yes, Russian science is different from ours in many particulars – what else could one expect after almost half a century of relative isolation? – but, as I hope to have shown, the Russian story is both comprehensible and intriguing in the same terms we use to understand our own science, and for the same reasons. Furthermore, understood in this way, Russian science can provide a helpful comparative perspective on American and European developments. It is time to go beyond our Cold War, Anglo-American, Western European biases and bring the history of Russian science again into the mainstream of our analysis.[29]

Finally, perhaps it is time to begin a systematic study of the structure, nature, and roles of the networks that have helped shape the history of science generally. We all know bits and pieces, relating to the people and materials we study. Much of this information is available, masked as biography, reminiscence, or anecdote. Yet I suspect its systematic importance will only become clear when we put the information together, and change our analytic focus from the individual to the network itself. Networks do not leave many public traces, so we must use the opportunities provided by living informants – genealogies, family archives, reminiscences, interviews, photographs, and private letters – before the information disappears. Yes, networks no doubt gained a special character and importance in Russia, given its political system and historical experience. But, upon reflection, can anyone active in science or the academy *anywhere* doubt that networks of the sort I have described also play pervasive and analogous roles in the West?

Curiously, it was my reconsideration of the "anomalous" case of Soviet genetics during the Khrushchev era that forced me to see what now seems obvious: when all the relevant disciplines and institutions had been seized, what other factor could account for the overthrow of Lysenko's "hegemony"? The anomaly *compelled* bringing networks to the fore. In retrospect, this is not surprising or unusual: pathology often provides insight into "normal" functioning by forcing us to pay attention to what, in the normal course of things, is hidden. Since coming to this realization of the centrality of networks, however, I have been increasingly impressed by its power as a tool in intellectual discourse and analysis more generally: I find myself seeing and understanding in new ways subjects I had already thoroughly studied, even those on which I had published. As I read articles in *Isis* and other journals, I am constantly surprised at the omnipresence of networks: sometimes they are actually talked about casually, elsewhere they are an essential part of the story – yet almost never are they explicit or analytically central.

What I propose is not especially new or radical: it is, as it were, a slight refocus, a five-degree shift, that lets us see in a new way – that can give new meaning and significance to things we all study, but which we have traditionally treated as "marginal" or "particularistic". Details that I now understand as being about networks are everywhere, hiding in a wide range of works, but dropping from view when the analysis and explanation begin. It often seems that, like Molière's character from "Le Médecin Malgré Lui", we have all been speaking prose without knowing it! That slight refocus can be remarkably illuminating, I have found, suddenly bringing into sharp relief patterns that had previously been only cryptic or implicit. What I propose, then, is that we liberate, reconfigure, systematize, and "center" the idea of networks in order to be able to see familiar things in a new way.

When I first entered the history of science under Everett Mendelsohn's tutelage, the "internalist-externalist" debate (between history of ideas and social history) was still raging and, in a curious sort of way, I think it still remains with us. The basic concern and indeed excitement of our field still centers, at least for me, on understanding the intricate interrelationships between the impressive and beautiful intellectual achievements of science on the one hand, and, on the other, its complex social, political, and cultural contexts – the influences of time and place read broadly. Over the intervening decades, the "social history" of science has made much progress, moving from polemical pronunciamentos to the sophisticated analyses of many case studies. From an almost Platonic focus on the "heroic scientist" or "unit ideas", and little attention to their social contexts, we have developed considerable sophistication in understanding how *disciplines* and *institutions* have mediated between science and its broader social milieu, one that, in the process, has greatly enriched our understanding of both lives and ideas in science.

It seems to me time to make it a troika. If we are to see more clearly both the unity and diversity of world science, we should consider bringing our analysis of *networks* to the fore – not as marginal, coincidental, personal, or particularistic details, but as a central focus of analysis, reflecting a real phenomenon, an influential (if often hidden) substructure in science that has helped to fashion, establish, protect, shape, and remake both *institutions* and *disciplines*. Doing so, I believe, may also enrich our understanding of those other subjects commanding such attention in our field: the moral economy of groups; the formation of schools; the varying place of women and ethnicities; the development of forms of laboratory and field practice; the origin of new specialties; the interdisciplinary and international spread of ideas, techniques, models, and metaphors; the appropriation of science to serve sociopolitical purposes; the conflicting agendas of public policy; and the ideologies, conventions, and popular images of science. By focusing on networks, we may find new commonalities in a wide range of work in our field, as well as new and better ways to bring "internal" and "external" together in our evolving understanding of the evolution of science itself.

Acknowledgements

Candor requires that I conclude this essay on networks with acknowledgment of the debt I owe to my own. I wish to thank my mentors, Everett Mendelsohn and Ernst Mayr, whose Soviet contacts greatly aided me during my early visits to Russia; A.E. Gaisinovich, Nikolai Vorontsov, Elena Liapunova, Daniel Aleksandrov, and the many others who welcomed me to Moscow and Petersburg, for graciously sharing their homes and friends with me; Manfred Heinemann, whose Halle conference provided the occasion for sharpening my thoughts on the subject; John Krige, for reading and commenting on a draft of the essay; and most of all, my colleagues Nikolai Krementsov and Dan Todes, for their candor, encouragement, and wisdom.

University of Pennsylvania

Notes

1 See my articles: "Biology in the Soviet Academy of Sciences, 1953–1965: A Case Study in Soviet Science Policy", in *Soviet Science and Technology: Domestic and Foreign Perspectives*, J.R. Thomas and U.M. Kruse-Vaucienne, eds. (Washington: NSF/George Washington University, 1977), pp. 161–188; and "Biology After Stalin: A Case Study", *Survey: A Journal of East and West Studies*, 102 (winter 197778), pp. 53–80.

2 See especially the monthly issues of *Priroda* [Nature], 1989–1998, which include heroic biographical treatments of more than half of the scientists in the listing above, each detailing, among their other accomplishments, their struggles to protect and cultivate genetics.

3 For a fine survey of actor-network theory, and criticisms of it, see Jan Golinski, *Making Natural Knowledge: Constructivism and the History of Science* (Cambridge: Cambridge University Press, 1999), pp. 37–43. Those familiar with the theory will be aware that the epistemic concerns of these approaches distinguish them from my own use of network, which refers rather to an informal,

voluntaristic, private, and fluid set of interlinking personal relations and associations based on ties of trust, family, friendship, "old school ties", shared concerns, common fascinations, and so forth. Here, I am consciously using "networks" and "networking" in the evocative way the word has come to be used in everyday speech, which I find simpler, looser, richer, and more metaphorical, useful, and suggestive than the technical ways various theoretical specialties – including our own – sometimes use the term. The same, of course, is also true of my use of the terms "discipline" and "institution", which have also been given technical meanings by sociologists and others. (Just to be clear, my use of "institution" here does not include either "marriage" or "the family", but refers to organizations that have names, administrators, and various ranks or positions, are usually housed in buildings, employ people, organize work or practice of some kind, and spend money that has to be gotten from somewhere.) If we are to rescue our own field from the fragmentation of subspecialization, and keep our own "theory and practice" together, there is much to be gained by using terms in ways that all historians can understand and relate to.

[4] See, for example, the memoirs of Richard Goldschmidt, *Portraits from Memory: Recollections of a Zoologist* (Seattle: University of Washington Press, 1956) for his vivid descriptions of the ways long-lasting associations were formed among students in Europe at the turn of the century, ranging from student drinking and dueling groups, to the "groupies" collecting around certain prominent professors, to the bondings that occurred at various marine zoological stations, such as those at Naples and Villefranche.

[5] In a pathbreaking article ("The Politics of Technology: Stalin and Technocratic Thinking among Soviet Engineers", *American Historical Review,* 79 (1974), pp. 445–469), for example, Kendall Bailes has demonstrated that the Shakhty Trial and the so-called "Industrial Party Affair" (which marked the "Great Break" and the advent of Stalinism) were not simply arbitrary in their victimization; however unfair the charges of "sabotage", the trials targeted a network of professionals who had been active in Russia's technocracy movement.

[6] This notion, proposed to me by Nikolai Krementsov, is quite suggestive and well worth exploring through further research.

[7] For example, see *Shkoly v nauke*, S.R. Mikulinskii *et al.*, ed. (Moscow: Nauka, 1977); and *Fiziologicheskie nauchnye shkoly v SSSR*, N.P. Bekhtereva, ed. (Leningrad: Nauka, 1988). The use of the term "school" in the Soviet context has sometimes been complicated by professional agendas: under a system in which various scientific "stars" were accorded almost saintly status, much reflected glory and unimpeachable legitimacy could come to those who could be seen as members of such a "school". This explains, in part, both the debates about which "pretender" was a great man's true heir (such as those which followed the deaths of Pavlov, Michurin, and Severtsov in the mid-1930s), and also the vague catch-all quality of "school" as employed in Soviet (and subsequent Russian) lexicon.

[8] For example, Daniel Alexandrov has been studying a number of intriguing circles that formed in Russia, including the so-called "Biological Circle" in pre-revolutionary St. Petersburg, which included not only biologists, but also humanists, including the butterfly taxonomist (and novelist!) Vladimir Nabokov.

[9] "*DrozSoor,*" short for "Sovmestnoe oranie drozofilshchikov"; many years ago, Theodosoius Dobzhansky and I.M. Lerner first translated it into English for me, using the phrase I have given. For details of the "society" and the arrest, see Mark B. Adams, "Sergei Chetverikov", *Dictionary of Scientific Biography*, 17, Supplement II (New York: Charles Scribner's Sons, 1990), pp. 155–165.

[10] For example (probably) spurious information about Nikolai Vavilov's "arrest" in 1936 was actually published in the *New York Times,* to be hotly contested in a subsequent interview given by Vavilov himself. One can also see from the obituaries of his former colleagues written by Theodosius Dobzhansky for various Western scientific journals in the 1940s and early 1950s that he was receiving information from a variety of networks; during this period, correspondence with émigrés and "unpersons" (as Dobzhansky was regarded at that time) was strictly forbidden. Given the isolation of Russia, what is remarkable is not how much of the information was faulty, but rather how much turned out to be true.

[11] For example, see my articles: "Science, Ideology, and Structure: The Kol'tsov Institute 1900–1970", in *The Social Context of Soviet Science*, Linda Lubrano and Susan Gross Solomon, eds. (Boulder: Westview Press, 1980), pp. 173–204; and "Sergei Chetverikov, the Kol'tsov Institute, and the Evolutionary Synthesis", in *The Evolutionary Synthesis: Perspectives on the Unification of Biology*, Ernst Mayr and William Provine, eds. (Cambridge: Harvard University Press, 1980), pp. 242–278.

[12] See "Iurii Aleksandrovich Filipchenko", *Dictionary of Scientific Biography*, 17, Supplement II (New York: Charles Scribner's Sons, 1990), pp. 297–303.

[13] See my articles: "Eugenics as Social Medicine: Prophets, Patrons, and the Dialectics of Discipline-Building in Revolutionary Russia", in *Health and Society in Revolutionary Russia*, Susan Gross Solomon and John E. Hutchinson, eds. (Bloomington: Indiana University Press, 1990); "Eugenics in Russia, 1900–1940", in *The Wellborn Science*, Mark B. Adams, ed. (Oxford University Press, 1990), pp. 153–216; and "The Politics of Human Heredity in the USSR, 1920–1940", *Genome*, 31, no. 2 (1989), pp. 879–884.
[14] Mark B. Adams, "The Soviet NatureNurture Debate", in *Science and the Soviet Social Order*, Loren Graham, ed. (Cambridge, Mass.: Harvard University Press, 1990), pp. 94–138.
[15] See my Ph.D. dissertation, Mark B. Adams, *Genetics and the Soviet Scientific Community, 1948–1965* (Ann Arbor: University Microfilms, 1972).
[16] My detailed knowledge of this I owe to Nikolai Vorontsov, his wife Elena Liapunova, and her sister, Natalya, all of whom attended the seminar. (As a student in Leningrad in 1956, Vorontsov was the person who brought the "Letter of 300", then making the rounds there, to the attention of his father-in-law and his Moscow colleagues.) Needless-to-say, I am immensely grateful not only for their friendship and help, but also for their willingness to "certify" my legitimacy to sometimes doubting members of their extensive, multiple networks. An overview of the circle and its impact has just been published by Natalya Liapunova and Nikolai Vorontsov: "Delo sester Liapunovykh", *Znanie–Sila*, 1998, pp. 34–47.
[17] For a masterful survey, see Alexander Vucinich, *Empire of Knowledge: The Academy of Sciences of the USSR (1917–1970)* (Berkeley: University of California Press, 1984).
[18] In the final months of his life, Stalin launched a campaign against a group of elite physicians (almost all of them Jewish) who, in their capacity as Kremlin doctors, were accused of being part of a conspiracy to murder the aging Soviet leadership. See Yakov Rapoport, *The Doctors' Plot of 1953: A Survivor's Memoir of Stalin's Last Act of Terror, Against Jews and Science* (Cambridge, Mass.: Harvard University Press, 1991).
[19] A synthetic organic chemist by profession, Academy President Alexander Nesmeianov seems to have been especially won over by the implications for new kinds of organic syntheses that DNA work would make possible; he later became famous in the Soviet Union for creating a commercial version of synthetic caviar from petroleum by-products.
[20] See my article, "Nikolai Ivanovich Vavilov", *Dictionary of Scientific Biography*, XV (Supplement I) (New York: Charles Scribner's Sons, 1978), pp. 505–513.
[21] See George Gamow, *My World Line: An Informal Autobiography* (New York: Viking, 1970).
[22] B.M. Zavadovsky had been an active participant in the genetics debates since the 1930s; he was the brother of the renowned experimental embryologist M.M. Zavadovsky, who was fired from his Moscow position in 1948 as a result of Lysenko's takeover. For much detailed information about Tamm and his role, as recalled by many friends and colleagues in his network (not all of them physicists!), see *Reminiscences About I. E. Tamm*, E.L. Feinberg, ed. (Moscow: Nauka, 1987).
[23] In the institutional edicts following the Lysenko meeting in 1948, that mainstay of genetics research – drosophila flies – had to be set free or, in the case of KGB and military research institutions, drowned in boiling oil. The physicist L.A. Artsimovich was bringing special strains of the flies back for Dubinin's new laboratory in the Institute of Biophysics.
[24] For the most authoritative treatment in English, see David Holloway, *Stalin and the Bomb: The Soviet Union and Atomic Energy 1939–1956* (New Haven: Yale University Press, 1994).
[25] See A.S. Sonin, *"Fizicheskii Idealizm": istoriia odnoi ideologicheskoi kampanii* (Moscow: "Fiziko-Matematicheskaia Literatura", 1994), pp. 114–160.
[26] Indeed, by contrast to what we hear about the comparable communities in the United States, Britain, or Germany, Russian accounts and reminiscences seem remarkably untroubled by the moral or political implications of their work. This is even glaringly apparent in Andrei Sakharov's remarkable *Memoirs* (New York: Knopf, 1990).
[27] Rumors rippled through the networks of Khrushchev's threat, and they were reported, for instance, by both Zhores Medvedev (*The Rise and Fall of T.D. Lysenko*, New York: Columbia University Press, 1969) and David Joravsky (*The Lysenko Affair*, Cambridge, Mass.: Harvard University Press, 1970). Judging by the recent publication of a previously unavailable transcript of Khrushchev's actual remarks (V. Iu. Afiani and S. S. Ilizarov, "... My razgonim k chertovoi materi akademiiu nauk", *Voprosy Istorii Estestvoznaniia i Tekhniki*, 1999, 1, pp. 167–173) the network information was generally accurate, but not entirely so – assuming, of course, that one chooses to credit the accuracy of the version that was transcribed, edited, and just recently published over the contemporary oral accounts provided by the networks. In any case, one suspects that it was the rumors of what he said, rather than the actual text, that contributed to his undoing.

[28] This view has been forcefully advanced, in much the same language, and with considerable documentation, in Nikolai Krementsov, *Stalinist Science* (Princeton: Princeton University Academic Publishers, 1997), pp. 777–794.
[29] Loren Graham's new book, *What Have We Learned About Science and Technology from the Russian Experience?* (Stanford: Stanford University Press, 1998), is a useful first step. See also his article, "Russian Science in the Twentieth Century", in *Science in the Twentieth Century* John Krige and Dominique Pestre, eds. (Amsterdam: Harwood , 1997).

GILBERT F. WHITTEMORE

THE MULTIDIMENSIONAL CHESS OF SCIENCE AND SOCIETY: A POSTWAR DEBATE OVER PLUTONIUM EXPOSURE

Abstract

This paper will examine the establishment by the AEC of limits for plutonium exposure shortly after World War II. Experimental data was only one of several factors at work. The AEC rejected as too low the limit established by an international conference and established its own higher level. An examination of this process reveals not only disagreements over the analysis of scientific data and the influence of Cold War priorities, but also the persistence of prewar attitudes towards worker safety and control of data gathered from the workplace.

Introduction

Every student of a renowned teacher gathers up slightly different melodies on which to play variations as the years pass. As one who now straddles the worlds of academia and law, I am struck by the utility in both worlds of two lessons learned from Everett Mendelsohn. First, when analyzing an apparently "purely scientific" dispute, look beneath the technical aspects to the underlying interests, but without losing sight of the science itself. Secondly, whatever insights one gleans from historical research should eventually be presented to a broad audience in plain language, avoiding terms that exclude the uninitiated, while still doing justice to the complexity.

This paper presents an example of these lessons by examining the debate, just after World War II, over what level of exposure to plutonium would be allowed in American weapons production plants. To apply the first lesson, the underlying organizing principle is the simple metaphor of multi-dimensional chess as a means of visualizing the variety of interests at work. The reader is left to judge whether I have been successful in applying Everett's second lesson, presenting a complex story in plain language.

G.E. Allen and R.M. MacLeod (eds.), Science, History and Social Activism: A Tribute to Everett Mendelsohn, 277–289.

The story begins with the puzzle posed by a single document, which serves as a window onto Cold War America at the end of the 1940s.[1] The document is a 1948 memo from the Chief of the Insurance Branch of the Atomic Energy Commission (AEC) to a Declassification Officer, also in the AEC. At issue was whether a report on blood changes due to chronic exposure to low level radiation should be declassified. All documents produced within the AEC's labs were "born classified" and therefore had to be officially declassified before publication. Here we have an instance of a scientific debate – the effect of low level radiation on blood – whose very existence some sought to keep hidden. Several aspects of the document intrigued me, by posing the following questions:

First, the author and addressee struck me as odd. The memo appears to be part of a routine consultation between the Insurance Division and the Declassification Office, not a unique memo resulting from an unusual situation (such as a large insurance claim, litigation, or, the nightmare of all bureaucrats, a public controversy). Apparently, the Insurance Division was formally and routinely involved in reviewing declassification of research reports. Why?

Secondly, judging from its title ("The Changes in the Blood of Humans Chronically Exposed to Low Level Gamma Radiation"), the research report under discussion contained no information related to weapons design or production. What were the arguments that a report such as this, unrelated to weapons, should remain classified? The Chief of the Insurance Branch describes the research report as indicating that "the tolerance levels for chronic exposure to gamma radiation which have been accepted both within the AEC and elsewhere may be too high". Later, he notes that "The results of the present study tend to cast greater doubt upon the validity of tolerance levels since the persons being studied had average exposure of only .2r/week, about 2/5 the tolerance levels". Had it been written in a university, such a report would have joined a growing flood of articles on the effects of radiation exposure. Why did the Insurance Division caution the Declassification Branch that "the question of making this document public should be given very careful study"?[2]

An initial answer begins to emerge as we look more closely at the content of the memo. The Insurance Branch officer lists three reasons for not declassifying the report. First, publication would weaken morale of employees at weapons plants. ("... the possibility of a shattering effect on the morale of the employees if they become aware that there was substantial reason to question the standards of safety under which they are working".) Secondly, the report could be used by labor unions as a basis for increasing wages. ("In the hands of labor unions, the results of this study would add substance to demands for extra-hazardous pay".) Third, the report could be used by plaintiffs' attorneys as a basis for workers' compensation claims. ("... the definite possibility that general knowledge of the results of this study might increase the number of claims of occupational injury due to radiation and place a powerful weapon in the hands of a plaintiff's attorney".)

Several aspects of this reasoning are worth noting. First, there is no claim that the results are invalid, either in experimental design, data collection or analysis. The argument is not that this is "junk" science, but that it is alarming science.

Second, none of the reasons for retaining classification bear any direct relation to foreign enemies obtaining knowledge about weapons. Instead, the Insurance Branch is concerned with its impact on the domestic workforce. Any link to national security can only be indirect, in the form of a fear that any workforce disruption would jeopardize weapons production.

Many of us, myself included, usually think of secrecy primarily as an aspect of the Cold War – the rippling outward of the mindset that developed within the closed, secret weapons facilities at Oak Ridge, Los Alamos, and Hanford. Classification was a means of maintaining a monopoly on atomic weapons. On viewing this document, I was tempted to jump to such an explanation. But one lesson learned from Everett was not to jump immediately to such a somewhat cynical conclusion, which usually is only a projection of one's own political inclinations. Inquiring more deeply usually reveals a web of relationships which are far more interesting than one's initial reading. As more documents emerge from this period, we are beginning to develop a more nuanced view of the uses of secrecy. Classification played a role, not only in the atomic arms race, but in much older political games.

The essence of my argument is that at times the classification of research was not always a struggle among superpowers, but at times also was a chess move in the game of domestic labor management relations. Classification, coupled with the authoritative voice of scientific expertise, helped maintain a government monopoly over atomic affairs in the domestic as well as in the international realm. This is by no means a novel argument, nor a startling one. However, historians today are challenged to penetrate beneath the glib conspiratorial view of the "X-files" to uncover the complexity of such games.

To attempt this, I would like to examine one of the industrial health standards related to this initial document: plutonium exposure as limited by the setting of a "maximum permissible body burden". A brief chronology of changes in the standard suggest interesting artifacts lie just beneath the surface of apparently objective quantifications.

World War II:	5.0 μg	(Manhattan Project)
September 1945:	1.0 μg	(Manhattan Project)
September 1949	0.1 μg	(Tripartite Conference)
January 1950	0.5 μg	(Division of Biology & Medicine, AEC)

For simplicity, visualize this as a game of three dimensional chess.

The First Level: The Complexity of the Science

First, although an industrial health standard – such as the maximum permissible body burden of plutonium – is simple in concept (how much plutonium should you carry about inside you?), its determination is a difficult scientific problem.

The essential scientific problem was, and still is, to determine the effects of radiation on humans. During World War II, the United States government sponsored both animal and human radiation experiments to provide data as a basis for industrial health measures. The subject of much controversy half a century later, the human experiments were designed to trace the metabolic pathway of plutonium. In theory, such data would allow those founding the new field of health physics/radiation biology to determine the rate at which the body naturally excretes plutonium. This in turn would allow for calculation of:

(a) the "body burden" – the total amount of plutonium in the body based upon excretion rates, and

(b) the "biological half-life" of plutonium as it decayed radioactively and the body gradually excreted some of it.

The scientific reasoning was elegant, but the human experiments fell far short of this goal. Aside from ethical concerns, which have been extensively discussed,[3] these experiments had serious shortcomings. Most of the subjects had serious illnesses which may have affected how they metabolized plutonium; doses were not carefully measured; the number of patients was very small. The scientific questions were complex, the human data was sparse, but the need for guidance was great. The atomic bomb was to be the wonder weapon to end the war; pressure was enormous to increase production of all its components, including plutonium.[4]

The response was to provide an authoritative voice by calibrating plutonium to the existing standard for radium. The radium standard itself is an example of a scientific "law" arising from a combination of scientific inquiry and institutional interests. Just prior to the entry of the United States into World War II, a radium standard had been established by an ad hoc committee at the request of the U.S. Navy. The Navy foresaw a dramatic increase in the use of radium-illuminated instrument dials in ships and planes, and sought expert guidance on industrial health standards. A small committee, headed by Robley Evans of MIT, met to set such a limit. Evans was skeptical of extrapolation from animals to humans. He had abandoned his own earlier work on rats and preferred human data, which he had begun to gather from dial painters and others who had suffered from radium poisoning. At the time the standard was set, his data were:

Radium Body Burden	Health Status
under 1.0 μg:	6 cases were symptom free after 25 years
1.0 to 2.0 μg:	2 of 5 cases had symptoms
2.0 to 9.0 μg:	2 of 5 cases had symptoms
over 10.0 μg:	2 of 5 were deceased (1 from leukemia; 1 from heart failure)

These data were sparse, but indicated that no symptoms appeared when the body burden was under 1.0 micrograms.[5] The committee eventually settled on a figure of one microgram, using as the criterion: "would you feel comfortable if your own wife or daughter were the subject?"[6] What is refreshing about such a criterion is its candor in admitting the paucity of data and relating its scientific conclusions to personal risk. But this candor was not published for over forty years. In the intervening period, the simple statement "1.0 microgram was the maximum permissible body burden for radium" was presented as a law of naure.

The "radium standard" became, and remains today, the baseline for all other standards for "internal emitters" – that is, radioactive substances which somehow find their way into the human body. The next problem was to calculate how much plutonium would produce the same biological effect as 1.0 microgram of radium.

During the war, the body burden for plutonium was initially set at 5 micrograms by simply comparing the amount of physical energy released by the decay of radium and plutonium.[7] This method did not account for any differences in how the body metabolizes the two elements. In the spring of 1945 Wright Langham advised lowering the limit by a factor of five, to one microgram. Animal data were indicating that plutonium was concentrated in the "organic matrix" of the bone, unlike radium, which appeared primarily in the mineralized bone. Concentration in or near the blood-forming portion of bone would be more hazardous than in the mineralized portion.[8]

Thus, beneath the apparent authority of precise numbers, lay a process of consensus driven by institutional needs and based upon sparse data which was analyzed using reasonable, but not unquestionable, assumptions about how the body metabolizes plutonium.

The Second Level: International Relations

Thus far in this story, the voice of scientific authority has been located within the United States government, especially that part of government concerned with weapons production – initially the Navy (concerning the radium standard) and later the Manhattan District (concerning the plutonium standard). At no time did these standards clash with the needs of the institutional authority which had sought and then applied the scientists' judgment. (Whether the lack of conflict is due to the level of the standards or to neglect in enforcing them is another story.[9])

In the post-war period, the game became more complex and illustrates how broad a view is needed to appreciate the "social context" of an apparently technical process of standard-setting. The future of atomic energy was seen as an international phenomenon requiring a worldwide vision. At one extreme were ambitious schemes for international control of atomic energy. At the other, was the vision, which eventually prevailed, of an attempt to maintain an American monopoly, a vision which later developed into thc atomic arms race.

International politics directly affected the process of setting exposure standards. Beneath the grand debates over the future world order, on a more mundane, daily level were the relationships among the wartime allies: Britain, Canada and the United States. One consequence of a desire to continue as an alliance in the development of atomic weapons was a series of meetings termed the "Tripartite Conferences". Attended by civil servants of the atomic programs in each country, these dealt with a variety of issues, among them standards for industrial health.

At the September, 1949 Tripartite Conference the participants discussed "permissible body burdens". National sovereignty clashed with international alliances and with the belief in the universality of natural laws. The meeting's agenda included a discussion of scientific issues, followed by an "agreement" on specific numbers. The underlining of the word "agreed" throughout the minutes indicates formal endorsements, though no record was made of any votes.[10] One such agreement, following a discussion of animal data, was that the prewar radium body burden of 1.0 microgram "might reasonably be regarded as the minimum amount of radium know (sic) to produce damage".[11]

The difficult question was to determine the amount of plutonium that would produce the same degree of biological damage as 1.0 microgram of radium. One problem was the uneven distribution of plutonium within bone: how close did it reside to the blood-forming portion of bone? This was an empirical question for which there was as yet no clear answer. This led to a second, non-empirical question: how large a safety factor should be introduced in recognition of the lack of sufficient data? Wright Langham of Los Alamos cautioned that "in assessing the equivalent quantities of isotopes known to be deposited non-uniformly, care be taken to avoid over-estimating the safety factors", adding some anecdotal evidence from his survey of Los Alamos workers. Also attending on behalf of the AEC, and more senior in its hierarchy than Langham, was Shields Warren, then Director of the Division of Biology and Medicine. Warren "summed up the discussion by saying that at the present time there is insufficient direct evidence for man, and that data from animal experiments has still to be relied on". At his suggestion, the group set the minimum damaging amount of plutonium at 1.0 microgram, the same as radium.[12] A workable industrial health standard would need to be less than a harmful amount – but how much less?

The next step for the Tripartite Conference was to derive a maximum permissible body burden from this minimum damaging amount. Essentially, this involved the choice of a safety factor. Dr. Joseph Hamilton noted that the present figure of 0.1 microgram for radium "corresponds to a safety factor of 10 and it was suggested by Dr. Hamilton that by comparison with the practice for other noxious agents, such a factor was reasonable".[13] One participant argued that if the relation between dose and damage was linear then 10 "does not represent a very great margin of safety", but the group as whole accepted ten as a safety factor.[14] Comparing the retrospective summary of this result in the official minutes of the Canadian and United States delegations reveals significant differences in attitude. The Canadian minutes describe the safety factor of 10, noting that, "by comparison with the practice for other noxious agents, such a factor was reasonable, "while the United States' minutes refer to it as "arbitrarily decided". The 0.1 body burden for plutonium is described by the Canadians as "maximum permissible body burden", while the United States delegation preferred the less authoritative term "best estimate".[15]

For the AEC, the net result was a lowering of the maximum permissible body burden for plutonium by a factor of ten, from the current 1.0 microgram to 0.1 microgram. Now, an authoritative voice of science from outside the AEC – the international Tripartite Conference – clashed with national interests – the AEC's production of atomic weapons. Wright Langham was greatly disturbed by the lower standard, and complained to Warren in the car to the airport.[16] He believed the standard was one that Los Alamos simply could not meet. Langham began a letter writing campaign to persuade Warren, his superior, to ignore the Chalk River standard. What followed was a series of correspondence among a closed circle of experts within the government.

The object of Langham's attack was not basic data (of which there was little), but the assumptions underlying the application of the data by the Tripartite Conference. "Safety factors" were the obvious point of vulnerability. Langham argued that the new lower value was "extremely conservative". This conservatism was contrasted with the potentially "drastic effect on the efficiency and productivity of the Los Alamos Laboratory".[17] Langham sought support from others, including Robley Evans at MIT. Evans agreed that the Chalk River conference results were conservative.

It was not difficult to mount such an attack, given the sparse data and the innate complexity of radiation effects. Langham, like a courtroom lawyer, attacked every weak point he could find in the derivation of the new standard. This included his own reconstruction of Chalk River's logic; a lack of hard data to support an additional biological toxicity safety factor of 15; differences between plutonium and radium decay products; observations that radium was more unevenly distributed within the body than plutonium; a lack of human data on chronic exposure; the inapplicability of acute toxicity animal experiments to chronic exposure; and comparisons with natural background radiation as a measure of safety. The supposedly authoritative voice of science was readily challenged by insiders who understood the shaky foundation on which such a deceptively simple number as "maximum permissible body burden" was built.

On 24 January 1950, Shields Warren met in his Washington, D.C. office with Austin Brues from Argonne National Laboratory, Karl Morgan from Oak Ridge National Laboratory, Wright Langham from Los Alamos National Laboratory, and Robley Evans from MIT.[18] In contrast to Chalk River, which was a formal international meeting recorded in minutes, no records of this meeting have yet been found. The outcome, not publicly reported until thirteen years later, was that Warren raised the maximum permissible body burden for plutonium from the Chalk River level of 0.1 microgram to 0.5 microgram. This was halfway between Chalk River and the level that Los Alamos had been using – not a surprising decision for a skilled administrator such as Warren.[19]

This remained the level in use as late as 1995, although the rationale for its derivation has changed over the years.[20]

How far have we come in understanding the document with which I started? Thus far, we see how the solidity of a statement of "natural law", such as "the permissible body burden for plutonium is 0.1 micrograms", could be undermined by insiders drawing attention to conflicting data or unproven assumptions. This was precisely what Langham had done to the Chalk River standard. Obviously, the power of insiders to mount such a challenge was enhanced by stamping "SECRET" on reports which questioned the AEC's own standards. But it is not enough to observe such tactics and explain them by simply saying "institutional self-interest". The challenge is to uncover how this "self-interest" worked, especially given the multiple "selves" involved within the AEC's workings. Why were such challenges seen as so threatening? To understand this, we must turn to the third dimension of the chess game.

The Third Level: The Politics of the Workplace

On this third level we have three major players: the government, the labor unions, and the private contractors. Each, in turn, had its own internal tensions.

"The" Atomic Energy Commission can be understood literally as the Commissioners themselves. But the AEC was also a large, complex system for research, development and production, with many components, some of which were in conflict with each other. One such conflict developed between the interests of the new profession of health physicists and the military desire for increased production of weapons-grade material. Health physicists waged a continuing campaign to develop their credibility and authority in the face of pressure to increase output.[21] Issuing a plutonium standard that would halt work at Los Alamos would undermine the credibility of health physics, especially given the uncertainty of the science and the growing strength of nationalism over internationalism as the driving force in America's atomic policy.

The second major player was the labor movement. Labor, too, had multiple parts: unions competing for new members by fighting to provide better benefits for their current members. At Oak Ridge, workers had been allowed to unionize towards the end of the war. This introduced union leaders as new players in the game. At the local level, their concern was with working conditions (which included benefits such as vacation and sick time as well as conditions themselves), job security and promotion, and wages. On the national level, any success at Oak Ridge could be used to promote organization at other locations throughout the country.

Management, in the form of the private corporations that managed AEC facilities, was the third player. But, here too, there were multiple parts in competition. A corporation would not lock out its workers in AEC facilities, but it could refuse to extend its contract and walk away from AEC work. DuPont did this at the end of the war, because of "the mistreatment to which they had been subjected".[22] In 1947 and 1948, labor disputes at AEC facilities were the direct result of such changes as a new contractor arrived and tried to impose its own policies for pay scales and benefits. Also, like the unions, the corporations played on a national level. At Oak Ridge, increased labor costs could be passed on to the customer, the AEC. But this was not true elsewhere. Management feared that concessions at Oak Ridge would then be demanded by the unions at its non-AEC plants elsewhere in the country. In the words of one witness before the Joint Committee on Atomic Energy (JCAE), there would be a "chain reaction", as pay scales and benefits were increased from plant to plant across the country.[23]

The politics of the workplace thus saw a confluence of three players – the government, the unions and management – each with its own internal tensions. In the late 1940s, the situation was complicated further by the fluid state of federal law governing labor-management relations generally. The Taft-Hartley Act governing labor-management relations was new and untested. Even within the JCAE hearings, Senators debated the role of the federal government. Was it fair to forbid unions to strike, without providing some restraint on management, such as compulsory arbitration? How active should the federal government be in negotiating solutions? Should these be imposed by the AEC? In December 1947 both labor and management at Oak Ridge wanted the AEC to step in and dictate a solution. The AEC, however, did not want to become drawn into direct management, and only at the last minute did the parties sit down to serious negotiation. By March 1948, the AEC staff was proposing a stronger role for the AEC, drawing up a proposal for basic elements of a contract, but avoiding details of wage rates and benefits. The AEC model was welcomed by labor, but rejected by management.[24]

A deeper current, prewar attitudes towards the role of government, management and unions, was also revealed. The AEC did not want to undertake direct management.[25] Already, some criticized the AEC as being an island of socialism in capitalist America. Unwilling and unable to be a government owned monopoly, and wishing to remain "only" a government procurement agency, the AEC appears to have been more concerned with retaining the loyalty of contractors than the good will of the unions. Thus, attempts to reduce labor tensions by developing a "common law" of employment contracts were not forced due to a fear of losing "our good contractors".[26]

The view which emerges steps from social context to cultural context – always less easily documented, but perhaps ultimately more important. In this instance, there was a half-hidden belief that decisions about industrial safety belonged exclusively to management. This attitude clearly preceded the war. The prewar history of the radium dial painters indicates how this belief applied to practice. Drinker's study was done in confidence for a private company. It was released only after the company had, in Drinker's eyes, released an edited version intended to deceive. The crusade for compensation and protection for the radium dial painters arose outside the management of the companies involved.[27]

Now that the war was over, the growing power of unions was another factor. Strikes were threatened at Oak Ridge in December 1947 and March 1948, the last avoided only when President Truman invoked the Taft-Hartley Act. In closed hearings before the JCAE, officials in December 1947 stated that if the diffusion plant was shut down even for only one day, "there was a possibility we might not get that plant started back into operation".[28] Halting the production of weapons grade uranium would be a "national emergency". Soon, the term "national emergency" became a flexible concept to broaden the legitimate scope of government intervention against any strike, not just those at the diffusion plant. Four months later, for example, the AEC's definition of national emergency had extended to *any* work stoppage at Oak Ridge. Two arguments were advanced. First, any halt, even in the production of radioisotopes for medical research, would eventually curtail research and hurt the national interest.[29] Second, labor unrest would spread. Managers and politicians spoke of labor unrest as a "chain reaction", a "cancer" that would spread beyond AEC facilities to other plants.[30] Only now do we find direct references to a foreign threat, in the person of "Uncle Joe".[31]

Conclusion

Can all these levels of play be assembled into an understanding of the declassification memo? Perhaps, but more usefully it demonstrates the questions to ask when examining the interaction of science, politics and law, presented as the result of scientific deliberations.

For the scientists, what were the consequences of admitting such questions? The Insurance Branch of the AEC believed that admitting the uncertainty of the science increased the political impact of a research report. Such a report might challenge pronouncements which suggested industrial health standards were a straightforward measurement of natural constants.

Moreover, the international importance of atomic weapons enhanced the political importance of industrial health standards as applied to weapons labs. Demand for increased weapons production, coupled with the desire of health physics to increase its status within the AEC, put the burden of proof upon those seeking safer standards.

Who has the tools to participate in such a debate? The story indicates how politically motivated challenges to scientific pronouncements can be mounted by those familiar with their actual basis. The complexity of nature almost always leaves room for criticism of any scientific analysis that has practical consequences. But effectively launching such criticism requires both the intellectual tools (first hand experience with how such standards were developed) and institutional access (entry to the office of Shields Warren). Restricting the availability of the tools, by retaining the secret status of a report questioning the workplace standards, effectively limited the participants in the debate. Historians can demonstrate how, in the past, such methods disenfranchised those effected by closed deliberations, with an eye toward alerting us to possible abuses today.

Finally, how did deeply rooted attitudes towards governance influence the use of scientists? Here, workplace politics added two further challenges to scientific authority. One was the prewar belief that data on workplace safety was owned by the employer – in this case by the AEC – which could restrict dissemination for reasons having nothing to do with the secret of atomic weapons. Moreover, the AEC struggled to find order within the complicated world of government, private contractors, and labor unions. Presented as a "scientific law", a "maximum permissible body burden" could endorse a common law of procurement contracts, at least in so far as industrial health was concerned, something which the AEC had been unable to achieve in the realm of wages and benefits. To weaken this law by declassifying AEC research that questioned the current standard would allow unions to do publicly what Langham had done privately: to politicize the process of establishing industrial health standards.

It is this last role of "science" as "law" which most intrigues me as an attorney. Having seen how uncertain and difficult it is to reach consensus on comparatively simple matters – such as a contract to purchase real estate – I can appreciate how tempting it is for administrators at all levels to treat science as a black box. You toss a rag bundle of difficult issues in at one end and receive a neatly woven solution at the other end. The last thing you want to do is open the box up and let the entire mess spill into the open. As someone once noted, two products one should never watch being made are sausages and laws. When used to establish standards for behavior ("law" in the attorney's sense), scientific rules may be produced by means which are equally unsightly. Using a single memo, we see specific lessons learned from Everett's influence. For me, however, the most valued aspect of his teaching is that he has not advocated a single abstract model or rigid methodology, but has instead nourished a spirit of inquiry through the thoughtful wording of vital questions. That spirit of inquiry takes us well beyond the analytical techniques of "science policy", which too often takes "the data" as a starting point, to the "polity of science", which sees "the data" as an end result of a process to be explored. As historians, we should seek not only to open the black box, but to give others the ability to do so as well.

Of Counsel, Stalter & Kennedy, LLP, Boston

Notes

"ACHRE No" refers to archival numbers assigned to documents by the Advisory Committee on Human Radiation Experiments. The documents now are part of the ACHRE collection in the National Archives.

[1] Memorandum, Clyde E. Wilson, Chief, Insurance Branch (AEC), to Anthony C. Vallado, Deputy Declassification Officer, Declassification Branch (AEC), "Review of Document by Knowlton", December 20, 1948, Oak Ridge Operations Record Holding Area, H-240, Building 2714-H, "Documents 1944–1994", File #969: "Communications and Records Section"; ACHRE No. DOE 120394-E-32.

[2] Memorandum, Clyde E. Wilson, Chief, Insurance Branch (AEC), to Anthony C. Vallado, Deputy Declassification Officer, Declassification Branch (AEC), "Review of Document by Knowlton", December 20, 1948, Oak Ridge Operations Record Holding Area, H-240, Building 2714-H, "Documents 1944–1994", File #969: "Communications and Records Section"; ACHRE No. DOE 120394-E-32.
[3] Advisory Committee on Human Radiation Experiments (ACHRE), *Final Report* (New York: Oxford University Press, 1996), Chapter 5: "Experiments with Plutonium, Uranium and Polonium", pp. 139–171.
[4] For a detailed study of the tension between productivity and safety, see Russell B. Olwell, "Atomic Workers, Atomic City: Labor and Community in Oak Ridge, Tennessee, 1942–1950" (Unpublished PhD dissertation, Program in Science, Technology and Society, Massachusetts Institute of Technology, 1997).
[5] In 1950, Evans summarized his data as follows: six cases under 1.0 microgram were symptom free after 25 years; two of five cases with body burdens of 1.0 to 2.0 micrograms had symptoms; two of five cases in the 2.0 to 9.0 microgram range had symptoms; two of five with body burdens over 10.0 micrograms were deceased, one from leukemia and one from heart failure. Letter, Robley Evans, MIT, to Wright Langham, Los Alamos Scientific Laboratory, January 13, 1950, DOE Human Studies Project # D-00628.
[6] Robley Evans, "Inception of Standards for Internal Emitters, Radon and Radium", *Health Physics*, 41 (1981), pp. 437–448, quoted in Advisory Committee on Human Radiation Experiments, *Final Report* (New York: Oxford University Press, 1996), p. 141.
[7] W.H. Langham, *et al.*, "The Los Alamos Scientific Laboratory's Experience with Plutonium in Man", *Health Physics*, 8 (1962), p. 753, quoted in ACHRE, *Final Report*), p. 141 (cit. n. 3).
[8] Memo, Wright Langham, Los Alamos Laboratory Scientific Division, to Hymer Friedell, May 21, 1945, ACHRE No. DOE-113094-B-7, p. 1 (cit. n. 3).
[9] ACHRE, *Final Report*, n. 4 (cit. 3).
[10] "Permissible Dose Conference Held at Chalk River, Ontario, September 1949", Atomic Energy of Canada Limited, Chalk River, Ontario, Canada.
[11] Permissible Dose Conference (cit. n. 10).
[12] Permissible Dose Conference (cit. n. 10), p. 16.
[13] Permissible Dose Conference (cit. n. 10), p. 17.
[14] Permissible Dose Conference (cit. n. 10).
[15] Permissible Dose Conference (cit. n. 10), p. 17. "Chalk River Permissible Dose Conference Notes", 24 October 1949, MIT Archives, Collection MC-80 (Robley Evans), Box 3, Folder "Chalk River", pp. 14, 16.
[16] "Langham upset on tolerance". Shields Warren, Diary for October 1, 1949, Shields Warren Papers, Boston University, ACHRE No. BU-121594-A.
[17] Wright Langham, Los Alamos Laboratory Health Division, Letter to Shields Warren, Director of AEC Division of Biology and Medicine, "Radiation Tolerance Proposed by the Chalk River Permissible Dose Conference of September 29–30, 1949", January 20, 1950, ACHRE No. DOE-020795-D-6, p. 1.
[18] Langham *et al.*, "The Los Alamos Scientific Laboratory's Experience ... " (cit. n. 7), p. 142.
[19] Langham *et al.* (cit. n. 7).
[20] Advisory Committee on Human Radiation Experiments, *Final Report* (New York: Oxford University Press, 1996), p. 142.
[21] For a detailed study of the tension between productivity and safety, see Olwell, "Atomic Workers" (cit. n. 4).
[22] U.S. Congress, Joint Committee on Atomic Energy, "Proceedings of an Executive Meeting of the Joint Committee on Atomic Energy", 6 May 1948, U.S. National Archives, NNLR - Record Group 128, Executive Transcripts, Box 1, JCAE document # 991, p. 225.
[23] 44 U.S. Congress, Joint Committee on Atomic Energy, "Proceedings of an Executive Meeting of the Joint Committee on Atomic Energy", 2 March, 1948, U.S. National Archives, NNLR - Record Group 128, Executive Transcripts, Box 1, JCAE document # 986, p. 77.
[24] U.S. Congress, Joint Committee on Atomic Energy, "Meeting of the Joint Committee on Atomic Energy", 2 March, 1948, p. 82 (cit. n. 23); JCAE, 6 May 1948, p. 221.
[25] JCAE 2 March, 1948 (cit. n. 23), p. 83.
[26] JCAE 2 March, 1948 (cit. n. 23), p. 83.
[27] For a detailed study, see Claudia Clark, *Radium Girls: Women and Industrial Health Reform, 1910–1935 (*Chapel Hill: University of North Carolina Press, 1997).

[28] U.S. Congress, Joint Committee on Atomic Energy, "Proceedings of an Executive Meeting of the Joint Committee on Atomic Energy", 17 December 1947, U.S. National Archives, NNLR - Record Group 128, Executive Transcripts, Box 1, JCAE document # 980, p. 255.

[29] Lilienthal, JCAE Hearing, 2 March, 1948 (cit. n. 23), p. 67, p. 87.

[30] Lilienthal, JCAE Hearing, 2 March, 1948 (cit. n. 23), p. 75.

[31] Lilienthal, JCAE Hearing, 2 March, 1948 (cit. n. 23), p. 81.

ELIZABETH S. WATKINS

RADIOACTIVE FALLOUT AND EMERGING ENVIRONMENTALISM: COLD WAR FEARS AND PUBLIC HEALTH CONCERNS, 1954–1963

Abstract

This essay considers the scientific, social, and political contexts of the debate over radioactive fallout. I contend that the growth of an environmental consensus in 1950s America was constrained both by the nature of the fallout debate and by the cultural climate in which it took place.

Introduction

In November 1962, a cartoon in the *Saturday Review* pictured a businessman at a bar confessing to the bartender, "I had just come to terms with fallout, and along comes Rachel Carson".[1] Cartoonist Herbert Goldberg recognized the similarities between chemical pesticides and radioactive fallout from nuclear weapons tests as environmental pollutants. So, too, did Rachel Carson. On the fourth page of *Silent Spring*, her book on the dangers of chemical pesticides, she drew a direct comparison between the effects of strontium-90, a radioactive isotope produced by nuclear explosions, and those of pesticides:

> Strontium-90, released through nuclear explosions into the air, comes to earth in rain or drifts down as fallout, lodges in soil, enters the grass or corn or wheat grown there, and in time takes up its abode in the bones of a human being, there to remain until his death. Similarly, chemicals sprayed on croplands or forests or gardens lie long in soil, entering into living organisms, passing from one to another in a chain of poisoning and death.[2]

Historians have interpreted this association to mean that controversy over radioactive fallout in the 1950s helped to "pre-educate" people in the 1960s to understand the problems presented by pesticides and other environmental contaminants, thus increasing awareness and concern about the environment.[3]

G.E. Allen and R.M. MacLeod (eds.), Science, History and Social Activism: A Tribute to Everett Mendelsohn, 291–306.

However, the health hazards of radioactive fallout were consistent with issues addressed by the public health movement since the early twentieth century, such as the filtration of the public water supply to protect citizens from contamination by unseen impurities. In the 1950s, the strontium-90 contamination of the food chain, with humans at the apex, alerted people to the interconnectedness of everything in nature. Both this ecological perspective, and the interest in protecting human health, contributed to the development of environmental thinking.[4] The seeds of the American environmental movement, which germinated with the publication of *Silent Spring* in 1962 and flowered in the late 1960s and 1970s, may have been sown in the 1950s with the radioactive fallout debate. Like so many other social movements of the 1960s, environmentalism can trace its roots to previous decades.

However, the radioactive fallout debate did not generate the broad-based grassroots environmental movement spawned by Carson's exposé of pesticides. The Cold War context in which nuclear weapons testing took place precluded the wholesale conversion of the fallout issue into a consensus on environmentalism. Advocates of nuclear weapons testing claimed that the tests were vital to the nation's defense and downplayed the effects of fallout as a minimal price to pay for national security. Critics of nuclear weapons tests who voiced concerns about the hazards of fallout were often portrayed as unpatriotic. In the anti-Communist climate of the 1950s, both peace proponents and emerging environmentalists found it difficult and unpopular to oppose the policies of the government, and in particular, the military.

By contrast, Rachel Carson attracted widespread support for her criticism of the chemical industry. Her portrayal of pesticide producers as greedy, insensitive, and destructive of the habitats of innocent animals (and people) fed into the American tradition of sympathizing with the "little guy" and siding against big business. Carson's status as an established author, along with a well-calculated publicity campaign, contributed to the attention garnered by *Silent Spring*.[5] The spirit of change that permeated the early 1960s also helped Carson draw support for her campaign. Although the concerns raised by radioactive fallout may have been identical to those brought about by chemical pesticides, and pre-dated them by several years, the circumstances in which the fallout debate took place differed in important ways from those surrounding the publication of Carson's treatise. This essay will consider the scientific, social, and political contexts of the debate over radioactive fallout. I contend that the growth of an environmental consensus in 1950s America was constrained both by the nature of the debate and by the cultural climate in which it took place.

Scientists were well aware of the dangers of radiation long before the first explosion of an atomic bomb at Alamogordo in 1945. In the 1920s, H.J. Muller, a geneticist at Indiana University, demonstrated that radiation caused genetic changes by using X-rays to induce mutations in fruit flies (work for which he won the Nobel Prize in 1946).[6] Thirty years later, as both the United States and the Soviet Union conducted dozens of above-ground nuclear tests, the U.S. National Academy of Sciences, the British Medical Research Council, and a scientific committee formed by the United Nations General Assembly all produced reports on radiation hazards. In February 1957, *Science* published the results of a study by a group of scientists at Columbia University, funded by the U.S. Atomic Energy Commission (AEC) on the effects of strontium-90 (three more articles appeared in the next three years).[7] In 1958, two prominent American scientists, Linus Pauling and Edward Teller debated fallout and disarmament on television; their arguments were popularized in books published that same year.[8] The titles – Pauling's *No More War*! and Teller's *Our Nuclear Future* – indicate the authors' diametrically opposed positions.

Debate over the hazards of radioactive fallout centered on three main issues: the process by which fallout descended to earth (its rate and uniformity), genetic hazards (damage to future generations resulting from mutations in an individual's sex cells), and somatic hazards (injury or illness caused to an individual exposed to radiation). Scientists often discussed radioactive fallout in connection with other forms of radiation, such as natural background radiation (from cosmic rays and naturally occurring radioactive decay on earth) and diagnostic medical and dental X-rays, to compare the relative risks to human health presented by each type. Also, the presentation and interpretation of statistics often depended upon a researcher's stance with respect to fallout hazards. Thus, for example, the low percentage of the world's population affected by fallout could be stressed, instead of the large number of individual cases of leukemia which would result, or vice versa. This problem of interpretation was compounded by the limited data available on the effects of radiation. Different groups of researchers disagreed over the validity of certain data; for example, some contested the extrapolation from animal experiments to humans. All these factors increased the difficulty of assessing the risk presented by nuclear fallout.

Nuclear explosions produce three kinds of radioactive fallout.[9] Local fallout, generated by both atomic and hydrogen bombs, consists of irradiated particles of surface soil and dust which are carried up into the air by the force of the explosion, and which then fall back to earth in the area around the site of detonation within hours. Wind-borne fallout, also generated by both types of nuclear bombs, is forced up into the troposphere where it is carried far away by the wind. This second type of fallout is brought down to earth by precipitation, usually within weeks of the explosion. Particles are tiny and invisible, compared with the larger, perceptible dust of local fallout. Radiation from this type of fallout was often detected in the Midwest and Northeast regions of the United States in the weeks following an atomic test at the Nevada proving grounds.

The third type, global fallout, is produced only by hydrogen bombs, or "superbombs". In these thermonuclear weapons, an atomic fission device is exploded to achieve the very high temperature necessary to trigger the fusion of hydrogen isotopes. The bomb is surrounded by uranium-238, which then undergoes fission as a result of the thermonuclear reaction. In other words, the hydrogen bomb is a fission-fusion-fission bomb. Radioactive isotopes produced from the fission of the uranium are catapulted up into the stratosphere, where they may circle the globe for years. This type of fallout, which contaminates the atmosphere for very long periods of time and which affects the entire globe, is a primary cause for environmental concern. After years in the upper atmosphere, much of the radioactivity dissipates, but those isotopes with long half-lives remain radioactive when they eventually descend to earth.

By 1959, new data on the process of fallout became available to the scientific community.[10] First, data from the Soviet test series in October, 1958 revealed that the rate of fallout to earth from the stratosphere, thought to take ten years, occurred at least twice as fast. Thus, a greater proportion of the fallout was still radioactive when it fell to earth. Second, the pattern of fallout was demonstrated to be non-uniform. That is, some areas received much higher levels of fallout than others, producing "hot spots" where the dosage of radiation might be considerably greater. Third, the densely populated Northern Temperate Zone received more fallout than anywhere else for two reasons: the United States and the Soviet Union tested their nuclear weapons only in the Northern hemisphere, and the pattern of circulation of air in the upper regions in the atmosphere caused much of the fallout to descend from the stratosphere into the troposphere over the temperate zone. Precipitation from the troposphere carried the fallout down to that region of the earth (including North America, Europe, and the Soviet Union).[11]

These new findings evoked both alarm – because the fallout reaching the Northern Temperate Zone was more concentrated and more radioactive than previously believed – and comfort, because the skies would be "cleared" of radioactive debris that much sooner, since fallout descended more rapidly than had been thought. The data was not strongly contested; rather, its interpretation varied according to one's position on nuclear testing.[12]

Scientists tended to agree on much of the data on radiation hazards to human inheritance, but differed on the relative importance given to radioactive fallout as a source of radiation. In 1956, the Committee on Genetic Effects of Atomic Radiation of the National Academy of Sciences (which included, among others, the prominent geneticists Muller, George Beadle, James Crow, A.H. Sturtevant, and Sewall Wright) drew the following conclusions:

> First, since any additional radiation is genetically undesirable the fallout dose is genetically undesirable. Second, the fallout dose to date (and its continuing value if it is assumed that the weapons testing program will not be substantially increased) is a small one as compared with the background radiation, or as compared with the average exposure in the United States to medical X-rays.[13]

The genetic risk from fallout was considered in comparison to the risk presented by other forms of radiation. By placing fallout in this context, the authors minimized its contribution to radiation exposure and thus refrained from making negative statements about the U.S. weapons testing program.

On the other hand, Linus Pauling estimated that the by-products from ten megatons of fission explosion-yield would cause an increase of one percent in the number of defective births. His position on weapons testing could not be mistaken:

> I believe that the national leader who gives the order to test the nuclear weapon and to release the fission products corresponding to 10 megatons of fission should know that in giving this order he is dooming 15,000 unborn children to a life of misery or early death.[14]

Although he agreed with the NAS report that radiation from fallout contributed only about ten percent as compared to the amount from background radiation, he argued that such a ten percent increase in radiation exposure would lead to a corresponding ten percent increase in the mutation rate, thus dramatically increasing the *number* of genetically affected children born each year.[15] For Pauling, the fact that radioactive fallout may contribute only a fraction of the total exposure to radiation was no comfort. Instead, any increase in exposure to radiation was bad, which added fuel to his arguments in favor of a nuclear test ban and disarmament.

Edward Teller, known as the father of the hydrogen bomb, downplayed the risk of genetic mutation caused by radiation.[16] He calculated that nuclear tests increased the number of mutations by only one-tenth of one percent, relatively insignificant compared with those caused by natural background radiation. People who lived at high altitudes (and therefore received more intense cosmic rays), such as the Tibetans, showed no demonstrable genetic differences. Teller applauded the study of radioactivity for "helping us to understand the strange process of life and of the curious substances which connect one substance to the next".[17]

In a 1959 report, the General Advisory Committee to the U.S. Atomic Energy Commission made the point that other people live in areas of the world with five times the background radiation of the U.S., "or more than one hundred times the average amount of radiation from fallout in the United States".[18] This neat statistic was derived from the estimate that fallout generated less than five percent as much radiation as that caused by natural background radiation. Whereas Pauling pointed to the increase in radiation exposure from fallout, Edward Teller and the Atomic Energy Commission, among others, sought to dismiss any apparent genetic hazard by focusing on fallout's relatively minor contribution to worldwide radiation.

There was less consensus within the scientific community on the somatic hazards of fallout. Concern about harm to the individual centered on the possibility of contracting cancer from exposure to radiation. Since the type of fallout associated with the bomb tests was primarily global (remaining in the upper atmosphere for years before descending to earth), exposure to radioactivity occurred not by inhalation or skin contact, but rather by ingestion. When radioactive isotopes fell to earth, they settled in the soil, in water, and on vegetation, thus entering the food chain. Scientists disputed "safe" limits to concentrations of these radioactive chemicals in the human body and whether a threshold existed below which no disease would result.

Nuclear bombs produce several radioactive isotopes with varying half-lives. Strontium-90, with a half-life of twenty-eight years, mimics the role of calcium in the body's metabolic pathways and is incorporated into bones. Medical researchers hypothesized that the presence of strontium-90 in bones would increase the risk of bone cancer and leukemia (cancer of the blood, which is manufactured in the bone marrow). Radioactive strontium is taken up by plants, and then by cows which eat the plants, and then by people who drink the cow's milk. Although both cows and humans absorb calcium in preference to strontium, the radioisotope was detected in the bones of people from all over the world.[19] Interestingly, vegetation grown in calcium-deficient soil absorbed more strontium, and people who relied upon plants instead of milk as their primary source of calcium incorporated even higher concentrations of strontium into their bones.[20] However, American studies of the strontium problem rarely included these non-western peoples.

In 1957, an AEC-funded study at Columbia University determined the world-wide average strontium content in man to be one ten-thousandth of the maximum permissible concentration, with levels in children three to four times greater than in adults (due to active bone growth).[21] A second report, one year later, confirmed that strontium-90 levels in bones would continue to increase, even if no more tests were conducted. However, even if testing did continue at the current rate, the concentration of strontium in bones would still be less than the maximum permissible concentration.[22] The maximum permissible concentration, arbitrarily set by the National Committee on Radiation Protection (a group of scientists, physicians, and government officials) in 1959, did not represent an absolute "safe" limit for human tolerance to strontium.[23] Levels of strontium-90 below the arbitrary maximum permissible concentration might still increase the risk of leukemia and bone cancer. The problem with determining causality between strontium-90 and cancer was the long lag time between exposure and onset of the disease, and the problem with setting maximum permissible concentrations was that nobody knew for certain the effects of low level exposure.

In *Our Nuclear Future*, Edward Teller chose to dismiss concerns about health hazards of fallout by demonstrating fallout effects to be insignificant in comparison with other sources of radiation: "we know enough to state positively that the danger from the world-wide fallout is less than many other radiation effects (e.g. natural cosmic rays, medical X-rays) which have not worried people and do not worry them now".[24] He compared the reduction in life-expectancy resulting from various habits and exposures and concluded that "the world-wide fallout is as dangerous as being an ounce overweight or smoking one cigarette every two months".[25] Teller estimated the reduction in life expectancy to be nine years for smoking a pack of cigarettes a day, one and a half years for being ten percent overweight, and one to two days for exposure to world-wide fallout.

Not unexpectedly, Linus Pauling's estimates of the somatic hazards resulting from radioactive fallout were significantly more alarming. He calculated that ten thousand deaths per year from leukemia and bone cancer would be caused if testing continued at the current rate.[26] He characterized strontium-90 as follows: "One teaspoonful of this poison, distributed equally among all the people in the world, would kill all of them within a few years".[27] This powerful image was repeated one year later in an article on fallout in *Newsweek*.[28] Not only were the data on the effects of strontium-90 debated and the use of statistics manipulated to suit different political positions on nuclear testing, but also the language in which the risk from exposure to fallout was described proved to be an effective tool for either diminishing or magnifying perceived health hazards.

Public concern about radiation exposure grew in response to the American nuclear weapons testing program in the 1950s. In May 1953, the fallout from a bomb tested in Nevada fell on the town of St. George, Utah, exposing residents to high levels of radiation. Another accident the following year heightened anxiety about the dangers of fallout from nuclear weapons tests. In March 1954, the U.S. government tested a hydrogen bomb at Bikini Atoll in the Marshall Islands. An unexpected wind shift carried the resultant radioactive cloud towards the islands, where 236 Marshallese and 31 Americans were exposed to radiation. In addition, a Japanese fishing boat trawling eighty miles from the test site was showered with radioactive fallout. Although the U.S. Atomic Energy Commission did not release an official report on the fallout from this test until February, 1955, the fate of the Lucky Dragon received great publicity in both Japan and America in newspapers, magazines, and newsreels.[29]

Magazine writers followed scientists in comparing the radiation exposure from fallout to that of natural background radiation and medical and dental X-rays. The depiction of fallout as contributing only a small fraction of the world-wide radiation served to allay fears about its effects. In the late 1950s, attention focused on the health hazards associated with contamination of food by radioactive fallout. Depending on the author's position (or that of the magazine), fallout was portrayed either as an invisible, insidious agent of disease and death, or as a minor side-effect of the U.S. nuclear testing program, essential to the nation's defense and security.

Two articles published within a year of each other – one in *Reader's Digest* (September, 1958) and the other in the *Saturday Evening Post* (August-September,1959) – presented very different approaches to the subject. The first was reassuring in tone: "The evidence thus far obtained [from the AEC's Project Sunshine] should lay to rest a sinister ghost of world terror of the much publicized "rain of radioactive death" which has led many people to call for an immediate end to nuclear testing".[30] The author reminded readers of the importance of American national security and blamed the nefarious Communists for instigating the panic over fallout:

> As we have seen there is some slight danger [from fallout]. But it must be remembered that the tests we have been conducting have been essential to the defense of this country and the small risk is deemed necessary from the standpoint of national security.

> There have also been dupes of Communist propaganda in neutral nations and among our own allies who have paraded and vociferated against U.S. atomic testing ... . It is now clear that a large part of the fallout scare was another of the Kremlin's big lies.[31]

The second article, by contrast, emphasized entirely different issues related to fallout and drew very different conclusions. The author expressed frustration over the lack of consensus among scientists and government authorities.

> This brings us to the core of the controversy that has kept the public in such a state of confusion. For the picture of fallout the public has had to rely upon the interpretation of fragmentary data by authorities with different viewpoints and policies.[32]

In other words, the confusion and controversy within the scientific community over radioactive fallout had filtered down to the public with neither answers nor resolutions.

The author also agreed with geneticist A.H. Sturtevant in taking issue with the AEC's method of risk assessment. The AEC claimed that the risk of fallout was significantly less than other risks presented by everyday life, such as riding in a car or swimming at the beach. But, as Sturtevant pointed out, one could exercise control over those everyday risks in ways not possible with fallout, and hence such analogies were invalid.[33] The author described fallout as "a silent killer which hides its poison among the more familiar causes of human illness and death and thus postpones positive identification".[34] He concluded that fallout presented a real and serious threat to humanity, and advocated peace negotiations rather than continued nuclear arming and testing. Unlike other contemporary articles, the *Saturday Evening Post* piece did not frame the discussion of fallout in terms of the importance of weapons testing to ensuring national security against the peril of Soviet supremacy.

A 1962 article in *Redbook* also took a more questioning approach to the fallout debate, focusing upon the contamination of milk by radioactive iodine. The authors asked why the U. S. government had not yet proposed concrete countermeasures (such as preparations for a conversion to milk treated to eliminate radioactivity).[35] They expressed frustration over the lack of consensus and decision-making among the various government agencies responsible for determining health policy. Given the lack of initiative and the practical difficulties of implementing countermeasures to radio-contamination of food, along with potential international criticism if the United States acted only to protect its own citizens from the worldwide effects of fallout, they hinted that the only feasible countermeasure might be the cessation of nuclear weapons testing.[36]

Earl Ubell, science editor of the *New York Herald Tribune* took a very different tack in an article written for *Parents' Magazine*. By emphasizing the importance of national defense and by downplaying the hazard of radiation from fallout, he reassured his readers that, at least for the time being, the health of their children was not in jeopardy. After explaining the processes by which radioactive fallout was generated and then entered the food chain, Ubell described both the genetic and somatic hazards of exposure to radiation. In comparing the risk presented by fallout with that of medical X-rays, he stated that "most scientists tend to concede" that "medical X-rays also are risky, actually riskier to the individual than fallout. (In most cases, however, the benefits far outweigh any possible harm.)"[37] Just as a sort of cost-benefit analysis was used to assess the risk of medical X-rays (success of treatment or diagnosis versus adverse radiation side-effects), Ubell applied the same evaluation to the relative risk of radioactive fallout with respect to national security.

> In the case of fallout, it is within the power of governments to eliminate radioactivity by a cessation of atom bomb tests. This raises questions of national power and security. How many additional leukemias and hereditary defects is it worth to remain militarily strong? ... These are moral questions each citizen must answer for himself with the greatest national good in mind as well as individual hazards.[38]

The author also compared the risk of fallout exposure to other risks of modern life, and concluded:

> The chances of being hurt by the rays appear to be far less than of being killed in an automobile accident. We know modern society produces risks. We try to estimate them, reduce them if possible and determine whether or not they are compatible with the benefits we supposedly receive ... One thing is clear: we are all bathing in radiation. We can only guess at the extent of the total risk, but we can be assured that the danger to any individual is small compared to the other dangers of modern life.[39]

Popular magazines often employed this sort of cost-benefit analysis to assess the relative risk posed by fallout. For example, *Newsweek* noted in 1962 that, of 1.7 million deaths in America that year, only 40 could be attributed to leukemia or bone cancer resulting from exposure to fallout. Furthermore, 110 children would be born with major birth defects caused by fallout radiation. The article commented that President Kennedy had to take into consideration the value of those 150 lives in making his decision to resume nuclear testing.[40]

The conservative *U.S. News and World Report*, which consistently supported the policies of the AEC and worked to allay public fears about fallout, ran a ten-page article in 1955 promising to give "factual answers to most of the questions now being asked" about atomic bomb tests and fallout. According to "top U.S. scientists ... the atom isn't generating monsters, isn't causing tornadoes, isn't endangering people, animals or crops".[41] Two years later, an article appeared claiming that the fallout "scare" was part of a Communist plot. The summary, in large type, on the first page, gives a good indication of the editorial stance:

> When U.S. or Britain sets off a nuclear test bomb, you hear loud cries of protest and warnings of grave danger to mankind. Let Soviet Russia explode a similar bomb, however, and you hear few protests. This is what Western officials find as they study the record. The conclusion that they reach is this: Much of the clamor about the danger of fall-out is inspired by Communists.[42]

In attempts to reduce concerns about the health hazards of fallout from nuclear tests, *U.S. News* printed two interviews with H.J. Muller, in 1955 and again in 1961. In the first, Muller noted that the small percentage of the population genetically affected by fallout radiation translated to a large number of disabled individuals, but the editors chose to print in large, boldface type at the top of one page: "Mutations from bomb tests: 'No significant deterioration'".[43] Similarly, editors extracted Muller's comment that the genetic risk from injudicious use of medical X-rays was probably greater than that of fallout from bomb tests for the summary at the beginning of the article, minus the word "probably". The summary also pointed out that there would be one thousand times as many naturally occurring mutations per generation as those generated by exposure to radioactive fallout.[44] The implicit answer to the question in the article's title, "What Will Radioactivity Do to our Children?" was: Very little indeed. The second interview with Muller in 1961 focused less on the genetic hazards of radiation and more on the relative dangers of nuclear testing and nuclear war. The take-home message: the health risk from testing was insignificant compared to the real danger of exposure to radioactive fallout in the event of nuclear war.[45] This Cold War logic underscored the value of nuclear weapons tests in maintaining a strong deterrent force, apparently critical to the prevention of nuclear war with the Soviet Union.

Articles in *Time* and *Newsweek* also linked radioactive fallout with the Cold War perception of the Soviet Union as the enemy. As early as 1954, *Newsweek* briefly reported the statement of geneticist A.H. Sturtevant that radiation was harmful to humans and that one source of exposure – bomb tests – could be avoided (by a test ban). However, one of the feature articles in that same issue, entitled, "Can We Win a War with Russia? Yes, But –" described the impressive strength of the Soviet military force.[46] In 1957, *Time* quoted AEC scientist Willard Libby as preferring the small risk from fallout to the "far greater risk, to freedom-loving people everywhere, of slackening our defenses against totalitarian forces".[47] The cover article of the 10 November 1961 issue of *Time* ended with the following paragraph:

> The realization that he [Glenn Seaborg, chairman of the AEC] must devote most of his energies to preparing for the possibility of war saddens Glenn Seaborg. But the choice is not his – or that of the U.S. For the Soviet Union has made it necessary for the U.S. to meet ruthlessness with strength.[48]

Although these news articles expressed contemporary concerns about the health hazards of radiation, they almost always placed the fallout issue in the larger, hostile context of the Cold War.

Consumer Reports, the monthly publication of the nonprofit Consumers' Union, ran its first article on radiation hazards from fallout and X-rays in September 1958, offering "a non-emotional, non-technical look at the essential facts".[49] After an explanation of the sources and biological effects of radiation, the article devoted four times as much space to X-rays as to fallout. The conclusion: "the facts indicate that there is cause for concern, but not panic".[50]

However, the focus shifted exclusively to fallout just six months later when *Consumer Reports* published a detailed report of its study of radioactive isotopes (particularly strontium-90) in milk:

> The fact is that fresh clean milk, which looks and tastes just as it always did, nevertheless contains (wherever you get it these days) an unseen contaminant, a toxic substance known to accumulate in human bone. There is an analogy: bacterial contamination. There, too, the presence of unseen contaminants can bring grave harm. Most of us, even if we are not bacteriologists, know by now the danger from germs, and yet know how to estimate it rationally, and not to fear it unreasoningly. It is exactly with this kind of reasoned concern that we seek to approach the more novel problem of radioactive fallout.[51]

The results of this study, which emphasized the adverse effects of food contamination to address concerns about the health hazards presented by radioactive fallout, were picked up by national news services and printed in newspapers around the country.

The journals of the health professions also considered the effects of radioactive fallout and the potential consequences of radiation in the aftermath of a nuclear war. Articles entitled "Contamination of Foods by Fallout" and "Survival in a Thermonuclear War" ran side by side in medical journals, where Cold War politics played a prominent role.[52]

Public health officials contended that the problems of radiation and fallout fell squarely within the boundaries of public health – just more contaminants in a long line of disease-producing agents found in the environment. In the early twentieth century, public health activists had directed their efforts to the filtration of water and the pasteurization of milk.[53] In the 1940s, several cities tried to rid the air of industrial pollutants through smoke control programs.[54] In 1958, the U.S. Public Health Service began to monitor levels of radioactivity in milk supplies. However, enacting effective controls required legislation. In regard to radioactive fallout, only the U.S. Congress had the authority to regulate the nuclear weapons testing program, and in the 1950s, few legislators were willing to do more than hold hearings on the matter.

It fell to activists to use fears about fallout to advocate a halt to nuclear weapons testing. The National Committee for a Sane Nuclear Policy (SANE) lobbied for a test ban.[55] In a full-page advertisement in the *New York Times* on 15 November 1957, SANE claimed that man "has the right to live and to grow, to breathe unpoisoned air, to work on uncontaminated soil".[56] The purpose of this ad was to jolt readers into realizing that a test ban would "eliminate at least one real and specific danger", that of radio-contamination. In his book on the test ban debate, historian Robert A. Divine quoted a SANE lobbyist who said in 1959, "The fallout issue has become our 'hottest' and we have a real chance to make a significant breakthrough which will affect our drive against testing".[57] In 1961, Women Strike for Peace (WSP) organized a nationwide strike of American housewives to protest against the arms race.[58] Both these groups, however, expressed their hopes not in terms of the environment *per se*, but rather in terms of achieving world peace.

Another group, the Greater St. Louis Committee for Nuclear Information (CNI, organized by Barry Commoner and other faculty members at Washington University), began to publish *Nuclear Information* in late 1958, "for the purpose of collecting and distributing to the public factual information about the effects of nuclear war, nuclear testing, and non-military uses of nuclear energy".[59] Although designed for a more limited readership, its articles often received attention from major newspapers, wire services, and television broadcasts, and thus reached a wide national audience. This monthly publication reported often on the levels of strontium-90 in the St. Louis milk supply and, in December 1958, announced the establishment of the "Baby Tooth Survey", an ambitious project to collect fifty thousand teeth per year from children in the greater St. Louis area to provide a record of strontium-90 absorption.[60] Nuclear Information reported periodically on the results of the study (which indicated an increase in the concentration of strontium-90 in the children's teeth). The survey received national publicity and, although not needed for the localized study, teeth were sent from many other states and countries.

SANE, WSP, and CNI directed their efforts toward a nuclear test ban. They met with success in August 1963, when the United States, the Soviet Union, and Great Britain signed the Limited Nuclear Test Ban Treaty, which prohibited testing in the atmosphere, under water, and in outer space. Although these groups used health and environmental concerns about fallout as means to achieve a specific end (the test ban), their popular appeal set a precedent for broader environmental causes.

Did conservation or nature organizations express concern about the environmental effects of radioactive fallout? Between 1954 and 1963, neither the *Sierra Club Bulletin* nor *The Living Wilderness* (the bulletin of the Wilderness Society) made any mention of the issue. *Audubon Magazine* printed one article in 1963 about elevated levels of strontium-90 and cesium-137 in Eskimos. Radioisotopes from nuclear test fallout had contaminated the simplified food chain (lichens to reindeer to man) of this population. Although the author commented on the effect of contaminants on the ecosystem (described as a "dynamic concept for 'environment', one which included the interactions of animals, plants, soil, water, air, and sunlight"), he emphasized the plight of the Eskimos and their right to proper representation as American citizens:

> The human population of the region is relatively sparse and, unfortunately, casts few votes in the world's halls of government. If a hot spot of similar magnitude were to develop in relation to peaches or beef or milk or tobacco

> [as opposed to lichens] it seems probable that no reason would be thought serious enough to allow contamination to continue.[61]

By contrast, the same periodical devoted many articles to the problem of the widespread use of pesticides (especially DDT) in the years 1958–1963. For the Audubon Society, pesticides were a greater danger than fallout to health and the environment.

In general, popular magazine journalists presented the fallout debate in the light of the relationship between the United States and the Soviet Union. Some did question the wisdom of nuclear testing, the escalating arms race, the prospect of nuclear proliferation, and, of course, the hazards of fallout. Most, however, supported nuclear testing on the grounds of national defense and security, and readers were reassured that fallout presented negligible risk.

This risk, however significant, was viewed as a human issue – of populations alive and yet to come. Rarely, if ever, were possible effects on other species considered. Similarly, there was little discussion of the contamination of the environment as a concern in and of itself. The emphasis on fallout as a human health issue rather than as a matter of ecological concern is illustrated in the pages of the specialized magazines surveyed: witness *Consumer Reports'* extensive studies of radio-contamination of food and milk and the almost nonexistent coverage of fallout by the conservation and nature periodicals. In the period between 1954 and 1963, "environmental thinking" about radioactive fallout focused mainly on human health and well-being.

However, a couple of letters to the editor of *Nature* hinted at a broader ecological perspective. One group of researchers reported findings of strontium-90 in deer antlers; another observed the cumulative effects of radioactivity in Pacific island plant species exposed to fallout from weapons tests.[62] Both communications revealed an increasing sensitivity toward ecological relationships. It was only a small step from the preoccupation with fallout contamination of food as a human health hazard, to the understanding that fallout polluted the very air, water, and land on which human life depended, to the realization that fallout could upset the delicate balance of nature.

Rachel Carson succeeded in combining these concerns about human health with an awareness of the surrounding environment. In 1959, she wrote her editor:

> As you know, it has always been my intention to give principal emphasis to the menace of human health, even though setting this within the general framework of disturbances of the basic ecology of all living things.[63]

Whereas fallout had been associated mainly with genetic mutation and somatic illness, Carson expanded the risk from pesticides to include damage to the rest of the biosphere. Her presentation of the pesticide issue combined all three aspects of environmental thinking – in historian Samuel P. Hays' terms: beauty, health, and permanence.[64] Carson eloquently demonstrated that pesticides jeopardized the aesthetic qualities of the environment, endangered human health and well-being, and threatened to upset the ecological balance of nature.

Radioactive fallout, by contrast, was largely portrayed as a health hazard. Scientists, journalists, and activists emphasized the health hazards of contaminated food and milk and the potential risk of disease and genetic effects. The pollution of the environment by fallout was a secondary issue, and even then, concern centered on how a polluted environment might impinge on human health and well-being. In this way, radioactive fallout was presented as more of a public health issue, than an environmental crisis.

The emphasis on health-related effects framed the debate over radioactive fallout in the 1950s and early 1960s. Initially, the social and political climate of the Cold War contextualized nuclear weapons testing as vital to national security. From this perspective, radioactive fallout was interpreted as an unfortunate side effect of American defense, and "Cold Warriors" (not only government officials, but also sympathetic scientists and journalists) manipulated scientific data to show that the adverse health effects of fallout were minimal anyway.

As the 1950s drew to a close, more and more people voiced their concerns about radioactive fallout and spoke out against the government's nuclear weapons testing policy. The work of organized groups such as SANE, CNI, and WSP received both media attention and popular support. The metaphors used to describe the dangers of radioactive fallout recalled earlier environmental toxins and set the stage for the reception of Rachel Carson's urgent message. Although the general public wavered between health and national security as the overriding factor in their appraisal of fallout, there was less ambivalence about the less political issue of pesticides.[65] The pesticide-induced silent spring foretold by Rachel Carson spurred an environmental movement that fallout failed to produce, so long as weapons tests were seen as indispensable to the prevention of nuclear war.

Carnegie Mellon University

Notes

1 Illustration in Linda Lear, *Rachel Carson: Witness for Nature* (New York: Henry Holt, 1997), after p. 366.

2 Rachel Carson, *Silent Spring* (New York, Fawcett Crest, 1962), p. 16.

3 Ralph Lutts, "Chemical Fallout: Rachel Carson's Silent Spring, Radioactive Fallout, and the Environmental Movement", *Environmental Review*, 9 (1985), p. 212; Samuel P. Hays, *Beauty, Health, and Permanence: Environmental Politics in the United States, 1955–1985* (Cambridge: Cambridge University Press, 1987), p. 28; Spencer R. Weart, *Nuclear Fear: A History of Images* (Cambridge, MA: Harvard University Press, 1988), p. 325; Lear, *Rachel Carson*, pp. 373–375 (cit. n. 2).

4 Hays, *Beauty, Health, and Permanence*, pp. 24, 28 (cit. n. 3).

5 Lear, *Rachel Carson*, pp. 397–427 (cit. n. 1). For an interesting analysis of Carson's use of anti-Communist rhetoric in her critique of the pesticide industry, see Jeff Ellis, "Redefining the 'Menace of Our Time': Rachel Carson and Barry Commoner as Environmental Patriots in Cold War

America", paper presented at the Organization of American Historians annual meeting, San Francisco, 1997.

6 Elof Axel Carlson, *Genes, Radiation, and Society: The Life and Work of H.J. Muller* (Ithaca, NY: Cornell University Press, 1981). For a biography of Linus Pauling, see Thomas Hager, *Force of Nature: The Life of Linus Pauling* (New York: Simon and Schuster, 1995).

7 The NAS report was published in *Science*, 123 (29 June 1956), pp. 1157–1164. The strontium-90 reports were published by J. Lawrence Kulp *et al.* in *Science*, 125 (8 February (1957) pp. 219–225; *Science*, 127 (7 February 1958), pp. 266–273; *Science*, 129 (8 May 1959), pp. 1249–1255; and *Science*, 132 (19 August 1960), pp. 449–454.

8 Linus Pauling, *No More War!* (New York: Dodd, Mead & Company, 1958); Edward Teller and Albert L. Latter, *Our Nuclear Future: Facts, Dangers, and Opportunities* (New York: Criterion Books, 1958). The transcript of the televised debate was published in *Daedalus*, 87 (December, 1958), pp. 147–163.

9 Lester Machta and Robert J. List, "The Global Pattern of Fallout", in *Fallout: A Study of Superbombs, Strontium-90, and Survival*, John M. Fowler, ed. (New York: Basic Books, Inc., 1960), pp. 26–27.

10 Ralph E. Lapp, "Fallout Hearings: Second Round", *Bulletin of the Atomic Scientists*, 15 (1959), p. 307.

11 Machta and List, "The Global Pattern of Fallout", pp. 27–29 (cit. n. 9).

12 See Carolyn Kopp, "The Origins of the American Scientific Debate over Fallout Hazards", *Social Studies of Science*, 9 (1979), pp. 403–422.

13 "Genetic Effects of Radiation", *Science*, 123 (29 June 1956), p. 1164.

14 Pauling, *No More War!*, p. 73 (cit. n. 8).

15 Pauling, *No More War!*, p. 67 (cit. n. 8).

16 Teller and Latter, *Our Nuclear Future*, pp. 130–133 (cit. n. 8).

17 Teller and Latter, *Our Nuclear Future*, p. 133 (cit. n. 8).

18 "Problems Presented by Radioactive Fallout", *Bulletin of the Atomic Scientists*, 15 (June 1959), p. 258.

19 J. Lawrence Kulp *et al.*, "Strontium-90 in Man", *Science*, 125 (1957), p. 219.

20 W.O. Caster, "From Bomb to Man" in John M. Fowler (ed.), *Fallout*, p. 44 (cit. n. 9).

21 Kulp, "Strontium-90 in Man", pp. 221–222 (cit. n. 19).

22 Walter R. Eckelmann *et al.*, "Strontium-90 in Man, II", *Science*, 127(7) (1958), p. 273.

23 Lapp, "Fallout Hearings: Second Round", p. 305 (cit. n. 10).

24 Teller and Latter, *Our Nuclear Future*, p. 124 (cit. n. 8).

25 Teller and Latter, *Our Nuclear Future*, p. 124 (cit. n. 8).

26 Pauling, No *More War!*, p. 105 (cit. n. 8).

27 Pauling, No *More War!*, p. 102 (cit. n. 8).

18 "Atom Fallout ... How Bad It Is ... What We Can Do About It Now", *Newsweek*, 53 (6 April 1959), p. 35.

29 Robert A. Divine, *Blowing on the Wind: The Nuclear Test Ban Debate*, 1954–1960 (New York: Oxford University Press, 1978), pp. 1–13; Allan M. Winkler, *Life Under a Cloud: American Anxiety About the Atom (*New York: Oxford University Press, 1993), pp. 93–94. See also the film, *Atomic Café* (1982).

30 David O. Woodbury, "Basic Facts About 'Fallout'", *Reader's Digest*, 73 (September, 1958), p. 51.

31 Woodbury, "Basic Facts About 'Fallout'", p. 55 (cit. n. 30).

32 Steven M. Spencer, "Fallout; The Silent Killer", *Saturday Evening Post*, 232 (29 August 1959), p. 87.

33 Spencer, "Fallout", 89 (cit. n. 32).

34 Steven M Spencer, "Fallout: The Silent Killer, How Soon is Too Late?", *Saturday Evening Post*, 232 (5 September 1959), p. 86.

35 Ruth and Edward Brecher, "What We Are Not Being Told About Fallout Hazards", *Redbook*, 119 (September 1962), p. 50.

36 Brecher, "What We Are Not Being Told", p. 106 (cit. n. 35).

37 Earl Ubell, "Will Radiation Harm Your Child?", *Parents' Magazine*, 35 (July, 1960), p. 35.

38 Ubell, "Will Radiation Harm Your Child?", p. 105 (cit. n. 37).

39 Ubell, "Will Radiation Harm Your Child?", p. 105 (cit. n. 37).

40 "Radiation: 110 Babies", *Newsweek*, 59 (11 June 1962), p. 62.

41 "All About A-Bombs, Fall-Out, Dangers in the Future", U.S. *News and World Report*, 38 (29 April 1955), p. 96.

42 "What's Back of the 'Fall-Out Scare'", *U.S. News and World Report*, 42 (7 June, 1957), p. 25.

[43] "What Will Radioactivity Do to our Children?", *U.S. News and World Report*, 38 (13 May 1955), p. 78.
[44] "What Will Radioactivity Do to our Children?", p. 72 (cit. n. 43).
[45] "What Fallout Really Means to You and your Children", *U.S. News and World Report*, 51 (13 November 1961), p. 41.
[46] "Warning", *Newsweek*, 44 (5 July, 1954), p. 73, and "Can We Win a War with Russia? Yes, But-", *Newsweek*, 44 (5 July, 1954), p. 30.
[47] "The Atom: The Peril of Strontium-90", *Time*, 69 (6 May,1957), p. 24.
[48] "The Atom", *Time*, 78 (10 November, 1961), p. 25.
[49] "Radiation Hazards from Fallout and X rays", *Consumer Reports*, 23 (September, 1958), p. 484.
[50] *Ibid.*
[51] "The Milk All of Us Drink and Fallout", *Consumer Reports*, 24 (March 1959), p. 103.
[52] R.H. Wasserman, "Contamination of Food by Fallout", *New York State Journal of Medicine*, 60 (1 December, 1960), pp. 3857–3862; S. Garb, "Survival in a Thermonuclear War" *New York State Journal of Medicine*, 60 (1 December 1960), pp. 3863–3866. See also Paul Boyer, "Physicians Confront the Apocalypse: The American Medical Profession and the Threat of Nuclear War", *Journal of the American Medical Association*, 253 (1985), pp. 633–643.
[53] Deborah Levine, "Typhoid Fever and Public Health in Pittsburgh", unpublished manuscript, Historical Society of Western Pennsylvania, 1997.
[54] Joel Tarr, "Changing Fuel Use Behavior and Energy Transitions: The Pittsburgh Smoke Control Movement, 1940–1950" *Journal of Social History* 14 (1981), pp. 561–588; Joel Tarr, "Railroad Smoke Control: The Regulation of a Mobile Pollution Source", in *Energy and Transport: Historical Perspectives on Policy Issues*, George H. Daniels and Mark H. Rose, eds. (London: Sage Publications, 1982), pp. 71–92.
[55] Divine, *Blowing on the Wind*, pp. 267–268; Winkler, *Life Under a Cloud*, pp. 105–106 (both cit. n. 9).
[56] Divine, *Blowing on the Wind* (cit. n. 9), p. 167.
[57] Divine, *Blowing on the Wind* (cit. n. 9), p. 267.
[58] See Amy Swerdlow, *Women Strike for Peace: Traditional Motherhood and Radical Politics in the 1960s* (Chicago: University of Chicago Press, 1993)
[59] *Information* (24 December, 1958). Initially titled *Information*, the name of the bulletin was changed to *Nuclear Information* in March 1959.
[60] *Information* (24 December, 1958), (cit. n. 59).
[61] William O. Pruitt, "High Radiation in Eskimos", *Audubon Magazine*, 65 (1963), pp. 284, 286.
[62] J. Hawthorn and R.B. Duckworth, "Fallout Radioactivity in a Deer's Antlers", *Nature*, 182 (1958), p. 1294; F.R. Fosberg, "Plants and Fallout", *Nature*, 183 (1959), p. 1448.
[63] Quoted in Paul Brooks, *The House of Life: Rachel Carson at Work* (Boston: G.K. Hall & Co., 1972), pp. 243–244.
[64] Hays, *Beauty, Health, and Permanence* (cit. n. 3), pp. 22–29.
[65] Eugene J. Rosi, "Mass and Attentive Opinion on Nuclear Weapons Tests and Fallout, 1954–1963", *Public Opinion Quarterly*, 29 (1965), pp. 280–297.

ROY MacLEOD

"STRICTLY FOR THE BIRDS": SCIENCE, THE MILITARY AND THE SMITHSONIAN'S PACIFIC OCEAN BIOLOGICAL SURVEY PROGRAM, 1963–1970[1]

ABSTRACT

Between 1963 and 1970, the Smithsonian Institution held a grant from the US Army to observe migratory patterns of pelagic birds in the Central Pacific. For six years, the Pacific Ocean Biological Survey Program (POBSP) collected a vast amount of data from a quarter of the globe little known to science, and difficult for civilian access. Its reports were (and remain) of great value to science. In 1969, however, the Program became embroiled in controversy. Some alleged that the Smithsonian, by accepting the military's coin, had violated its own rules governing the receipt of government funds and the publication of research. Recent investigations have pointed to a number of unexplained relationships between the POBSP and the Army, during a period of intense activity in chemical and biological weapons testing. The controversy marked a watershed in Smithsonian-military relations. As yet, its history is incomplete. What is known, however, suggests that the POBSP involved a highly problematic *mésalliance* between science and secrecy during the height of the Cold War. Its gradual unfolding prompts questions of contemporary relevance, that await contemporary answers.

"You have got to keep an eye on the military at all times, and it doesn't matter whether it's the birds from the Pentagon or the birds in the CIA". Thus Harry Truman cautioned an enquiring interviewer in 1962, long after leaving office, but not too long to have forgotten power.[2] Today, this message might be read as the Old Testament lesson for a sermon about real birds, as revealed in the history of the Pacific Ocean Biological Survey Program. The corresponding New Testament lesson is found in the Gospel of St. John, chapter 8, verse 12: "Ye shall know the truth, and the truth shall make you free" – words etched in marble in the foyer of the original headquarters of the Central Intelligence Agency in Langley, Virginia. To some, they seem ironic, amidst the passing

G.E. Allen and R.M. MacLeod (eds.), Science, History and Social Activism: A Tribute to Everett Mendelsohn, 307–337.

trade in secrecy that has characterized relations between academic science, the military and the intelligence services since the Second World War. More secret still – at least until the revelations of the last decade – have been the intentions prompting military sponsorship of civilian research projects. However, it is in precisely this context that a story unfolds, involving one of the nation's most cherished establishments, the Smithsonian Institution, during one of the most difficult periods in its 150-year history. It is a story of changing meanings and misgivings, of confusion between means and ends, of ambition and of conscience. While familiar to many, it remains studied by few.[3] Twenty-five years on, it combines elements of fact and speculation, prompting conclusions that remain conjectural and open-ended.

Students of the Cold War have made us familiar with many moments in recent science where academic and military interests have coincided.[4] In America, the overwhelming importance of military investment in basic research is a matter of public record. Equally, the efforts of the armed services to cultivate scientific knowledge for potential strategic use, are well documented. This phenomenon, we also know, was hardly exclusive to the United States. On the contrary, the creation of what has become known as the "military-academic-industrial complex" has found parallels in most advanced industrial states, both East and West. Thanks, however, to changing rules of access, and to the spirit of openness and reflection gaining ground in many American institutions, it is now possible to study this phenomenon – at least in its American aspect – more intensively and dispassionately. It is also possible to extend our understanding of the military-academic partnership from the list of well known cases involving the physical and social sciences, to less well known cases arising in the biological and environmental sciences; and to consider the consequences of such relationships not only for national laboratories and universities, but also for hospitals, museums, and private research institutions.[5] For this reason, the participation of Smithsonian staff in secret military-oriented research projects came as a shock to the national psyche, raising fears more quickly and more sharply, than any number of standard military contracts to university researchers.

Pacific Birds and the National Interest

In outline, the facts of the case are briefly told: conclusions and verifications will take longer. In the autumn of 1962, representatives of the Army and Navy approached the Smithsonian Institution, with a request to conduct a survey of bird migration patterns over a vast, almost uninhabited, and scientifically little-known area of the Central Pacific (see Fig. 1). What the *intentions* of this military project were, remain conjectural.[6] To the ornithologists at the Museum of Natural History, the project was a godsend. The area to be studied included the Hawaiian, Line and Phoenix chains, the Gilbert and Marshall groups, Baker and Howland islands, Wake Island, French Frigate Shoals, and Sand Island on Johnston atoll, today the site of America's principal chemical weapons incinerator – an area expensive to reach and restricted in access to all

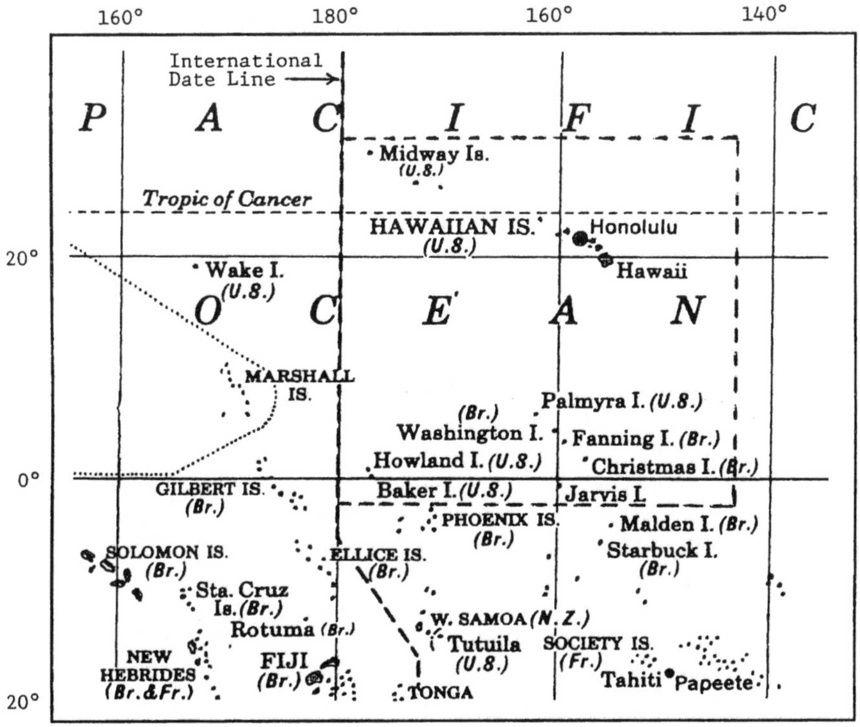

Figure 1. Area of the Central Pacific where the Pacific Ocean Biological Survey Program took place (main study area shown inside dashed line box). From: Hammonds Historical Atlas. 1960. Maplewood, NJ: C.S. Hammond & Co. H-39.

but the most determined commercial and diplomatic interests.[7] To even get to the area, a substantial degree of military and naval cooperation was, and is, essential.

If military motives were needed to justify research in the Pacific, this was nothing new. The European presence in the Pacific had always reflected strategic intentions. Since the mid-eighteenth century, moreover, scientific expeditions invariably combined military and scientific motives.[8] And not for the first time, would the Smithsonian be a willing partner. The Institution prided itself on a long history of service to government agencies, including the military, that preceded the American Civil War and continued well beyond the Second World War.[9] There was nothing especially novel about a military project wishing to involve Smithsonian experts. Perhaps it was unusual to have a contract with the Army Chemical Center at Fort Detrick, Maryland – but nothing more.

As it was, the Army project was welcomed by the Smithsonian Castle. Leonard Carmichael, the Secretary of the Smithsonian between 1953 and 1964, favored the appropriate application of science to national defense. I can find no resistance to the suggestion that the Smithsonian submit a proposal,[10] and no dissent to the decision, on 9 October 1962, to accept a $208,000 contract for a project that was expected to last between two and five years. So began the Pacific Ocean Biological Survey Program (POBSP).[11] The principal applicant, and principal investigator, was Dr. Philip S. Humphrey, the newly-appointed Curator of Birds and Chairman of the Department of Vertebrate Zoology at the Natural History Museum. Between 1963 and 1968, the annual agreement was repeatedly renewed, eventually costing $2.7 million – at least until that time, possibly the largest single contract, from any source, received by the Natural History Museum.

That the program had an obvious military importance, was easily conceded. Overtly, there was a military interest in forming a comprehensive environmental picture about an area which, from the beginning of the Pacific war, had acquired special military importance. During the war, American forces had built a huge forward base on Canton, in the Phoenix chain, and after 1945, the area experienced tests at Christmas and Malden Islands, and later became part of the Pacific Missile Test Range. (see Fig. 2) To establish benchmarks for continuing studies of environmental change in the region was clearly of importance.[12] Moreover, Pacific birds had a significant physical presence. On Midway and other islands, aircraft taking off from and landing on military air bases were routinely threatened, and sometimes damaged by flights of birds. Proverbially huge albatrosses were known to kill themselves by flying into guy wires and communications antennae. The military had, in short, a practical problem.

From the scientific perspective, the survey was enormously appealing. Little was known about the region's migratory birds, or of their seasonal movements. Such a project in descriptive natural history was entirely suited to the Smithsonian's traditional portfolio, and fitted easily into its mission – to better serve the "increase and diffusion of knowledge". For the ornithologists of the Natural History Museum, moreover, it was the chance of a lifetime. From the survey would come direct observation of a number of pelagic species, their breeding, food habits and behavior. Scientists would band birds by place of origin, and trace their movements with the aid of volunteer reporters throughout the region. Avian blood samples and stomach contents would be analyzed to detect changes in environmental conditions. And ship-borne oceanographic and hydrographic observations would track the direction and speed of avian movements in response to ocean conditions and wind patterns.

If, in its scope, the study was well worthy of the Smithsonian, so its task seemed safely routine. The Army appeared to want merely a survey. In the contract documents, no hypotheses were framed, no theories were to be tested. As far as the Smithsonian was concerned, staff were simply to write up their observations, copies of which were to go routinely to the Army, as sponsor. The right to publish was assumed; there was no suggestion that any scientific information would be suppressed.

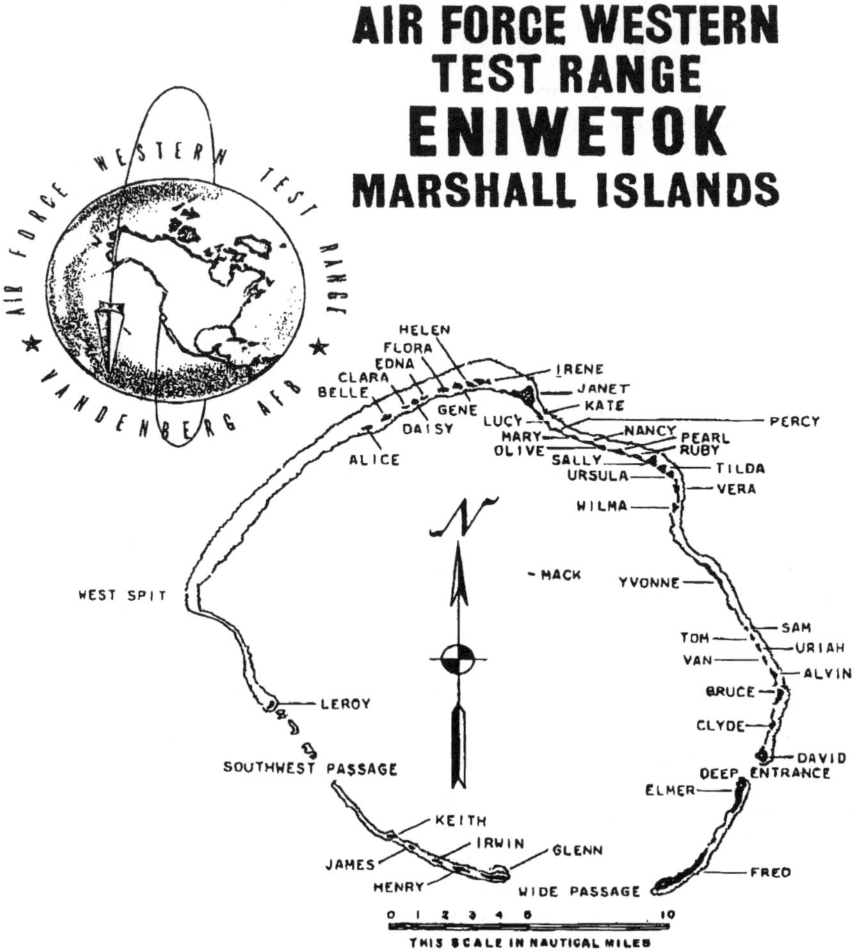

Figure 2. The POBSP selected certain atolls at Eniwetok for study. From: R. Crossin to J. Bushman and Robert Pyle, map attached to letter 1/29/69. Box RU 245, Smithsonian Archives.

At the outset, the program was based at Humphrey's office in the Natural History Museum in Washington, DC. Humphrey was often away on other work, including a Rockefeller-funded Smithsonian project studying viruses in bird populations located at Belem, in Brazil. In 1967, he left the Smithsonian to become Director of the Natural History Museum and Chairman of the Department of Zoology at the University of Kansas, but remained on paper the principal investigator. In his absence, Smithsonian naturalist Charles Ely ran the program, hired and promoted staff, and kept work underway.[13] In 1964, a "field office" was set up in Honolulu, under Richard Pyle, in association with the University of Hawaii, to provide logistic liaison with the Navy. The scientists had continuous use of three Navy auxiliary vessels and other small ships. "There is nothing really particularly hush-hush about our program", Humphrey assured the Bishop Museum in Honolulu, which generously lent its facilities.[14] Nonetheless, all Smithsonian personnel were required to have security clearances. The reason was never formally stated, but most assumed that it merely reflected the regulations surrounding the restricted military and naval areas they would inevitably visit.

The Program's newly-recruited field staff, fresh from graduate school, took to their work with enthusiasm. On their first operation beginning in August 1963, the Navy took them on a series of 15-day cruises from Pearl Harbor, enabling them to survey all bird life visible to radar and the human eye, and to collect oceanographic data over a 50,000 square mile area around the Leeward Islands. There was no explanation given for this exercise – code-named Project STARBRIGHT – beyond a reference to its importance for tracking "the seasonally changing conditions both physical and biological of the oceanic environment".[15] Records suggest that the naturalists sought – in vain – to be informed of the military motive in collecting this data.[16] Beginning in December 1963, and continuing for six years, staff kept up weekly accounts in an internal diary, called "Droppings from the Eagles Roost". But, staff were reminded of the need for security.[17] In a memorandum, Charles Ely cautioned all concerned to "be careful in discussing the project with outside people. No one wants to be branded a security risk as the result of an idle conversation". He continued:

> It should be enough for our men to know that they are securing data for the Division of Birds, SI, and that the military is interested in learning the ecology of areas in which they may someday be committed. Military and ecology are both nice vague terms. Use of military facilities and ships is easily explained because they are the only ones available for these areas.

In any case, it was not the scientists who were the worry, Ely insisted, but rather "other groups", whose presence or interest in the matter remained unspecified. However, no one was afterwards to mention the live bird shipments, and everyone was to forget the word "STARBRIGHT".[18] From April 1964, staff were called to Ft. Detrick and given special inoculations, ostensibly to avoid contracting diseases in test areas. Apparently, however, no one was told, *which* diseases. In fact, staff were told as little as possible, and then only on a "need to know" basis. From her office at the Smithsonian, program secretary Jane Church – an "ardent bander", and by common consent, "den-mother to the Pacific fieldmen"[19] – complained about having to keep classified documents, which had to be inspected every four months.[20] But what these documents contained, is not known.

In 1964, at Humphrey's request, the Army agreed to widen the survey the following year to include all wildlife on all islands in the region at all times of the year; and eventually, botanical studies as well. The survey then extended beyond the Hawaiian sector, to include Johnston Island, the islands in the Phoenix Group, and the Aleutian Islands of Alaska. Reports were sent routinely to the sponsoring Army Chemical Laboratories at Fort Detrick, and also to the Army's Deseret Testing Center, at Fort Douglas, Utah. In 1964, senior staff from both Ft. Detrick and the Smithsonian met in Salt Lake, and apparently also visited Deseret, to discuss field methods and outcomes.[21] In certain respects, the Smithsonian seems to have provided "resident staff" for the Army in the Pacific.

Not all the going was easy. Each year, the work of banding and description was much the same – indeed, too much the same, some complained. Civilian and military priorities were not always identical; communications between Washington, Honolulu and the field were not always easy; and life on board a small Navy fleet tender could be monotonous and uncomfortable. To be kept for weeks on a lonely, sparsely vegetated island, in searing temperatures, without shade, banding thousands of sooty terns – the most abundant tropical bird – alive with ticks and lice, was not the exotic life featured in Pacific holiday brochures.[22] Sometimes, given the circumstances, tempers flared.[23] But practical drawbacks aside, staff bore their task stoically and morale was generally high. They were reminded that the project existed by courtesy of the Army, which included some who thought there should be no publication at all.[24] To most, the burden of secrecy – whatever its justification – remained a distinct nuisance, especially where the purpose of a given exercise was not obvious.

In the Kaneohe area, for example, staff were asked to feed mosquitoes upon captured birds, and collect sera samples, which were sent to Ft. Detrick. But no explanation was given for this, and no details were reported in the scientific press. The "funding agency", Ely reported, "felt that the project is secret and that any publication will provide information to our enemies (real or imagined)".[25] Secrecy was particularly tiresome when the Army requested staff to sail out to study birds on Baker Island – a one-square mile atoll 2000 miles southwest of Hawaii, where, as it transpired, there were many cats, but no birds at all.[26] But Ely took issue with his critics. "We are constantly being criticized for our secrecy", he wrote a colleague,

> but some SECRET clearance is necessary for many of the islands where we work and for many of the ships and military groups with which we are necessarily associated. I am willing to discuss any non-classified parts of the project with any interested biologist. However, I do not feel that we need to apologise for accomplishing work that other groups have not or could not accomplish. Our at sea study, for example, is absolutely unparalleled in the history of ornithology.[27]

By this view, the end justified the means. "I am going to see if we can get rid of the two secret documents we have", wrote Jane Church, the secretary, chaffing at Ft. Detrick's routine visit to her office: "If we can, this joker will only have to come to visit us once a year instead of every four months".[28] But protest was futile.

From 1965, the Program issued a series of designated Army Reports (reaching No. 91 by 1969), as well as scientific reports intended for the open literature. Some reports, sent to the U.S. Army Materiel Command, were marked "Secret;" all were marked "Confidential".[29] The director's secretary also kept quarterly status reports, which amount to an official history of the Program. Beginning in September 1965, the staff also issued a small, popular newsletter – the *Pacific Bird Observer* – edited (until issue No. 6, February 1967) by Tina Clapp. Inspired by Roger Clapp, one of the Program's earliest and most resourceful members,[30] this regular publication was intended for volunteer bird watchers, among others, and carried items of ornithological interest, distribution maps, and photographs of the Program in action. An early editorial made a point of emphasizing the economic value of bird-banding for the Pacific tuna industry, which owed its survival to the careful monitoring of seabirds as indicators of fish movements.

Banding was the Program's most visible achievement. Over 400,000 birds were banded by 1965, and a further 750,000 by 1968. These included Masked Boobies, Great Frigatebirds, Shearwaters and several other major species – specimens of which can be seen in the Natural History Museum today. Many of these magnificent birds showed a Coleridgean resilience to human contact. In the Marshalls, for example, they seemed less affected by nuclear and missile tests than expected. On Midway, human beings were far more dangerous to the birds than vice-versa. Navy personnel clubbed sooty terns for sport, and brought dogs from Sand Island to run through colonies of incubating albatrosses.[31] On Sand Island (Johnston Atoll), where there were 300,000 indigenous sooty terns, naval personnel drove jeeps on joy rides through the colonies during their incubation period.[32] In the Phoenix group, the U.S. Air Force, which cultivated a more "ecological image",[33] invited a consultant from the Fisheries and Wildlife Service to set out guidelines for the treatment of birds around its base on Canton. On Howland and Baker, both administered by the Department of the Interior, cats, which were introduced to keep rats at bay, had predictable consequences for bird life. Scientific interests eventually helped shape environmental concerns; but in the early stages, military interests remained paramount.

In 1966, the Army agreed to continue the project to the end of 1967, but with a changed emphasis – linking the survey to marine mammals, deep sea fish, and the wider ecology of atolls and islands – in the Eastern Pacific, including the Californian and Mexican shorelines. By 1966 staff were no longer required to be immunized against unspecified diseases, for reasons that also remained unexplained. As to the military significance of it all, Humphrey, by then on his way to Kansas, says he was kept in ignorance.[34] When asked to take a study group to Wake, he asked Jane Church to write the officer in charge, in hopes that "maybe then he would give us a little more information about what he is up to".[35] There is no reply on record.

As 1967 drew to a close, Humphrey reported on the Program's progress. From 2500 man-hours spent on thirty islands had come data for thirty publications, transforming scientific knowledge of the region. Looking forward to a renewal of the grant, he promised a major publication on seabird distribution.[36] Indeed, by this time, the Program's emphasis had shifted from what Humphrey described as an exploratory and descriptive phase, into a "far more complicated one, in which we were more concerned about correlations and interrelationships, and in which we had to deal with vast quantities of data".[37] At the end of 1967, the project was renewed for 1968, and extended to include the observation of radiation effects on the reproductive behavior of sooty terns on Eniwetok, in the Marshalls. Again the purpose of these correlations was not explained.

Between January 1967 and June 1968, the Program added another twelve survey cruises, amounting to over 10,000 miles of observations. The contract's administration was transferred from the Chemical Corps to the newly-created Army Research Office, also at Ft. Detrick. However, from early 1968, the "funding agency" seems to have expressed diminishing interest. Apparently, its curiosity had been satisfied. But research continued, and when Army funding ended in June 1969, a no-cost extension was granted to 30 June 1970, to give time to write up. Publication was destined to continue for years afterwards.[38] Samples and specimens remained with the Deseret Test Center. What became of them – or what eventual use was made of them – has not been disclosed.

The Survey and its Successes

In the history of Pacific ecology and ornithology, nothing of this kind, or of this scale, had ever been attempted – nor, it is said, will ever be attempted again.[39] Beginning with only four naturalists, the project eventually employed over two dozen, many of whom went on to senior positions in academic and government science. By August 1967, in over 100 cruises, they traversed more than four million square miles of ocean, as far north as the Pribilof Islands – an area almost the size of the continental United States – made over 200 visits to islands and atolls, and banded over a million sooty terns. By 1969, they had sent hundreds of bird specimens to the Smithsonian; and an unknown (but very large) number of blood and stomach content samples to laboratories at Ft. Detrick and Deseret. Even on a marginal cost basis, the Program's expense to the Army, above and beyond the Smithsonian's contract, must have been colossal.

At a descriptive level, the Program was a blockbuster success. It amounted, in the words of one knowledgeable observer, to "a quantum leap forward in scientific knowledge of this enormous area".[40] From 1969 onwards, the *Atoll Research Bulletin* published a series of reports on the natural history of the vast region, as well as on seabird migration and behavior.[41] By 1972, Program staff had published over forty-five papers, including reports on French Frigate Shoals, Pearl and Hermes Reef, Gardner Pinnacles, Necker, Nihoa, Lisianski and Laysan Islands in the Northwestern Hawaiian chain, as well as on Johnston, some of the Phoenix chain, Kwajalein in the Marshalls, and American Samoa.

There were, perhaps, some surprising gaps. Among the Leeward Hawaiian Islands, for example, Midway, located in a nuclear test area, was notably missing from the published reports, although there may be unpublished material still in the Smithsonian. Of the equatorial islands in the south Central Pacific, reports on only Caroline and Vostok were published. For some gaps, there are perhaps easy explanations. On the Phoenix chain, for example, the grant expired before the report could be finished. But the wealth of data collected on Wake Island, a nuclear storage site, merited more than the limited report that emerged.[42] The natural history of Johnston was described at length,[43] but while surveys were also prepared for Baker and Howland,[44] it appears that these were not published.[45]

In February 1969, secretary Jane Church, chose to remove from her files all reference to the Program's military contacts, and encouraged her Honolulu station to do likewise.[46] Many aspects of the Program could be pursued more easily if Ms Church's advice had not been so carefully followed. As it was, her directive succeeded; nothing of the military connection remains in the Program's Smithsonian archives. To track correspondence through military records is extremely difficult, not least because of the administrative reorganizations that took place in the Army during the mid-1960s, which have severely interrupted the archival record.

Unless evidence emerges to the contrary, however, certain assumptions can be made. First, we can surmise that the contract – DA 18-064 – was in fact initiated by the Army Chemical Corps at Fort Detrick, which was involved in biological weapons testing from at least 1942. We can further surmise that, somewhere at issue, was the mode of transmission of rare diseases, carried and incubated in the blood of birds and mammals, and transmitted by mosquitoes and ticks. Blood samples were taken to determine what diseases were carried, while banding, used in conjunction with incidence data of diseases in different localities, would help determine movements of disease carriers. That was the "defensive" aspect. For the "offensive" aspect, there was the possibility – remote as it may seem – that migratory birds could deliver toxins to enemy territories. Scientific evidence that this could be done – or done reliably – remains problematic. But the close relationship between "defensive" and "offensive" research was an established fact. After the Second World War, Ft. Detrick co-sponsored studies of possible "germ weapons" with the Department of Agriculture, the Public Health Service and the Army Medical Corps.[47] In 1959, this work assumed an offensive profile, when the Chief Chemical Officer was instructed to prepare a five-year program for the development of anti-crop weapons. Thereafter, the Kennedy administration devoted $4 billion to basic research on a range of offensive chemical and biological weapons.[48] Annual expenditure on CBW increased from $10 million in the 1950s to $352 million in 1969. By the mid-1960s, precisely at the time the Smithsonian contract began, Ft. Detrick was at the center of a web of CBW research and development contracts placed with 300 universities, research institutes, and corporations.[49]

After August 1962, when the Chemical Corps was discontinued, the administration of the Smithsonian contract was passed to the Army Biological

Laboratories at the Army Chemical Center and under the Army's new Materiel Command. At neighboring Edgewood Arsenal, the former Chemical Corps laboratories were at the time folded into a new Chemical Biological Radiological Agency and put under Munitions Command. They also formed part of the Army Chemical Center.[50] The Smithsonian sent its reports to the Materiel Command and to this Center. The archives of this Center, and of the new Chemical and Biological Defense Command that inherited the laboratories, are restricted. What follows must, subject to further documentation, remain in the realm of the speculative.

The Army, having commissioned the Smithsonian Program, and renewed its contract, may have had the support or assistance of the Special Operations Division (SOD) of the Chemical Corps' research center at Ft. Detrick. From 1952 onwards, this Division worked closely with the Technical Services Section of the CIA, establishing what it called a special security environment for "the maintenance, assessment and evaluation of a designated balance of biological and chemical disseminating systems for operational readiness".[51] This work produced stockpiles of lethal agents, including animal viruses. Designated MK-NAOMI, the program continued the work of MK-ULTRA, the secret enterprise begun by Allen Dulles during the Korean War for the investigation of behavior modification drugs for possible offensive use.[52] There is no confirmed evidence that any of these drugs were ever deployed. However, from the late 1950s onwards, CIA-funded chemists worked on studies of poisons found in natural products in the South Pacific and the Amazon, and added these to their biological arsenal. By the 1960s, the U.S. had at least ten different biological weapons available for use, reportedly including stocks of botulinus, staph. enterotoxin, and brucellosis, as well as the Venezuelan equine encephalomyelitis (VEE) virus, which can immobilize a person for two-to-five days.[53]

During the same period, studies of dissemination – "delivery systems" – of biological weapons are known to have formed part of the research portfolio of both the CIA and the Chemical Corps. Among the more notorious projects were those to develop so-called "nondiscernible microbioinoculators" – the "dart gun" was one example – and aerosol sprays; along with the potential use of insects (including ticks) and other organic agents. Scientists working for MK-ULTRA also trained animals – guided animals – for possible offensive purposes, continuing a line of research that dated from at least the Second World War, and which also featured in the postwar history of behavioral psychology.[54] Collaboration between the Army and the CIA offered a window of "easy passage from defensive to offensive applications".[55] The potential use of birds as carriers of toxins could easily have been among the many so-called "what if?" projects – global, encompassing, descriptive studies – conducted by private foundations and other institutions with CIA and military support.[56] To "know everything" about the world – at whatever cost – was a luxury, but one prized by governments and intelligence agencies since the Ptolemies endowed the ancient Library of Alexandria.

On April 9, 1973 – the day after the White House Counsel, James Dean, began cooperating with the Watergate Prosecutor – President Richard Nixon told his aide, Bob Haldeman, that it was time to erase the White House tapes. The tapes survived, but the Technical Services records did not. On the orders of the CIA's Director, Richard Helms, the Technical Services Section shredded many of its files in 1973. Only a small fraction of the CIA's interest in this field may ever be made public. What links – if any – existed between the CIA, the Chemical Corps, and the POBSP are unknown. At the time, accusations were rife. Today, a quarter century later – and until an authoritative history can be written – we can only suggest a range of possibilities. But whatever those possibilities, we do know that before the Program ended, it was destined to be tested by public opinion, and so raise questions that concern us to this day.

"Disclosure"

The serenity, if not the secrecy, that had surrounded the operations of the POBSP for over six years was abruptly dispelled on Tuesday, February 5, 1969. The occasion was an NBC-TV broadcast, the second of a new series of "First Tuesday" documentaries, a "magazine format" contemporary affairs program. That evening, the discussion included a segment on secrecy and science, and focused upon chemical and biological weapons research.

Coming at the end of the 1960s – a decade of violence at home and abroad – the Program coincided with protests against the military-industrial complex, and classified links between academic institutions, the military, and the intelligence services. Coupled with increasing concern about America's use of herbicides in Vietnam, the Program's very existence rendered the Smithsonian a vulnerable target.

On December 9, 1968, Washington correspondent William E. Small, writing in *Scientific Research*, alleged that the Smithsonian was deeply committed to studies in Brazil and the Pacific, dealing with the mechanisms by which rare viruses and blood parasites are transmitted from birds, mammals and insects to man. Sidney Galler, whom Small interviewed, admitted that such studies existed, but insisted that Smithsonian researchers were "free from pressure by the military, conducting research of their own choosing, just as any scientist would under a similar agreement".[57] "What they [the DOD] do with the data", Philip Humphrey disarmingly added, "I don't have any idea. We just send them copies of our results". A subsequent letter to the editor by Philip Siekevitz of the Rockefeller University condemned these "ingenuous" replies, and saw the issue as one of "adventuristic, imperialistic American military policy, the same as is causing the disaster of Vietnam"."I for one", he added, "consider it about time that we scientistsget down to really thinking about our roles in society".[58]

At a time when the war in Vietnam was being fought on evening television, it was perhaps just a matter of time before television also brought the Pacific birds into the living room. NBC, after well over six months' preparation, opened its documentary with compelling footage of the University of Utah at Salt Lake, only eighty miles from the Army's Dugway Proving Ground, which it cited as one of fifty academic institutions taking part in secret nerve gas experiments. Spokesmen for the University explained that its research was prompted by local fears of pathogenic leakages from the nearby weapons testing site. However, they also admitted that, between 1952 and 1969, the University had received Army contracts worth $480,000 a year, and had employed forty scientists to study the role of "vectors" – carriers of pathogens – and patterns of transmission in a range of diseases, including tularemia, rocky mountain spotted fever, bubonic plague, anthrax and parrot fever. Some of these diseases were endemic to local wildlife, but others were suggested to the Utah researchers by the Army Chemical Corps. The University claimed that such work, while classified, was not secret; the Army retained vetting rights over publication – but these, the University insisted, were rarely if ever exercised.[59]

Such reassurances were not reassuring. On Sunday, March 17, 1968, the University relayed to Dugway reports from a local livestock producer that 3000 of his sheep had died in Skull Valley, adjacent to the Proving Ground. The cause of their death was never ascertained, or at least never announced; but popular opinion held that nerve agents had been used, and had drifted in the wind from the test area. For the Army, the ensuing publicity was disastrous. Almost immediately, Congress began tightening controls. Open air testing of bio-weapons was restricted, and even sea dumping of chemical munitions was stopped.[60]

Appearing only eleven months after these events, the NBC broadcast revived memories of a similar kind.[61] In constructing a catalogue of conspirators, NBC's Tom Pettit reported that the Smithsonian had for years been a "cover" for Army tests, both in the Pacific and also in Brazil, where Humphrey also directed a project. On air, Pettit interviewed Robert L. Standen, a school teacher who had worked with the POBSP as a field biologist between February 1964 and July 1965. During this time he had taken part in studies on Baker Island. On nation-wide TV, Standen said a test had taken place involving a "biological carrier"; but refused to say where. The NBC supplied the information that this had taken place on Baker during six weeks in the northern Spring of 1965 – its purpose being to enable Army, Navy and Air Force personnel to see how animal vectors would behave in a tropical environment. "No germs were involved.[62] In effect it was a checkout of an animal delivery system for CBW", Pettit said. The allegation was unconfirmed, but stunning. Seizing the political moment, Senator Joseph Clark (D-Penn.) claimed that the Smithsonian had been a screen for Army efforts to locate a satisfactory CBW test site and a cover for "ultra-secret" tests on potential animal vectors.

On the face of it, the evidence was suggestive. If the Army were looking for a remote place to conduct experiments (including aerosol tests) without fear of wider contamination, a birdless island in the Central Pacific in an established test area. far from anywhere else, was ideal. Some months earlier, scientists had speculated on the possibility of finding such an island, uninhabited by man and unvisited by birds. Uninhabited, unvisited Baker and Howland Islands, might be places to start.[63] If there were bird-life, or any other life in the vicinity, it was important to learn about it. At the same time, if "guided birds" could be developed and tested there, that would clearly be of interest to the Pentagon, as well as to a potential enemy.

The Army admitted conducting tests using birds at Baker.[64] There is also speculation that birds from Johnston were tested at Ft Detrick, using particle aerosols of VEE virus, although whether there were also field tests is not known. Whether Smithsonian staff were complicit in these experiments; or if so, whether they, their director, or anyone else at the Smithsonian were informed of the significance of such experiments, are altogether different questions. To all intents and purposes, the Smithsonian had embarked upon a project of bird-banding and oceanographic recording in the interests of science. It had not asked, nor was it informed, of any military intentions. Had it been duped?

Such nuances were lost in the subsequent furor in the press. On February 5, 1969, *The Washington Post* announced that "Smithsonian Bird Research [was] tied to Germ Warfare Study",[65] and quoted noted author and CBW expert, Seymour Hersch, in citing the need of the Defense Department for an isolated island free of birds to conduct tests, which would otherwise spread diseases around the world. On the same day, *The New York Times* announced that the Senate Foreign Relations Committee had "obtained information suggesting that the Army, under the guise of a bird study by the Smithsonian Institution, is looking for a remote Pacific site to conduct experiments in chemical-biological warfare".[66] Overnight, the "Pacific Project" was transformed from an innocent file title into a suspicious code-word.[67] The press announced that the matter was to be reviewed by the Senate Subcommittee on Disarmament, soon to begin hearings on bacteriological weapons.

Meanwhile, several lobby groups, including the Scientists' Committee on Chemical and Biological Warfare, were holding demonstrations against the use of herbicides in Vietnam. The Federation of American Scientists called for a total ban on bacteriological warfare, and 1400 scientists at MIT heard Noam Chomsky and others call for a nation-wide, anti-war "research strike" on March 4, 1969.[68] Within the scientific community, there were angry divisions. At MIT, George Wald, the Harvard Nobel Prize-winning biologist, savaged the American Institute of Biological Sciences for co-sponsoring a conference at Ft. Detrick in 1968.[69] On the other hand, a conference at Stanford (billed as "convocation, not confrontation") – which coincided with the MIT meeting – heard Merrill J. Snyder, a microbiologist from the University of Maryland, praise Ft. Detrick for its good work.

Whatever the merits of Ft Detrick and its work, the publicity was damaging for the Smithsonian. From the day after the broadcast, the NBC's allegations produced a flurry of denials. The Army denied any ulterior motive behind its sponsorship of the six-year, \$2.7 million study. But what, then, were its motives? Confusion was further compounded. Studies at Baker, the Pentagon said, were to investigate the natural distribution of diseases among migratory birds. No, the DOD Public Information office said they were to investigate the problem of dealing with birds flying on military landing strips. In either case, they had nothing to do with chemical or biological weapons.[70]

George Watson, a member of the Program's staff who also held research funds from the Army for other studies of migratory birds – and who perhaps spoke for some of his colleagues – retreated into the bliss of ignorance, when he quixotically assured reporters that "I am naive enough *not* to know where my money comes from".[71] But rumors spread and implicated the Smithsonian birdmen in a sinister conspiracy. Why else should bird watching, whether in Brazil or the Pacific, be financed by Fort Detrick?

When approached by the press, the Castle categorically denied any connection. "The Smithsonian has never engaged in any chemical or biological research for the armed forces or anyone else", Sidney Galler insisted.[72] It emerged that Galler, Assistant Secretary for Science, had come to the Smithsonian in 1965 from the Office of Naval Research (ONR), where he had been head of the Biology Branch (1950–1965). At the ONR, Galler had overseen projects related to "environmental warfare", and had even helped Humphrey get a contract for work similar to that of the Pacific Program. "I wasn't interested in the germs", Galler was reported as saying; "I was interested in the animals and their behavior that could be utilized by an enemy to carry the germs". "Some Pacific birds", he added, can "migrate tremendous distances and reach target areas with about 97% accuracy". At the same time, he said, he was not at first aware that the Pacific Program was classified.[73] In August, 1968, when the contract was renegotiated, he obtained the removal of the clause that required prior military approval before publication.[74] It is not clear whether this modification – vital, according to Sidney Galler, for "developing a successful relationship between a scientific organization and a sponsoring agency"[75] – made a significant difference to the outcome.

For two weeks, the issue festered and fermented, and became the talk of Washington's conversation culture. Within the Smithsonian, two urgent meetings of the Senate of Scientists took place in the Natural History Museum. The Senate had been established in 1963 to represent the professional concerns of the research staff. Loosely modeled on the idea of an American university Senate, it reflected discontent with poor communications across the Mall, and was intended to function, in its own words, as a "trouble shooter and source of collective opinion outside normal administrative channels".[76] For some time, it had limited its recommendations to such matters as parking, library services, data processing, and publications. Now, for the first time, there was an issue that transcended parking space.

Clifford Evans, as Chairman of the Senate (1968–1970), called upon Galler to explain the sequence of events that had occurred since the article in *Scientific Research*. It transpired that Pettit had apparently telephoned the Smithsonian some weeks before the television program was aired, and had been sent a copy of the contract. The issue in Evans' mind, was whether there were any circumstances in which the Smithsonian should abridge a researcher's right to seek funds.[77]

The answer was of great importance to staff and secretariat alike. On 10 February 1969, Leonard Carmichael, by then retired from the Smithsonian, received a call from S. Dillon Ripley, the distinguished ornithologist and wartime Far East OSS analyst, who had succeeded him as Secretary in 1964. In an *aide-memoire* prepared after the event, Carmichael recalled that the Army contract had stipulated only that reports be submitted to the Army; scientific papers were to be submitted to the Army before publication only to ensure that "they did not provide non-public information concerning military installations".[78] To the best of his memory, no report was ever challenged or paper altered. By implication, the Program was not "secret".

This was not enough to satisfy the Senate of Scientists. A second meeting of the Senate heard Galler deny he had ever seen the contract and its clause concerning classified research, and then say that he had since removed the offending clause in the last renewal. He further announced that neither he nor Richard Cowan, Director of the Natural History Museum "had ever been given full access to information about the Pacific Bird Project ... ".

Opinion was divided on whether Defense funds, however badly needed, should be *ipso facto* excluded, or whether such funds should be accepted on a case by case basis. The Senate debated the issue, resolving to "keep its cool for the present", and to deplore the lack of communication between staff and administration that had allowed "this type of thing" to happen.[79] A statement was prepared for Ripley emphasizing freedom of publication and rejecting involvement with classified research; this, Galler later qualified, to distinguish between institutional and individual engagements.[80] On February 9, Humphrey was called before the Senate of Scientists to explain his work at Belem and in the Pacific. At the Senate's insistence, Humphrey agreed to terminate the contract in July 1969, and to explore ways of removing DOD funds from Belem. A tape-recording of the conversation was reportedly kept for Ripley.[81] The following week, Nora Beloff, writing for *The Observer* in London reported that the case would surely now go to Capitol Hill and be reviewed by the Senate Sub-committee on disarmament, which was to open hearings in early March.[82]

Against the allegations aired on NBC, the Castle issued a press statement explaining that the Survey was a "basic research program consistent with the Institution's traditional scientific pursuits".[83] Its purpose was simply to accumulate data on the plants and animals of the region and on climatic and oceanographic variables "to permit broad ecological conclusions to be drawn". Its funds, it explained, came from the Army Research Office. Ripley, who claimed to have been previously uninformed about the Survey, defended it as fascinating – a "wonderful project from the scientific point of view – the fulfillment of a dream".[84] No mention was made of Fort Detrick, the Chemical Corps, or for that matter, the distribution of diseases.

The issue came to a head on February 21, when Philip Boffey, a respected reporter for *Science*, killed the story – or so the Smithsonian thought – in a major feature article.[85] The Castle unequivocally denied that it had been "an unwitting dupe or cloak" for biological warfare research, and Boffey found no evidence to the contrary. Boffey dismissed the NBC's allegations as "marred by the use of loaded words and guilt-by-association reasoning".[86] Standen, and everyone else on the Program team, had been routinely excluded from military discussions, whether in the field or in Washington. Senator Clark's testimony was dismissed as factually unsound. Worse, *The New York Times* (of which Boffey later became Science Editor) had apparently not confirmed its sources. No one had taken the precaution of interviewing senior staff at the Smithsonian. The attack was, apparently, just a press beat-up. Boffey easily demolished it, and indirectly endorsed the Castle's plea of innocence. He limited his own criticisms to the observation that, in such heady times, "an institution that wishes to maintain an unblemished reputation can't merely follow its traditional *mores* – it must consider the changing values of the public as well".[87]

On February 26, the Senate of Scientists heard Ripley say that the project had "no more scientific merit", and should be discontinued, and confirmed the extent of Humphrey's role in it.[88]

In its leader on March 3, 1969, *Barron's* believed the Smithsonian had been "cleared" of any guilt, even by association.[89] And there the story might have ended. But it did not. On the contrary, it became a *cause célèbre*.

On March 5, 1969, Congressmen in closed session heard reports that the United States was spending $300–350 million a year on CBW research and development – a figure too low in the estimation of some members of the House Appropriations Committee, but too high for Rep. McCarthy (R – New York), who demanded less secrecy surrounding such research.[90] From the Army agency that commissioned the work, came no comment at all. Instead, the Defense Department re-issued a statement, re-stating that the Survey was designed, innocently enough, to study the natural distribution of diseases, as they might affect the health of servicemen and civilians abroad; to study the impact of U.S. installations on bird populations; and to study the persistent problem of birds colliding with aircraft on tropical airstrips. These were valid reasons, even if no one believed that they exhausted the list.

Nowhere in the original Smithsonian contract was the study of animal diseases mentioned. However, in a statement to Senator William Fulbright and the Senate Foreign Relations Committee, and copied to the Secretary of the Smithsonian, the Defense Department explained that it was one of "many agencies and governments" interested in the problem of diseases carried by migrating birds and animals. Similar studies were being supported by the Rockefeller Foundation in Brazil, by the United States Public Health Service in Mexico, and by the Gorgas Memorial Laboratory in Panama. These sought data on the susceptibility of hosts, and on the life history and migratory habits of such animals. The American military presence in the Pacific had, the Department stated, raised concerns about bird populations in affected areas; and the Smithsonian had submitted a proposal for basic research to supply the data. After six years, the Program had shown that U.S. military activity had not in fact reduced bird populations. Its studies had thrown new light on migratory habits and, intriguingly, had found that certain species were susceptible to certain diseases, and to at least one human disease.[91]

Into the melee thus came the suggestion that the initiative for the Program might have actually come from the Smithsonian, and that the distribution of disease was a key element in its inception. But the Smithsonian had stated repeatedly (and correctly) that it did not do research on disease; the Program merely allocated blood samples and specimens. What the Army did with them, was the Army's concern. However, the Defense Department had stated unequivocally that it "had not been studying birds as potential carriers of biological warfare agents".[92] But what, then, were these studies for? Someone was being economic with the truth.

The Smithsonian's Albatross

Biological vectors come in two categories – communicable (e.g. smallpox) and non-communicable (e.g. anthrax and tularemia). Whether the Army conducted field tests of such biological vectors is difficult, if not impossible, to confirm.[93] There is speculation that the Army did conduct tests on Baker with uninfected mosquitoes and that cats were also put on Baker for experimental purposes. But the results, if any, of such alleged tests are unknown. However, in at least one case we know the Central Pacific was used as a biological testing area. It has been revealed that in the summer of 1968, open air tests were conducted downwind of Johnston Atoll, 1600 km south west of Hawaii. These were part of a series that reportedly began as early as 1964,[94] and would therefore have overlapped with the Smithsonian contract. Following the model of nuclear tests, the tests involved a fleet of ships – in this case, anchored upwind from a set of barges loaded with Rhesus monkeys. The tests consisted in the dissemination and monitoring of small amounts of a fine "bio-powder" (the nature of which has not been disclosed). Following the "laydown", over half the monkeys died. These tests were so "successful" that William Patrick, a noted "bio-weaponeer", has concluded that tactical use of similar "laydowns" could devastate whole cities – more effectively, it was said, than ten-megaton hydrogen bombs.[95]

At the time, all this was secret. However, even a rumor of secret "tests" was enough to alarm the region and awaken local concern. On March 6, 1969, the Hon. Patsy Mink (D – Hawaii), a leading voice in the House of Representatives, asked the Secretary of the Smithsonian to clarify the project's implications for the Hawaiian Islands. Ripley assured her that "accusations made to the effect that the Smithsonian is conducting germ warfare research or even that we are serving as a cover for the Department of the Army in such research are unwarranted".[96] Behind the scenes, however, the Castle went into damage control. Ripley issued a sharp notice, emphasizing that "responsibility for public information rests with the director of the Office of Public Affairs. All information issued to the press and other communications media should be issued by or through this Office".[97] Richard Cowan reined in the Program by reducing Humphrey's role, curtailing fieldwork, and seeking to wind up the writing of final reports as quickly as possible. Humphrey was then acting for the Smithsonian in three capacities: as a technical advisor for a rain-forest exhibit, as principal investigator of the POBSP, and as liaison with the Belem project. When his status as an "absentee landlord" was criticized, his role was redefined, and he continued as research associate in the Department of Vertebrate Zoology for the completion of the final reports. After July 1, responsibilities for Program data was transferred to George Watson, who was required to inform the Secretary and the Museum of any further work or grants received.[98] Cowan also asked that the administration of Humphrey's Brazilian research be moved from the Smithsonian to Kansas.[99] Some suggested that either the National Science Foundation should take over the work,[100] or that it should just be separated from the Smithsonian.[101] There was the example of Utah's Department of Ecology and Epizoology, which formed a private company (in this case, Ecodynamics, Inc.) to bid for contracts from the Deseret Test Center.[102] The abrupt about-turn on the part of the Castle was a victory, of sorts, for the Senate of Scientists. Never again would intra-Smithsonian commitments be so poorly coordinated. And the Senate, thus empowered, went on to defend staff who took up positions of conscience during the Vietnam moratorium.[103] Humphrey was apparently eager to continue the project under the umbrella of the Smithsonian, but only on his terms, and with continued support from the "funding agency".[104] However, he must have known that this would require Ripley's agreement – which, since the NBC program, was at best an open question.[105]

Further military support of the Program was, in any case, becoming less likely, as the political climate chilled. Ten years earlier, in 1959, the House of Representatives had been assured that biological agents were "nothing new and mysterious", and that micro-organisms that might be used as weapons were the "same diseases which public health measures and personal hygiene traditionally aim to thwart".[106] A decade later, such attempts to domesticate the question had lost their appeal. Bio-weapons research was intended "to minimize the possibility of technological surprise".[107] Yet, it had become increasingly clear that other countries, including the poorer countries of the world, could also develop "germ warfare" techniques, leaving the United States at an unprecedented disadvantage. In April 1969, the Secretary of Defense was asked to explain to the American people the benefits of the policy of Nixon's Administration. Continued support for CBW research was certain to meet resistance in Congress.

As far as is known, the Smithsonian Archives do not reveal whether the Institution wished the project to continue, or if so, whether the military would have agreed; but in December 1969, John Bushman, the Smithsonian's Army liaison, reported that his Command had said "no" to further funding.[108] In 1969, pressured by public opinion, the Senate's Foreign Relations Committee had begun a year-long investigation into the Nixon Administration's chemical, biological and radiological warfare (BR) program, and into its practice of co-opting academic scientists into classified research. In November, President Nixon reaffirmed a "no first use" principle for chemical and incapacitating weapons, and renounced American offensive research for biological weapons.[109] By Executive Order, U.S. Government agencies were required to destroy all bio-weapons technology, including "any specifically developed organism or toxin and information which would assist in developing such weapons".[110] This reportedly reduced the Pentagon's $22 billion R&D and procurement budget by $2 billion.[111] At the same time, a "thorough review" of "all reports, studies and reference material was conducted", and less than 20% of archival material was retained.[112] Absence of evidence is not evidence for absence, but without documentation, much of the story will forever remain unclear.

There, at present, the historical matter rests. The POBSP was officially wound up in June 1970. The last year included an extension to write up final reports, and to catalogue specimens – some perhaps destined for Utah, but most for the Smithsonian. During this final year, Jane Church, the Program's ever-efficient secretary, kept to the letter of the law, insisting upon security clearances for all staff, whilst hoping that soon "it will all just go away".[113] In the end, it did all go away, if more with a whimper than a bang.

In the end and despite the huge cost, a final report proved difficult to produce. In what remains the best overview, published halfway through the project in the Smithsonian's *Annual Report* for 1965, Humphrey envisaged the Program would establish a descriptive baseline, which would be invaluable to future generations, conducting comparative studies of man-induced changes in the environment.[114] The need for such, he said, was urgent; and all the Smithsonian scientists who have written on the subject have agreed. Given global warming and other effects of environmental change upon animal life, it is even more urgent today. In 1970, Humphrey reported that he had assigned individual staff to write up separate parts of the project – e.g. island accounts, life histories, migrations, and taxonomic revisions – within five years' time.[115] In 1973, Warren King published a summary of the ecological side of the Survey,[116] but a quarter century later, we still lack an integrated analysis of the inter-relationships between seabird migration, habitat and ecology of the principal species of the region.

Nevertheless, in scientific terms, the Program's achievements were monumental. Its many papers form a major contribution to ornithology.[117] For the first time, science had a grasp of the ecology of the Northwestern Hawaiian islands, as well as population estimates for the great seabird colonies of the Phoenix group.[118] Thanks to the Program, Sand Island ranks among the best studied seabird colonies in the world.[119] Many of the Program's staff went on to do significant research afterwards.[120] Not least in this respect, the Smithsonian faithfully observed its continuing role in the long history of scientific expeditions, dating from the Wilkes Expedition and the *Challenger* survey of the nineteenth century, to which both science and the national interest remained deeply indebted.[121]

Conclusion

At the end of the day, the Smithsonian described the Program as "one of the most successful modern day field studies ever done".[122] Its scientific value, not least to the study of biodiversity, is abundantly clear. Yet, it had begun without a theoretical model to test, or set of questions to answer. It was, in fact, an exemplary exercise in nineteenth century natural history, funded with twentieth century objectives. Its scientific history properly lies in its extensive collection of data and reports. But, as this essay has argued, that history is in certain respects incomplete. The POBSP remains not so much a significant memorial, as a problematic moment in the history of the Smithsonian Institution.

Today, it remains unclear what part "bird studies" played, and whether (or if so, how) they featured within the larger American program of biological weapons tests. The public information office at the Deseret Center reports that it holds no administrative papers about the "birds project;" the Archivist of the U.S. Army states that no papers relating to the project can be found. None have as yet been located in the CIA papers deposited in the National Archives; and key Department of Defense records for this period have been destroyed. Documents recently obtained under the Freedom of Information Act by Professor Richard Steiner of the University of Alaska, who is collaborating on a study of the Program with Mr. Mark Rauzon, suggest that, far from being unique, it was only one of a series of projects on disease vectors undertaken during this period by military and civilian agencies – including the Office of Naval Research and Naval Medical Research Units (NAMRU) – based at various points on the globe, including Cairo, Taipei and Jakarta.[123] These may also have, from time to time, employed or hosted Smithsonian staff.[124]

In the fullness of time, before memories fade, the Program may yet yield information that will clarify the alleged military purpose, one way or another. There remains a serious scientific interest in the results of the ca. 5000 blood samples and specimens ostensibly still held by the Army, and of unique importance to the ecological movement.[125] There is also a legitimate historical (if no other) interest in allegations concerning Cold War research on animals as carriers of disease, or as monitors of biological activity in potentially enemy localities. If such were among the "real" issues underlying military sponsorship, then a continuing agenda cannot be ruled out. Research can presumably give answers to some questions, such as: can migratory birds carry biological weapons ("diseases") reliably; and if so, which, how, and how far? Can the analysis of migratory birds actually reveal the location and nature of biological and chemical weapons tests? The recovery from Vladivostok of a Frigatebird banded on one of the Phoenix Islands was enough to suggest to visiting academics a likely connection between the Program and a keen military interest in such questions.[126] Locating birds carrying traces of incubated diseases could have serious intelligence implications.

Since 1969, research on biological weapons has greatly "advanced", and the POBSP is no longer "news". But the fundamental historical and procedural questions it once posed have refused to go away.[127] In 1985, sixteen years after NBC's "disclosures" *The Washington Post Magazine* featured an account of the Program as "one of the largest and most mysterious undertakings" in the Institution's history.[128] By this time, it was fairly clear that the Program had been, in fact, two projects, not one. The Smithsonian had pursued its commitment to the increase of knowledge. The Army – the "funding agency" – obtained its data, and made a use of it that remains obscure. A fair division of labor?

Perhaps, despite the elapse of so much time, it is still too early to rush to judgement. A new and potentially controversial study of American biological weapons testing has recently appeared,[129] and public concern is not likely to diminish.[130] For the present, on the basis of what we know, we may offer three conclusions and a speculation.

First, in relation to the allegations made against the Smithsonian, it may be argued that field staff were almost certainly not knowingly complicit in bio-weapons testing. At the same time, while press allegations of "conspiracy" were wide of the mark, senior members of the Program did come across as being selective in their memory of events, and naive in their interpretation. Stephen Jay Gould, writing from Harvard, deplored the Smithsonian's apparent willingness to forego research ethics for the sake of federal dollars.[131] Surely, few who took part in systematic naval cruises under conditions of high secrecy, or who met with staff from the Deseret Test Center, or who were involved in the collection and dispatch of blood samples to Army laboratories could have discounted a serious military motive in what they did. Arguably, of course such matters were not made clear to the Castle. If not, the question remains, why not? And if they had been, what would the Castle have done? The issue remained to haunt the Smithsonian, and sowed distrust between the Natural History Museum's staff and the Smithsonian's management. "If there was any doubt of the unity of natural history scientists in the Senate of Scientists", recalled Clifford Evans, "that's when it really got unified, over that flap". Humphrey, in Evans' view, had "made a deal with the White House to get involved in germ warfare. He went and told Ripley and Galler, and those guys kept it over there and didn't tell him, 'Look, man, you're doing the wrong thing because we don't work in this kind of activity'".[132] Perhaps it is true that "Ripley got his fingers burnt".[133] I have found no evidence that the Castle was in any sense complicit in bio-weapons testing. But what seems indisputable, is that better communications within the Institution were urgently needed.

Second, the Smithsonian's technical defense of its right to open publication was correct, well justified and, apparently, uncontested. Its publications offer ample proof of its capacity to publish. However, it is not clear whether the stated clearance limitations, actually inhibited *complete* publication. There do seem important gaps in the record – notably in relation to Baker, Howland and perhaps less mysteriously, Christmas Island (whose absence may be more easily explained in military terms). However, even the absence of a report on Christmas is interesting, considering the amount that has been published on other test sites, including Bikini, Eniwetok and Mururoa.[134] More disconcerting, was the fact that the "restrictions" were not reciprocal – that is, there was no authority given the Smithsonian to restrict its client's use of data. The question here went, again, to the motives of the sponsor. If, in fact, scientifically invaluable specimens collected by the Program and sent to Army laboratories were actually destroyed without consultation – in 1969 or afterwards, when the operational files were destroyed – then there remain questions that are still worth asking.

Thirdly, the events of 1969 reveal the Smithsonian as a benign if somewhat bifocal agency, caught off guard and finding itself, in the midst of the Vietnam war, in an uncomfortable relationship to military sponsorship. Before the 1960s, no one questioned the Smithsonian's patriotism, or its place in the spectrum of government-sponsored research. By the late 1960s, however, the public mood had shifted, and as the Institution discovered, the Castle had not kept pace. At one level, the Pacific Program was a perfect manifestation of the Institution's desire to be of service to government, coupled with an understandable keenness for government funds and facilities. But the Smithsonian enjoyed a special, almost unique position of public trust – exemplified in the public's affection for its museums, its zoo, and its Secretary, S. Dillon Ripley, a man of high scholarship and integrity, widely respected by the scientific community and Congress. It was this sense of public trust that, some argued, had been betrayed by what seemed like an unseemly co-habitation with interests that were then waging a chemical, if not also a biological war in South East Asia.

In a larger context, and in a way representative of the Cold War at its deepest, the Smithsonian's experience reveals a growing division between public and academic perceptions of fundamental research.[135] The Smithsonian's acceptance of military work was held by a growing number of citizens, congressmen and practicing scientists to be incompatible with the ethos of free enquiry. The Director of the Program felt safe in seeking refuge behind a long-accepted distinction between basic research and its later application (or misapplication). But by the late-1960s, such a distinction appeared tenuous, if not untenable. Academics committed to open publication also relied upon government support, much of which was military, and some of which might be classified. For a closeness between academic and military interests, there was ample rhetorical justification of the most patriotic and democratic kind. "My observation", noted Sidney Galler, "over many years of experience is

> that an important by-product of the association of federal agencies with the academic community has been an extremely valuable balance in viewpoint which has resulted in the strong reliance of many agencies on the advice received from the university scientists through councils, committees, task forces, etc.

"I am convinced", he added, "that this interaction is a major force in the evolution of a fundamental research capability in this country which is second to none". By the same token, he argued, "it would be a great mistake to build high fences between the agencies and the academic community which would separate these two important elements and reduce the wholesome influence of our scholars and scientists on the decisions and courses of the federal establishment".[136] Whether such "high fences" were ever built, remains a matter for further investigation, although after the events of 1969, one could forgive the Smithsonian for being less keen to contract for "Defense-related, applied research". In negotiating a contract with the Castle in 1971 for further studies of the Phoenix Islands, the Air Force representative expressed fears of a growing association between serious scientific researchers and a lunatic fringe ... who have gone off the deep end in their anti-military activities." He hoped the Smithsonian would see reason – and

> recognise that in the national interest some broad wording of an anti-military nature could seriously damage American national interests, since at some point we may need, as we have in the Phoenix Islands, to establish tracking facilities or other capability and do it without any appreciable injury to the environment. We will certainly appreciate your help and that of the Institution to combat this kind of erroneous thinking.[137]

In the event, a compromise was reached. As the relevant Smithsonian officer sensibly put it, "We are always ready to continue our firm relationship with the Air Force on this matter".[138] In fact, the Smithsonian continued to accept defense contracts – between 1975–85, over $10 million in research funds – but after 1970, none, I believe, for work whose publication was classified.

What was the "Birds Project" really about? Clearly, there were both scientific and military motives, and perhaps an intelligence motive as well. Whether funds came ultimately from the Army or from the CIA, perhaps made little difference to the research outcome; but even the remotest suggestion of CIA involvement raised deep sensitivities. In 1985, Ted Gup called attention to the connection between the CIA and Leonard Carmichael's participation in the Human Ecology Fund, a well-known CIA conduit for research on behavioral control. It is known that the CIA routinely used academic institutions to conduct research. Carmichael's connection with the HEF was personal, not institutional, but there is evidence that he met HEF staff at his Smithsonian office. In 1977, Ripley wrote to the CIA, disassociating the Smithsonian from any CIA work on behavior modification. By that time, there was no longer any association between the POBSP and the military. But by that time, perhaps it was no longer necessary – the CIA was apparently working on a "bird study related to biological weapons" of its own.[139]

We conclude with a speculation. Thanks to the thirty year rule, we know that three days after Kennedy took the oath of office, his newly appointed Secretary of Defense, Robert McNamara, celebrated for his application of management techniques at the Ford motor company, embarked upon a massive review of American defense operations and spending. It is argued that among some 150 separate projects, there was at least one, Project 112, that concerned chemical and biological warfare. In the project, staff were asked to examine all possible applications of CBW, including uses as an alternative to nuclear weapons; and to "prepare for the development of an adequate biological and chemical deterrent capability, to include cost estimates, and appraisal of domestic and international political consequences".[140]

The military, it is said, held that such weapons already existed, but were of limited use; what was useful, was their unique ability to produce "controlled temporary incapacitation", rather than mass destruction of property. The project recommended a four year program of research on CBW weapons systems and biological agent manufacture. The program was to focus on sprays rather than bombs, and on incapacitating rather than lethal agents; all three services were to share the costs; and all tests would take place outside the continental U.S., and preferably in climates that would replicate likely target areas in the Soviet Union and Vietnam. These conditions were, coincidentally to be found in the Aleutian Islands, and in the Central Pacific.

In September 1970, the Smithsonian came under renewed scrutiny by the House Committee on Administration, which produced the first comprehensive congressional report on the Institution in over a hundred years. Whilst "patting them on the back" for their good work, the chairman agreed that the Institution had become, in the words of *Science*, a "far more complex and diversified organization than most people, including most members of Congress, realize".[141] Its staff, Sidney Galler observed, had nothing in common with the popular image of "guys with green eye shades or with pith helmets and butterfly nets".[142] Thanks to the collections of the Natural History Museum, Richard Cowan observed, researchers could "determine such things as the natural radiation levels found in planktonic organisms collected before nuclear weapons tests, the load of metallic compounds carried by organisms before air pollution reached its present levels, and the biochemical makeup of organisms collected before DDT and other persistent pesticides came into use".[143] The Institution's budget, despite having risen from $46 million to $55 million in three years, was nevertheless suffering slow starvation brought about by the demands of science. The Department of Vertebrate Zoology was singled out as especially "undernourished". Given such needs and opportunities, funding from the military seemed merely natural.

Whether the Pacific Ocean Birds Survey Program was one of the series of scientific projects initiated by the military during the first months of Kennedy's Administration, whether it played a role in biological and chemical weapons testing, and whether the Castle was conscious of taking on this obligation deserve answers this paper has not been able to give. Requests for information under the Freedom of Information Act await reply. Without them, we will not know why the birds of the Central Pacific came to play a long-running role in the theatre of the Cold War. We do know that they were principal characters in one of the most damaging episodes in the history of one of America's best-loved institutions. We do not know what military uses arose from their study. But we suspect that these are not questions that we can safely ignore. The history of the Program is a matter that is no longer, to say the least, strictly for the birds.

University of Sydney

Notes

[1] Research for this paper was made possible by a Short Term Travel Grant from the Smithsonian Institution, to whose tradition of supporting free enquiry I am greatly indebted. Special thanks are owed Dr Pamela Henson, Director of the Institutional History Division of the Smithsonian Archives and Mr William E. Cox of the Smithsonian Archives; Dr Roberta Rubinoff, Head of the Office of Fellowships and Grants; and Mr Brian LeMay, of the International Office. For many details about the POBSP, I wish to thank Mr Roger Clapp, an early member of its field staff. For responding to requests for information, I am grateful to Dr Jeffery K. Smart, Command Historian, U.S. Army Chemical and Biological Defense Command and Mr Norman M. Covert, Chief, Public Affairs, Fort Detrick, MD; and also to Mr Philip Boffey; Dr David Challinor; Mr Ted Gup; Dr Philip Humphrey; Mr Mark Rauzon; and Professor Richard Steiner; and to the staff of the National Security Archive, Washington, DC. For helpful comments on earlier drafts, I am grateful to the late Professor Fritz Rehbock of the University of Hawaii; Professor David Stoddart of the University of California at Berkeley; and Mr Daniel Greenberg, editor of *Science and Government Report*. A full account of the scientific achievements of the Program remains to be written.

[2] Cited in M. Miller, *Plain Speaking: An Oral Biography of Harry S. Truman* (London: Victor Gollancz, 1974), p. 392.

[3] T. Gup, "The Smithsonian's Secret Contract: The Link between Birds and Biological Warfare", in *The Washington Post Magazine*, 12 May 1985, pp. 8–14.

[4] See for example, Peter Galison, Bruce Hevly and Rebecca Lowen, *Controlling the Monster: Stanford and the Growth of Physics Research, 1935–1962* (Stanford: Stanford University Press, 1992); Stuart W. Leslie, "Profit and Loss: The Military and MIT in the Postwar Era", *Historical Studies in the Physical and Biological Sciences* 21 (1990), pp. 59–85; Rebecca S. Lowen, "Exploiting a Wonderful Opportunity: The Patronage of Scientific Research at Stanford University, 1937–1965", *Minerva* 30 (1992), pp. 391–421; and Rebecca S. Lowen, "Transforming the University: Administrators, Physicists, and Industrial and Federal Patronage at Stanford, 1935–49", *History of Education Quarterly* 31(3) (1991), pp. 365–388.

[5] See Christopher Simpson, *Universities and Empire: Money and Politics in the Social Sciences during the Cold War* (New York: Norton, 1998).

[6] It is stated (without confirming reference) by E. Regis, in *The Biology of Doom: America's Secret Germ Warfare Project* (New York: Harry Holt and Co, 1999), p. 188, that the military delegation met with Remington Kellogg, Director of the U.S. National Museum (1948–1962), who also held the title of Assistant Secretary of the Smithsonian (1958–1962). A native of Iowa, Kellogg was trained at the University of Kansas, and became a specialist in fossil marine mammals. During the First World War, he served in the Army Medical Department in France, where he studied rodent-carried diseases. During the Second World War, he led the development of Field Studies in Brazil concerned with the study of mammals as carriers of disease (*Year Book, American Philosophical Society*, 1972, pp. 205–210). By inclination, training and administrative position, Kellogg appears to have been an ideal contact for the military.

[7] D.N. Leff, *Uncle Sam's Pacific Islets* (Stanford: Stanford University Press, 1940).

[8] R. MacLeod and P.F. Rehbock, eds., *Darwin's Laboratory: Evolutionary Theory and Natural History in the Pacific* (Honolulu: University of Hawaii Press, 1994); J. Gascoigne, *Science in the Service of Empire: Joseph Banks, the British State and the Uses of Science in the Age of Revolution* (Cambridge: Cambridge University Press, 1998).

[9] P.M. Henson, "The Smithsonian Goes to War: The Increase and Diffusion of Scientific Knowledge in the Pacific", in *Science and the Pacific War, 1939–1945*, R.M. MacLeod, ed. (Dordrecht: Kluwer, 1999), pp. 27–50.

[10] J.M. Paley (U.S. Army Chemical Corps) to Humphrey, Biological Laboratory notes 9/25/62, Box 8, RU 245, Smithsonian Institution Archives, Washington, DC.

[11] POBSP Contract, 1964, Box 8, RU 245, Smithsonian Archives.

[12] The POBSP indeed, selected certain atolls at Eniwetok for study. R. Crossin to J. Bushman and Robert Pyle, letter 1/29/69, Box 5, RU 245, Smithsonian Archives.

[13] C. Ely, The Pelagic Program of the Pacific Ocean Biological Survey Program (nd).

[14] P.S. Humphrey to Edwin Bryan, Pacific Scientific Information Center, Bishop Museum, letter 9/4/63, Box 15, RU 245, Smithsonian Archives (cit. n. 11).

[15] POBSP Contract, June 1964.

[16] W. Banko to C. Ely, letter 11/26/63, Box 5, RU 245, Smithsonian Archives.

[17] Although, in time-honored fashion, it did afford a moment for disgruntled verse: "Starbright, Starbright/First Project came to Light/Wish what I may/That I might/ Had started the darned/

Thing off Right". [The remaining lines of the verse are too salty to print]. Natural History Museum, *Droppings from the Eagle's Roost*. Entry 12/13/63.

[18] C. Ely to Research Curators, letter 4/20/64, Box 15, RU 245, Smithsonian Archives.

[19] *Droppings from the Eagle's Roost*. Entry, 4/18/66 and 1/3/64.

[20] J. Church to C. Ely, letter 10/5/65, Box 12, RU 245, Smithsonian Archives. See also L. Cole, *The Eleventh Plague: The Politics of Biological and Chemical Warfare* (New York: W.H. Freeman, 1997).

[21] W.S. Miller (Chief, Test Chamber Branch, Technical Evaluation Division, Chemical Corps Biological Laboratories, Ft. Detrick) to C. Ely, letter 4/10/64, Box 15, RU 245, Smithsonian Archives.

[22] Patrick J. Gould, "Sooty tern" (*Sterna fuscata*). Paper No. 68, Pacific Ocean Biological Survey Program, in W.B. King, *Pelagic Studies of Seabirds in the Central and Eastern Pacific Ocean*, Smithsonian Contributions to Zoology No. 158 (Washington, DC: Smithsonian Institution Press, 1974).

[23] C. Ely to P. Humphrey, letter 3/5/65; C. Ely to W. Banko, letter 4/20/65; C. Ely to P. Humphrey, letter 5/12/65; C. Ely to J. Church, letter 11/4/65 Box 12, RU 245, Smithsonian Archives.

[24] C. Ely to J. Church, letter 12/9/65, Box 12, RU 245, Smithsonian Archives.

[25] C. Ely to P. Humphrey, letter 12/9/65, Box 12, RU 245, Smithsonian Archives. A procedure was apparently devised to clear reports through Charles Ely, which included vetting by John Bushman, the military liaison officer. It is not known, however, whether this procedure was used, or if it was, whether it discouraged publication, or actually restricted the publication of reports on certain localities.

[26] C. Ely, memo 4/23/65 and 4/28/65, Box 12, RU 245, Smithsonian Archives.

[27] C. Ely to Neil Walker, letter 9/15/64, Box 13, RU 245, Smithsonian Archives.

[28] J. Church to C. Ely, 10/5/65, Box 12, RU 245, Smithsonian Archives.

[29] J. Church to staff, memo 1966, Box 6, RU 245, Smithsonian Archives.

[30] cf. R.B. Clapp, "Frigatebirds of the Pacific", *Pacific Bird Observer* 1 (1965), p. 3; "Ruddy Turnstone", *Pacific Bird Observer* 2 (1965), p. 9; "Sibley to study Condors", *Pacific Bird Observer* 3 (1965): 4; "Blue-faced Booby: Demon or Dunce", *Pacific Bird Observer* 3 (1966), pp. 9–11; "Additional New Records of Birds from the Phoenix and Line Islands", *Ibis* 110(4) (1968), pp. 574–575; "Three Unusual Shorebirds from Midway Atoll, Pacific Ocean", *Elepaio* 28(9) (1968), pp. 76–77; "The Birds of Swain's Island, South Central Pacific", *Notornis* 15(3) (1968), pp. 198–206.

[31] W.B. King, "Conservation Status of the Birds of Central Pacific Islands", *Wilson Bulletin* 85(1) (1973), pp. 89–103, p. 93.

[32] *Ibid.*, p. 95.

[33] *Ibid.*, p. 98.

[34] P.S. Humphrey to author, personal communication 5/12/96.

[35] P.S. Humphrey to J. Church, letter 3/2/67, Box 13, RU 245, Smithsonian Archives.

[36] P.S. Humphrey to I.E. Wallen, letter 8/8/67, Box 5, RU 245, Smithsonian Archives.

[37] P.S. Humphrey to R. Pyle, letter 5/7/69, Box 13, RU 245, Smithsonian Archives

[38] cf. A. Binion Amerson, "Ornithology of the Marshall and Gilbert Islands", *Atoll Research Bulletin (ARB)* 127 (1969), pp. 1–348; "The Natural History of French Frigate Shoals, Northwestern Hawaiian Islands", *ARB* 150 (1971), pp. 1–383; A. Binion Amerson and K.C. Emerson, "Records of Mallophaga from Pacific Birds", *ARB* 146 (1971), pp. 1–30; A. Binion Amerson, R.B. Clapp and W.O. Wirtz, "The Natural History of Pearl and Hermes Reef, Northwestern Hawaiian Islands", *ARB* 174 (1974), pp. 1–306; R.B. Clapp, "A Specimen of Jouanin's Petrel from Lisianski Island, Northwestern Hawaiian Islands", *Condor* 73(4) (1971), p. 490; "The Natural History of Gardner Pinnacles, Northwestern Hawaiian Islands", *ARB* 163 (1972), pp. 1–25; "Albinism in the Black Noddy (*Anous tenuirostris)*", *Condor* 76(4) (1974), pp. 464–465; "Specimens of three species of *Pterodroma* from the Pacific Ocean", *Ardea* 62(3/4) (1974), pp. 246–247; "Additional Records of Lizards from the Tokelau Islands, Polynesia", *J. Herpetol* 9(4) (1975), p. 369; "Gray-backed Tern eats Lizards", *Wilson Bull.* 88(2) (1976), p. 354; "Notes on the Vertebrate Fauna of Tongareva Atoll", *ARB* 198 (1977), pp. 1–7; "Review of Buff-Breasted Sandpiper Records in the Central Pacific", *Elepaio* 40(6) (1979), p. 93; "Occurrence of Central Pacific Blue-faced Booby and Sooty Tern in Southeast Asia", *Tori* 29(2/3) (1980), pp. 95–96; "A Summary of Alcid Records from Hawaii", *Colonial Waterbirds* 9(1) (1986), pp. 104–107; "A Specimen Record of the Fork-tailed Swift from the Marshall Islands", *Elepaio* 49(1) (1989), pp. 1–2; "Notes on the Birds of Kwajalein Atoll, Marshall Islands", *ARB* 342 (1990), pp. 1–94; R.B. Clapp and C.D. Hackman, "Longevity Record for a Breeding Great Frigatebird", *Bird-Banding* 40(1) (1969), p. 47; R.B. Clapp and L.N. Huber, "Imperfect Albinism in a Red-Tailed Tropicbird", *Condor* 73(1) (1971), p. 123; R.B. Clapp

and W.B. King, "Status of the Kokikokiko *Acrocephalus aequinoctialis*",*Bull. Brit. Ornithol. Club* 95(1) (1975), pp. 2–3; R.B. Clapp, V.B. Kleen and D.L. Olsen, "First Records of Emperor Geese in the Northwestern Hawaiian Islands", *Elepaio* 30(6) (1969), pp. 51–52; R.B. Clapp, M.K. Klimkiewicz and J.H. Kennard, "Longevity Records of North American Birds: Gaviidae through Alcidae", *J. Field Ornithol.* 53(2) (1982), pp. 81–124; R.B. Clapp and E. Kridler, "The Natural History of Necker Island, Northwestern Hawaiian Islands", *ARB* 206 (1977), pp. 1–102; R.B. Clapp, E. Kridler and R.R. Fleet, "The Natural History of Nihoa Island, Northwestern Hawaiian Islands", *ARB* 207 (1977), pp. 1–147; R.B. Clapp and R.L. Pyle, "Noteworthy Records of Waterbirds in Oahu", *Elepaio* 29(5) (1968), pp. 37–39; R.B. Clapp, C.S. Robbins and K.W. Kenyon, "Additional Records of Black-legged Kittiwakes from the Northwestern Hawaiian Islands", *Elepaio* 40(7) (1980), pp. 98–99; R.B. Clapp and F.C. Sibley, "Longevity Records of some Central Pacific Seabirds", *Bird-Banding* 37(3) (1966):, pp. 193–197; "Notes on the Birds of Tutuila, American Samoa", *Notornis* 13(3) (1966), pp. 157–164; "New Records of Birds from the Phoenix and Line Islands", *Ibis* 109(1) (1967), pp. 122–125; "The Vascular Flora and Terrestrial Vertebrates of Vostok Island, South Central Pacific", *ARB* 144 (1971), pp. 1–9; "Notes on the Vascular Flora and Terrestrial Vertebrates of Caroline Atoll, Southern Line Islands", *ARB* 145 (1971), pp. 1–18; R.B. Clapp and P.W. Woodward, "New Records of Birds from the Hawaiian Leeward Islands", *Proc. U.S. National Museum* 124(3640) (1968), pp. 1–39; R.B. Clapp and W.O. Wirtz, "The Natural History of Lisianski Island, Northwestern Hawaiian Islands", *ARB* 186 (1975), pp. 1–196; C. Ely and R.B. Clapp, "The Natural History of Laysan Island, Northwestern Hawaiian Islands", *ARB* 171 (1973). pp. 1–361; J.S. Marks et al., "Notes on Longevity and Flightlessness in Bristle-thighed Curlews", *Auk* 107(4) (1990), pp. 779–781; M.J. Rauzon, C.S. Harrison and R.B. Clapp, "Breeding Biology of the Blue-gray Noddy", *J. Field Ornithol.* 55(3) (1984), pp. 309–321; Paul W. Woodward, "The Natural History of Kure Atoll, Northwestern Hawaiian Islands", *ARB* 163/164 (1972); P.W. Woodward and R.B. Clapp, "First Record of Baird's Sandpiper from the Central Pacific", *Elepaio* 30(3) (1969), p. 25.

39 Interoffice Records, Sera Samples, 1962, Box 6, RU 245, Smithsonian Archives.

40 D.R. Stoddart to author, personal communication 11/3/99.

41 Amerson, 1969, 1971 (cit. n. 38)

42 Stoddart to author, 11/3/99 (cit. n. 40).

43 A. Binion Amerson and Philip Shelton, "The Natural History of Johnston Atoll, Central Pacific Ocean", *ARB* 192 (1976), pp. 1–479.

44 Army Reports, F. Sibley and R. Clapp, "Biological Survey of Baker Island", Natural History Museum, Smithsonian Institution, 1965; R. Clapp, "Howland and Baker Islands", Natural History Museum Smithsonian Institution, April 1967.

45 Records of the Program's studies of Baker are found in Box 172, RU 245, Smithsonian Archives.

46 J. Church to R.L. Pyle, letter 2/7/69, Box 12, RU 245, Smithsonian Archives.

47 L. Cole, *Cloud of Secrecy. The Army's Germ Warfare Tests over Populated Areas* (Totowa, NJ: Rowman and Littlefield, 1988), p. 33.

48 S. Wright, ed., *Preventing a Biological Arms Race* (Cambridge, Mass: MIT Press, 1990), p. 33.

49 *Ibid.*, p. 35.

50 N. Covert, *A History of Fort Detrick, Maryland* (Fort Detrick: U.S. Army, 3rd ed.,1997). See also J.K. Smart, Historical Highlights. U.S. Army Chemical and Biological Defense Command, Special Study No. 1, 1994.

51 D.F. Chamberlain (Inspector General, CIA) to DCI, CIA, memo 5/20/75, National Security Archive, Washington, DC.

52 J. Marks, *The Search for the Manchurian Candidate: The CIA and Mind Control* (London: Allen Lane, 1979), p. 57.

53 U.S. Biological Warfare Programs, "U.S. Army Activity in the U.S. Biological Warfare Programs, 24 February 1977", in *U.S. Senate, Committee on Human Resources, Hearings: Biological Testing Involving Human Subjects by the Department of Defense*, Annex A, 1977, pp. 23–234, pp. 40–51 et passim.

54 J.H. Capshew, "Engineering Behavior: Project Pigeon, World War II and the Conditioning of B.F. Skinner", *Technology and Culture* 34 (4), (1993), pp. 835–857.

55 Wright, *Preventing a Biological Arms Race* (cit. n. 48), p. 36.

56 J. Ranelagh, *The Agency: The Rise and Decline of the CIA* (London: Weidenfeld and Nicolson, 1986); P. Agee, *Inside the Company: CIA Diary* (Harmondsworth: Penguin, 1975); V. Marchetti and J.D. Marks, *The CIA and the Cult of Intelligence* (New York: Laurel, 1980).

57 William E. Small, "DOD Supporting Bird Studies in Pacific, Brazil", *Scientific Research*, December 9, 1968, p. 27.

58 Philip Siekevitz, letter to *Scientific Research*, January 6, 1969, pp. 6–7.

59 University of Utah, "Germ War Film Sparks U Debate", *University of Utah Chronicle*, February 10, 1969, pp. 1 and 3.

60 J.K. Smart, *Chemical Warfare: Lessons Learned from the Past*, U.S. Army Historical Division, Aberdeen Proving Ground, Special Study No. 8, March 1991, p. 24.

61 "Under Guise of Pacific Isle Bird Study: Army Plan for Biological Warfare Tests Reported", *Honolulu Star Bulletin*, 2/5/69, p. 1; "Pacific Bird Study Not Linked to Warfare Tests", *Honolulu Star Bulletin*, 2/5/69, p. 16; "Poison in the Pacific", *Honolulu Star Bulletin*, 2/7/69, Box 6, RU 245, Smithsonian Archives.

62 "Bird Project a Cover-up for War Test, NBC Says", in *Evening Star*, 2/4/69, p. 1; "Smithsonian Denies Military Aid in Tests", in *Washington Star*, 2/4/69, p. 6; P.M. Boffey, "Biological Warfare: Is the Smithsonian Really a 'Cover'?" *Science*, 163 (February 21, 1969), pp. 791–796, p. 794.

63 V.B. Scheffer (U.S. Dept of Interior, Fish and Wildlife Service, Seattle) to W.R.Norwood (High Commissioner of the Trust Territory, Saipan, Mariana Islands), letter 1/13/69, Box 6, RU 245, Smithsonian Archives.

64 C. Ely, memo 4/23/65, Box 12, RU 245, Smithsonian Archives.

65 Stuart Auerbach, "Smithsonian Bird Research Tied to Germ Warfare Study", *The Washington Post*, February 5, 1969, p. 7.

66 "News Feature", *New York Times*, 2/5/69.

67 G.C. Wright to President L. Johnson, letter 12/30/68, Box 6, RU 245, Smithsonian Archives; Department of Defense to G. Wright, letter 1/18/69, Box 6, RU 245, Smithsonian Archives.

68 N. Beloff, "Are Bird Watchers Helping Germ Warfare Research?" *The Advocate News*, 2/19/69, Box 6, RU 245, Smithsonian Archives. The same article claimed that, although the Defense Department was currently receiving between five and eight times more applications for grants from academics than it could fund, relations with the universities had become "far less amiable and productive than they were".

69 B. Nelson, "MIT's March 4: Scientists Discuss Renouncing Military Research", *Science*, 163 (March 14, 1969): 1175–1178; F.H. Tschirley, "Defoliation in Vietnam", *Science*, 163 (February 21, 1969), pp. 779–786.

70 Department of Defense, "Pacific Project" (nd 1969), Box 6, RU 245, Smithsonian Archives.

71 J. Church to P. Humphrey, letter 2/14/69, Box 13, RU 245, Smithsonian Archives.

72 "Under Guise of Pacific Isle Bird Study: Army Plan for Biological Warfare Tests Reported", *Honolulu Star Bulletin*, 2/5/69, p. 1.

73 Gup, "Smithsonian's Secret Contract ..." (cit. n. 3), p. 14.

74 S. Galler to J. Sugar, letter 4/1/69, Box 6, RU 245, Smithsonian Archives.

75 S. Galler to J. Church, memo 1/13/69, Box 6, RU 245, Smithsonian Archives.

76 Introduction, RU 429, Smithsonian Archives.

77 See Clifford Evans, Memorandum 2/5/69, Box 4, Folder 2, RU 429, Smithsonian Archives.

78 Carmichael Papers, Smithsonian file, 2/11/69, American Philosophical Society, Philadelphia.

79 Third Council Meeting: Senate of Scientists, 2/6/69, Box 4, RU 429, Smithsonian Archives.

80 S. Galler to Ripley, 2/7/69, Folder 2, Box 4, RU 429, Smithsonian Archives.

81 Clifford Evans to Officers of Senate of Scientists, 2/9/69, Folder 2, Box 4, RU 429, Smithsonian Archives.

82 Nora Beloff, "Row Breaks over Birds and Germ Warfare Mystery", *The Observer (London)*, 2/16/69.

83 Press Release, Folder 2, Box 4, RU 429, Smithsonian Archives.

84 P.M. Boffey, "Biological Warfare: Is the Smithsonian Really a 'Cover'?" *Science* 163 (September 21, 1969): 791–796, p. 791.

85 *Ibid.*

86 *Ibid.*, p. 793.

87 *Ibid.*, p. 796.

88 Evans' notes of meeting, 2/26/69, Folder 2, Box 4, RU 429, Smithsonian Archives.

89 *Barron's Weekly*, 3/3/69.

90 "New Chemical Warfare Superiority Cited by Army in Congress Briefing", *The Washington Post*, 3/5/69, Box 6, RU 245, Smithsonian Archives.

91 "Pacific Project" (nd), Box 6, RU 245, Smithsonian Archives.

92 Boffey, "Biological Warfare ..." (cit. n. 84), p. 792.

93 DISA (Defense Information Systems Agency), 1998, Bibliographic Search.

94 R. Preston, "The Bioweaponeers", *The New Yorker*, 3/9/98, 52 65, p. 60.

95 *Ibid.*

[96] P. Mink to S.D. Ripley, letter 3/6/69, Box 6, RU 245, Smithsonian Archives; S.D. Ripley to Mink, letter 3/18/69, Box 6, RU 245, Smithsonian Archives.
[97] S.D. Ripley to Staff, Folder 2, Box 4, RU 429, Smithsonian Archives; see also entry, *Droppings from the Eagle's Roost*, 2/19/69.
[98] Clifford Evans to "Whom it may Concern", 11/17/69, Box 4, RU 429, Smithsonian Archives.
[99] R. Cowan to P. Humphrey, letter 3/5/69, Box 13, RU 245, Smithsonian Archives.
[100] J. Church to G. Watson, letter 3/19/69, Box 5, RU 245, Smithsonian Archives.
[101] P.S. Humphrey to J. Church, letter 4/15/69, Box 13, RU 245, Smithsonian Archives.
[102] D. Parker to J. Church, letter 7/15/69, Box 5, RU 245, Smithsonian Archives.
[103] Clifford Evans to General Counsel, 10/16/69, Box 4, RU 429, Smithsonian Archives.
[104] P.S. Humphrey to R. Cowan, letter 3/26/69, Box 13, RU 245, Smithsonian Archives; P.S. Humphrey to J. Church, letter 9/9/69, Box 13, RU 245, Smithsonian Archives.
[105] P.S. Humphrey to R. Pyle, letter 3/26/69, Box 13, RU 245, Smithsonian Archives.
[106] "Research in CBR" (Chemical, Biological and Radiological Warfare), Report of the Committee of Science and Astronautics, 86th Congress, House of Representatives, Washington, DC., Report 815 (1969), p. 7.
[107] SIPRI (Stockholm International Peace Research Institute), *The Problem of Chemical and Biological Warfare*, Vol II: *CB Weapons Today* (Stockholm: Almavist and Wiksell/New York: Humanities Press, 1973), p. 193.
[108] J. Church to P. Humphrey, letter 12/16/69, Box 13, RU 245, Smithsonian Archives.
[109] Smart, "Chemical Warfare ..." (cit. n. 60), p. 25.
[110] P. Cilladi-Rehrer (Command Historian, U.S. Army Medical Research and Materiel Command) to Professor R. Steiner, letter 10/2/98, Smithsonian Archives.
[111] *The New York Times*, 7/4/69.
[112] N. Covert (Command Historian) to M. Rauzon, letter 7/17/98.
[113] J. Church to P. Humphrey, letter 4/17/69, Box 13, RU 245, Smithsonian Archives.
[114] P.S. Humphrey, "An Ecological Survey of the Central Pacific", *Smithsonian Year* (Washington, DC: Smithsonian Institution, 1965), pp. 21–30.
[115] P.S. Humphrey to G.E. Watson (Department of Vertebrate Zoology), letter 2/9/70, Box 13, RU 245, Smithsonian Archives.
[116] W.B. King, "Conservation Status ..." (cit. n. 31).
[117] R.B. Clapp, M.D.F. Udvardy and A.K. Kepler, *An Annotated Bibliography of Laysan Island, Northwestern Hawaiian Islands*, Atoll Research Bulletin No. 434, 1996.
[118] Stoddart to author, 11/3/99 (cit. n. 40).
[119] R.W. Schreiber and and J.L. Chovan, "Roosting by Pacific Seabirds: Energetic, Populational and Social Considerations", *The Condor*, 88 (1986), pp. 487–492, p. 487.
[120] E.A. and R.W. Schreiber, "Great Frigatebird Size Dimorphism on Two Central Pacific Atolls", *The Condor*, 90 (1988), pp. 90–99, p. 90; J. Burger, E.A.E. Schreiber and M. Gochfeld, "Lead, Cadmium, Selenium and Mercury in Seabird Feathers from the Tropical Mid-Pacific", *Environmental Toxicology and Chemistry*, 11 (1992), pp. 815–822.
[121] For the Wilkes Expedition, see Herman J. Viola and Carolyn Margolis, eds., *Magnificent Voyagers: The U.S. Exploring Expedition, 1838–1842* (Washington, D.C.: Smithsonian Institution Press, 1985). For the *Challenger* survey see See Margaret Deacon, *Scientists and the Sea, 1650–1900: Study of Marine* Science (New York: Academic, 1971); Herbert Swire, *The Voyage of the Challenger: A Personal Narrative of the Historic Circumnavigation of the Globe in the Years 1872–1876*, 2 vols. (London: Golden Cockerel, 1938); Eric Linklater, *The Voyage of the Challenger* (London: John Murray, 1972); Susan Schlee, *The Edge of an Unfamiliar World: A History of Oceanography* (New York: E.P. Dutton, 1973); and Philip F. Rehbock, ed., *At Sea with the Scientifics: The "Challenger" Letters of Joseph Matkin* (Honolulu: University of Hawaii Press, 1992).
[122] Gup, "Smithsonian's Secret Contract ..." (cit. n. 3), p. 9.
[123] K.D. Akers (FOIA) to R.G. Steiner, letter 8/12/98.
[124] D. Challinor to author, personal communication 11/30/98.
[125] *Providing Scientific Information About Birds*, Ornithological Council, 1997.
[126] D.R. Stoddart to author, personal communication 8/14/96; Scheffer to W.R. Norwood (cit. n. 63).
[127] S. Aftergood, *Project in Government Secrecy. Federation of American Scientists* (Washington, DC: website <www.for.org>, 1999).
[128] Gup, "Smithsonian's Secret Contract ..." (cit. n. 3), pp. 8–14.
[129] Regis, *The Biology of Doom* (cit. n. 6).
[130] J.T. Richelson, "Scientists in Black", *Scientific American*, 278(2) (February 1998), pp. 38–45.
[131] Stephen Jay Gould, "Smithsonian's Albatross", *Science* 164 (May 2, 1969), p. 497.

[132] Clifford Evans, Oral History Project Interview, Senate of Scientists Project, 5/28/75, Archives and Special Collections, Smithsonian Institution, p. 66.
[133] Evans, "Oral History Project" (cit. n. 132), p. 67.
[134] Stoddart, Personal communciation, 1999 (cit. n. 126).
[135] Frances Stornor Saunders, *Who Paid the Piper? The CIA and the Cultural Cold War* (London: Granta, 1999).
[136] Galler to J. Sugar, letter 4/1/69, Box 6, RU 245, Smithsonian Archives.
[137] P.F. Hilbert (Deputy Undersecretary of the Air Force) to Ripley, letter 9/3/71, Box 528, RU 99, Smithsonian Archives.
[138] D. Challinor to P.F. Hilbert (Deputy Undersecretary of the Air Force), letter 9/22/71, Box 528, RU 99, Smithsonian Archives.
[139] Gup, "Smithsonian's Secret Contract ..." (cit. n. 3), p. 13.
[140] Regis, *The Biology of Doom* ... (cit. n. 6), p. 185.
[141] Luther J. Carter, "Smithsonian: Natural History is Undernourished, Panel Finds", *Science*, 169 (Sept 4, 1970), pp. 960–963.
[142] *Ibid.*
[143] *Ibid.*

MICHAEL A. FORTUN

THE HUMAN GENOME PROJECT: PAST, PRESENT, AND FUTURE ANTERIOR

Abstract

This paper re-reads several important events that have been historicized as "origins" of the Human Genome Project (HGP), arguing that historians of the HGP have deployed methods and logics that are homologous to those of contemporary genetics. Each relies on sequences that can be faithfully reproduced, and privileges nuclear control while marginalizing complex networks and interactions. While the writing of both history and biology seem to demand and display unidirectional causality, in the end they each have to rely, indirectly, on a future that has yet to arrive yet, already structures the past and present.

Let's Do the Time Warp Again

"The Human Genome Project has just entered warp speed".[1] So begins an article in the March 1999 issue of *Science*, describing a new effort by a recombinant public/private consortium to produce at least a "first draft", ninety-percent-complete sequence of the human genome by March 2000 – approximately five years ahead of original projected timetables. The consortium was organized by the NIH's National Human Genome Research Institute and the Wellcome Trust, but is funded in large part by such pharmaceutical companies as Bayer, Novartis, Glaxo Wellcome, Pfizer, Roche, and SmithKline Beecham. The new quasi-public effort responds to corporate high-speed sequencing initiatives being undertaken by companies like Celera Genomics and Incyte Pharmaceuticals, and the likelihood that much of the sequence-based information being produced in such venues will be privatized.

In the accelerated worlds of genomics, the public/private distinction isn't the only thing to blur under hybridizing forces. As any science fiction fan will tell you, when large technoscientific machines enter warp speed, time itself is likely to be distorted – dilated, looped, forked, wormholed, or otherwise warped. Intrepid scientists encounter previous versions of themselves or alternate

G.E. Allen and R.M. MacLeod (eds.), Science, History and Social Activism: A Tribute to Everett Mendelsohn, 339–362.

histories altogether, and causality and change no longer seem to operate unidirectionally.

This essay re-reads several events in the early history of the Human Genome Project that are almost always presented as "origins" of today's high-throughput HGP. Now that economics is apparently shuddering under the force of its own warp drives, these origin-events might be re-read for signs of other kinds of history, other kinds of time, and causal chains other than the straight ones that we generally take for granted. Only through establishing some kind of rapport – however tentative, experimental, and counterintuitive – with the untimeliness of genomics will we be able to respond better to its overwhelming scientific, social, and ethical complexity. If there's one thing that organisms, genomics, and history have in common, it's this: they're always out ahead of themselves. And that is not an easy position to be in when it comes to those proliferating social and ethical matters involving genomics where something bearing some resemblance to "control" is called for. My hope is that by interrogating the various ways in which "the future" is called up in the structure of our thought today, we may be able to meet it with some better appreciation of how its warp has already been woven into us, into the histories that we write, and into the sciences that we practice.

One of the reasons I decided to undertake a history of the Human Genome Project as my Ph.D. thesis (with Everett as chair of my committee) was that, even before its "official beginning" (this declaration is discussed below), stories of its origins had begun to appear in print. Scientists promoting the project, journalists, science writers, and even a historian or two seemed to find it imperative to uncover a foundation for the nascent endeavor, and to narrate an orderly history on top of that foundation. It was as if the question of the "origins" of the fast-moving Human Genome Project prompted such an immediate mixture of anxiety and curiosity that the work of history could not wait.

Perhaps as a result of this impatience, historiography comes to mime biology in a hunt for origins: what gave the Human Genome Project "life"? From what ancestor(s) did it evolve? What socio-historical code was present "in the beginning", from which, even in the presence of much developmental noise, the body of the Human Genome Project unfolded? Who wrote the program for these events? And above all: how can we sequence these historical events as fast as we possibly can, so that we might finally understand what it means to be humans who undertake massive accelerated DNA sequencing projects?

One could round up the usual suspects, all of which have appeared in the line-up at one time or another, and have been identified by the appropriate authorities. The Human Genome Project originated with: great men, great ideas, great meetings, great technology, great government reports, and other great things. The origins multiply, as we will shortly see, which only makes the hunt for them more intense. Thus, the early events have been narrated over and over again, each iteration intensifying the effects of origin, order, control, and rational decision-making.[2]

Surely it's no coincidence that historians of genetics and genomics can be remarkably like molecular geneticists: prone to fixating on dogmatic command structures, glossing or erasing the complexities of developmental events, and especially in their privileging and even essentializing of some center or core as the founding "origin" against which all other forces pale in significance. Life scientists and their chroniclers might be said to share the same discursive space, and to deploy similar machines. The logic is inescapably a reproductive one: everything is coded in a way that ends up producing the same origin stories, one more time.

A different reading of these putative origins is required, something along the lines of "cultural critique" as articulated by George Marcus:

> The strategy of engagement with the disciplinary mainstream that I have in mind rests on finding and intellectually probing effective oppositional space within mainstream discourses. This means finding where the "fissures" are – that is, finding those concepts, methods, ideas, practices, and life experiences within the culture of the mainstream, about which there is self-doubt and uncertainty ... This in turn means understanding these potentially self-critical cultural formations ... *ethnographically*, in their own terms and expressions.[3]

The fissures of mainstream historical accounts of the Human Genome Project will be analyzed here: the figures, rhetorics, and events within or around these constantly re-iterated origins that are often marginalized, hurriedly passed over, or patched up. I'm particularly fascinated by minor moments that undermine or rattle the stability of the dominant origin motifs in the standard histories, and the dominant conceptualizations of linear time that structure them. This essay will focus on several instances when mainstream origin discourse of genomics begins to crack or warp – whether the subject under question is a social project, a history, or an organism.

In other words, which both are and are not metaphorical: this essay is less interested in history's genome than in its cytoplasm, swarming with active substances and structures, open to multiple forces and contingencies, and exhibiting characteristics that challenge us to re-think the question of time and "sequence".

Hence the odd figure of the "future anterior". My title is a playful supplement to James Watson's 1990 article in *Science* magazine, "The Human Genome Project: Past, Present, and Future", the publication of which marked the "official" beginning of the Human Genome Project, and attempted to put aside the disagreements and debates that had troubled the scientific community over the previous four or five years – to relegate difference safely to the past.[4] The simultaneity that the printed page provides for those four words in Watson's subtitle already begins to shake the sequential organization of time: past, present, *and* future are "always already" *present*, in some sense. If time did not warp in this way, if it were not always subject to the subtle disjunctive force of the "and", it would not be such a perennial subject of inquiry, puzzlement, and delight.

When I undertook a history and sociology of the Human Genome Project, it was to interrogate historiography and, therefore, time. The Ph.D. thesis was framed in terms of "the uses of history".[5] Restless, I began to recast that history in terms of speed: the development of and desire for ever-faster machines, sciences, and economies.[6] In that effort – and I should have seen this coming – I have been far too slow. Everyone has a theory about speed these days, and when *Science* starts referring to warp speed, and a genomics company takes its name from the Latin for speed (Celera), the trope begins to seem a little *pass*[3]. I am still in hot pursuit of speed genomics, but that approach needs a supplement. In my latest bid to get ahead of myself, and ahead of the warp-driven wavefront that *is* genomics, I have come to approach genomics in terms of the future anterior and the promise.

The future anterior is an odd and difficult trope, a full account of which will not be attempted here. (Indeed, a full account is forbidden by its very structure.)[7] I can only beg the reader's patience, for it is best approached indirectly, and over time. Like many odd and difficult things, one can find it articulated in French under the heading of the *futur anterieur*, that tense which signifies *what will have been*. The (il)logic of the operations of the future anterior will become apparent, as I hope to show that if the future anterior is in some sense ungraspable, it is in another sense unavoidable. It is, in any event, intimately linked to the rhetorical figure of the promise.

The promise is that "rhetorical figure of confounding a future with a present, which is at work in every dialectic of presupposition" – an "epistemologically illegitimate rhetorical figure" that nevertheless provides the "ground of meaning, the ground of understanding, the ground of the science of literature".[8] In promising, one gets out ahead of oneself through a kind of unstable, ungainly, but unavoidable boot-strapping operation. Accounting for the promise, and specifically accounting for the Human Genome Project as a series of promises, can only happen on the installment plan. The present essay is an early payment into an account of "promising genomics", an account that, like my debt to Everett, will never be fully paid off.

"Let the Games Begin!": San Diego 1989

Since sometime around the publication of J.L. Austin's *How to Do Things with Words*, we have been better able to appreciate the ways in which at least some enunciations accomplish or perform something rather than simply describe or represent it.[9] Performative speech acts are never simply in opposition to constative speech acts, but always exhibit some complex and subtle entanglements that, indirectly, involve the future anterior and the promise. The "We the people" that opens the Declaration of Independence is one much-discussed example. This "We the people" who author and authorize the document *will only have existed* in the future – the future produced through the signing of the document. The "founding fathers" are founded, through the promise, in the future (anterior).

"Let the games begin!" is another example, perhaps more in keeping with the particular kind of performative effect that historical origin stories often provoke. They do not simply document a beginning, but rather become part of the historiographical apparatus that *produces* that beginning. A simple illustration comes from the story of the "official" beginning to the Human Genome Project.

At San Diego in October 1989, a few thousand scientists gathered for the conference "Human Genome I". The conference (sponsored by *Science*) was, as its name suggests, the first collective gathering on a grand scale of the molecular biologists, human geneticists, biochemists, computer scientists, biotech salespeople, and others involved in fully mapping and sequencing the genomes of humans and other problematic organisms. Also present were a swarm of journalists, a number of people writing books on the Human Genome Project, at least two bona fide historians of science, and at least one anthropologist. It was, as are most events with an "I" in the title, an historic occasion.

Among the many distinguished speakers was James Watson, whose scientific fame originated with his demonstrated capacity for getting out ahead of others. By 1989, he had already established himself as the most important initiator of the Human Genome Project; he had taken an active part on various expert panels, had testified in public hearings before the U.S. Congress and lobbied its members in more private settings, had harangued who knows how many recalcitrant scientists who opposed one aspect of the project or another, and had become the first director of the Office of Human Genome Research at the National Institutes of Health (NIH). In his address at Human Genome I, Watson thought it advisable to clear up potential confusion:

> ... NIH and DOE [Department of Energy] are working together to see if we can get it done as fast as possible ... because if it takes forever then we're not very interested, so we've said fifteen years, and I guess we have to declare a date when the fifteen years start [laughter], and since we want to achieve success we want to put that date off as far as possible [more laughter], so we've now declared this date is the [beginning of the] next fiscal year, October 1, 1990 ... So if people ask, when does the fifteen years start, that's the answer.[10]

Knowing a good joke when they heard one, the gathered scientists shared Watson's sense that origins and calendars did not form a natural fit. Yet they also knew that they would all be tested someday on the "success" of that fit by their patrons in the U.S. Congress. The origin would have to be performed, and Watson was just the man to do it. But even as Watson was performing the fit, invoking the institutions and time-tables of the State to settle some of the arbitrariness and indeterminacy about which one could only "guess", and "declaring" the answer to this question of origins, perhaps the gathered scientists were also laughing at the even better joke: the reason why we're all here in uncomfortable chairs in a cavernous, characterless hotel ballroom trying to digest mediocre institutional food, *is located in the future*. The "real" start of the Human Genome Project had yet to arrive; genomicists were already out ahead of themselves, authorized by the future they had already begun to promise.

In the Beginning Was the Table

Every history of or historically-inclined statement concerning the Human Genome Project will exhibit some performative quality of this sort, even if on a more subtle register than Watson's "declaration". By *saying* an origin, historians or historically-inclined scientists will in some measure, like Captain Jean-Luc Picard on the decks of the warp-driven *Enterprise*, "make it so".

For example, at the beginning of the essay "Out of Eugenics: The Historical Politics of the Human Genome" – itself the essay which begins Daniel J. Kevles and Leroy Hood's book *The Code of Codes* – is the opening sentence: "The scientific search for the 'Holy Grail' of biology dates back to the rediscovery, in 1900, of Gregor Mendel's laws of inheritance".[11] Assuming that neither de Vries, Correns, or Tschermak directly invoked Arthurian legend, the apparently constative statement of the historian harbors some performative elements as well: it realizes the grand continuity of the "scientific search", it re-affirms the kinship among the great men of that tradition, and it helps allay any anxiety a reader might have about what the Human Genome Project announced about the future: stay calm, we've been doing this for a hundred years.

Just a few pages later, the candidates for "origin" have multiplied, when we read that "the human genome project originated largely from initiatives taken in the mid-1980s by Robert Sinsheimer and Charles DeLisi".[12] This more immediate origin still bears the reassuring stamp of great men, whose names serve as prominent hooks on which to hang the historical narrative. Like radioactively-tagged markers of DNA on a dense and writhing chromosome, the names "Sinsheimer" and "DeLisi" serve as stable, luminous landmarks in a dim and shifting landscape. And like the DNA markers that are routinely stored in, bought by, sold to, or shared between genomics laboratories today, these names of people and events are easily archived and easily circulated, ensuring their faithful reproduction and future re-circulation.

The effect is unavoidable, of course. I could not have written a history of the Human Genome Project in the United States without such a handy apparatus, nor would I expect anyone else to do so. But if we are to avoid oversimplifying our narratives of how such simultaneously massive and disseminated events come to be, it is necessary to denaturalize such ready-to-hand tools, and connect them to the social and material technologies on which they rely.

In my thesis, I named such effects "BOGSAT history", after the phrase and its quicker acronym that are widely employed in government and other bureaucratic circles: "bunch of guys sitting around tables". Among the many valuable materials archived by Robert Cook-Deegan (who headed the U.S. Office of Technology Assessment project that analyzed proposals for a Human Genome Project, and who later wrote *The Gene Wars*, the best book available on the formation of the HGP), one can find the actual seating charts from many of the prominent meetings in these origin stories of the Human Genome Project.[13] Or if there is no seating chart, there are lists of participants that guarantee a person's place in history. But rather than fostering inquiry into (or at least speculation on) what kind of notes might have been passed between two people, or what might have been murmured between bowed heads, these lists and charts merely BOGSATiate history. The lists of meetings and attendees are archived and then re-iterated in a series of origin stories. The monstrously complex phenomena subsumed under the name "the Human Genome Project" are figured as the straightforward outcome of plans and decisions made around a piece of furniture.

The following sections dwell upon some more marginal events and figures that usually escape the BOGSAT grid. They show how BOGSAT accounts reproduce themselves, and provide exploded and disseminated accounts of these "initiatives" that are all too easily and often collapsed into the names of Charles DeLisi (a 1984 meeting at Alta, Utah, connected via a text to a 1986 meeting at Santa Fe, New Mexico) and Robert Sinsheimer (a 1985 meeting at Santa Cruz). They also try to tune into the ways in which the future might be said to be as worthy of consideration as anything that happened in 1900.

AT THE SUMMIT: ALTA 1984

A number of scientists came together at the Utah ski resort of Alta in December 1984, funded by the U.S. Department of Energy, to "discuss direct analysis of DNA". Although their conclusion was that current "methods of direct DNA analysis were inadequate to detect the expected increase in mutation frequency from radiation exposure at Hiroshima and Nagasaki", the conference nevertheless is frequently cited as an origin of the Human Genome Project, since it "brought together a welter of related ideas that would grow into the DOE genome project".[14]

The primary vehicles for the promotion and continuation of this BOGSAT story have been DeLisi's own historical account for the *American Scientist* and a 1989 article in the journal *Genomics*.[15] In his historical retrospective for *Science* magazine, Charles Cantor, then director of the Human Genome Center at the DOE's Lawrence Berkeley Laboratory, cites the *Genomics* article to further stabilize this meeting as one of "several independent origins" of "the Human Genome Project".[16] The chemist and writer Robert Shapiro used the *Genomics* article and interviews with Cantor and several others to again ground this past.[17] The Department of Energy's first published annual report on its human genome activities also highlighted this event, and even included the 1989 article in *Genomics* in an appendix to the report.[18]

The *Genomics* article thus circulates and re-circulates in other texts as the main reference point for historicizing the Alta meeting. So it is fitting that the central image of this BOGSAT event, repeated throughout these texts, is not of the Alta meeting itself, but of the biophysicist Charles DeLisi reading a text.

The story runs like this: A staff member of the U.S. Office of Technology Assessment (OTA), Michael Gough, had attended the 1984 Alta meeting and returned to Washington to work on what would become the OTA report *Technologies for Detecting Heritable Mutations in Human Beings.* The report "had been requested by Congress in anticipation that controversies over Agent Orange, radiation exposure during atmospheric testing in the 1950s, and exposure to mutagenic chemicals might find their way to court ...".[19] (These words and phrases are ports to other toxic origin stories, too frequently closed off with the name "Hiroshima", that could further complicate our understanding of the multiple desires and demands running through the origins of the HGP.) It was a draft of this report that is the text in question, providing the link between the 1984 Alta meeting and the 1986 meeting in Santa Fe.

Some of the accounts are straightforward, deploying tropes of initiation, firsts, and seeded messages:

> An early draft of that report, given to me by DOE's David Smith in October 1985, shortly after I [Charles DeLisi] arrived at DOE, was the initial stimulus that led to the Santa Fe workshop.[20]
>
> Charles DeLisi, then newly appointed director of the Office of Health and Environmental Research at DOE, read a draft of this report in October 1985, and while reading it first had the idea for a dedicated human genome project.[21]
>
> Among those who read the report of the Alta conference and got the message was Charles DeLisi.[22]

Other accounts got the same "message" and, while they exhibited a few flourishes in its evolving expression, reproduced it with a high degree of fidelity:

> A rumpled former physicist, DeLisi recalls looking up from a government report on biotechnology one day and thinking that if you could only compare the genome of a child with those of its parents, DNA base by DNA base, you would have an unparalleled medical and research tool for studying mutations.[23]

> DeLisi had thought hard about how such data [DNA sequence information] might be analyzed to reveal the genetic bases of human disease. In October 1985, he found himself thinking hard about that problem again while reading a draft report on the detection of heritable mutations in human beings. He later recalled that he suddenly looked up from the report with the thought of a marvelous way to expose mutations: compare the genome of a child with that of its parents, DNA base pair by base pair. The thought led DeLisi to consider whether it might be feasible to obtain the base-pair sequence in an entire human genome.[24]

With slight variations and embellishments, the basic origin story persists.

Why? Because it *really* happened and *really* is an origin of the Human Genome Project? Because it points to the DOE's "historical mission" of monitoring and studying mutation rates, legitimating its involvement in the Human Genome Project at a time when DOE scientists were often portrayed as mere technicians who knew little about real biology? From a certain perspective these are certainly good reasons for its inclusion in histories. But the story also persists because it is grounded in a BOGSAT event, the memories organized around it, and the texts produced from it, all facilitating the reproduction of certain histories.

The reproduction of the "looking up" trope in the last pair of citations is especially noteworthy. It is a reproduction that proceeds almost exactly in the manner of "DNA base pair by DNA base pair", a textual trait preserved perhaps for its evocation of evolutionary promises: humans are forever "looking up", away from the gritty materiality of texts, machines, and mutations, and toward those elevated realms where "thinking hard" (and "thinking hard again") is the noble path continuing upward, toward the future.

The stability of the origin story is always under threat, however – by unconscious tropologies, by forgetting, by the perpetual openness of all events to multiple reinterpretations. Hence the necessity that BOGSAT archival technologies and their spin-offs; they provide the machinery to counteract the inescapable warps of language, the failures of memory, and future forces of recombination.

There are other ways to re-visit origins, however, as allegorized by the purpose of the Alta meeting itself: employ a more sensitive technology for registering small but frequent narrative mutations. One such historical technology is the "raw" interview transcript.

I interviewed the organizer of the Alta meeting, Ray White, one of the leading scientists in the construction of genetic linkage maps.[25] In the exchange it is not always clear who is providing information to whom, whose memory is being tested, who the authority on this history is supposed to be:

> MF: So, I know you were at, I guess it was the 1984 Alta meeting that the DOE sponsored?
>
> RW: ... Some people track the genome project – I had rather thought that it probably was Sinsheimer's meeting that was really the initiating factor. And I was invited to that, heard about and was invited to it, but it sounded pretty silly to me. So I didn't go to that. More than silly ... it wasn't really relevant to what we were doing, and it did seem to be a bit far-fetched at that point.
>
> MF: Because of the emphasis on sequencing?
>
> RW: Yeah. And it was just – it was clear that Sinsheimer wanted to set up an institute at – whatever that place is ...
>
> MF: Santa Cruz.

RW: Santa Cruz, right. It just seemed a bit grandiose ... Anyway, at the Alta meeting – it was interesting in some ways. It was instigated by a DOE guy – what was his name?

MF: Mort Mendelsohn?

RW: Mendelsohn, right. He had called me out of the blue one afternoon, and said that there had been a discussion stimulated at a Japanese meeting, a Hiroshima meeting, where they were trying to measure mutations in Japanese Hiroshima survivors, and that the prospect had been raised that you could do this with restriction enzymes. That caught my interest – I didn't believe that for a minute [laughs]. Which is, you know, always the sort of thing that gets you engaged in doing stuff. So at any rate, we talked a little bit and it was clear that there was, from his point of view, that the rapidly developing DNA technologies were potentially interesting with respect to this issue, measuring human mutation rates. So he asked if I would help him identify the people and sort of set up and structure a meeting. Alta seemed like a good idea to both of us ...

I don't know that anything really came from it. Is it your impression that anything came from it? Can I ask you occasional questions?

MF: Oh, absolutely. Um – depending on who you talk to, certainly ...

RW: I became convinced that you couldn't use restriction enzymes to measure mutations.

MF: ... certainly within DOE and the people who sort of write the history from that perspective, they say that it was a very important meeting.

RW: So they have it salutary in developing DOE policy over ...

MF: Yeah, more in the policy development, as this is something that could be done. Now I still get conflicting stories as to what DOE's sort of real motivation was behind this. You know, some people say well, it's because the weapons work was running out and they didn't know what else to do, but when I interviewed Charles DeLisi, he was like, this was the furthest thing from my mind and hadn't really come up as an issue.

RW: It was really him, wasn't it? Wasn't this a personality-driven thing?

MF: He did push it a lot.

[Long pause.]

RW: Where are we?[26]

Alta as "origin" of the Human Genome Project is better seen as an emergent effect, the outcome of multiple contingent interactions such as these between historian and respondent. Its stabilization into a textual form that is easily and reliably iterable comes at the price (which is not to say we shouldn't sometimes *pay*) of complexity and alternative interpretations. Given the fluctuations of memory, the variable emphases placed on personalities and events, the subtle negotiations of authority among participants, texts, and historians, White's final off-hand question is always worth returning to: where are we? And even: when are we?

WHY NOT NOW?: THE "IMMEDIATE IMPETUS" AT SANTA CRUZ, 1985

Even the organizer of the "originary" Alta meeting, then, considers the meeting at Santa Cruz to be the real "initiating factor" of the Human Genome Project. James Watson has characterized this May 1985 meeting called by Robert Sinsheimer and Robert Edgar at the University of California at Santa Cruz as the "first serious proposal" for a centralized effort to map and sequence the human genome.[27] Santa Cruz has a very secure place in all origin stories of the Human Genome Project, with virtually every historical account mentioning the Santa Cruz meeting in some way or another.

The first of these came from Sinsheimer himself. Readers of the journal *Genomics*, should they have tired of reading about the latest developments in physical mapping or sequencing strategies, could have read Sinsheimer's own account of this meeting, which carefully lists the participants in this BOGSAT event.[28] A "Nova" documentary that broadcast this history of the Human Genome Project to "viewers like you" of public television beautifully reiterated this BOGSAT quality when it staged a re-enactment of texts being set around a table, with Sinsheimer and a few "stand-ins" shown taking their seats.[29]

The significance attributed to this event is given in terms of *sequencing* the human genome, and in terms of *priority. The Gene Wars* makes the point simply and clearly: "The first meeting focused specifically on sequencing the human genome was convened in 1985 by Robert Sinsheimer of the University of California at Santa Cruz. While the genome project did not grow out of the meeting, or even emerge as a topic of discussion, the 1985 Santa Cruz gathering did plant the seed". In keeping with the nature of seeds, the idea of sequencing the entire human genome "acquired a life of its own" (to continue with *The Gene Wars*' phrasing), and the future that had been packed into this tiny germ thus began to unfold.[30]

Reading Sinsheimer's article itself allows us to see ways in which the future appears to operate, indirectly, *on* these originary events, rather than merely sprouting *from* them.

First, Sinsheimer settles the technical question of whether the human genome could and should be sequenced with a most interesting logic, that of the future anterior. Since "the human genome surely would someday be sequenced, once and for all time", Sinsheimer wrote, "why not now?"[31] Once again, the reasoning here is a futural one: there will come a "someday" when *it will have been done* – when the timing of the event will be irrelevant, when there is no future in terms of DNA sequence to be read, only a "now" – therefore, why not just submit and do it now?

The future exerts its logic in still more subtle ways here as well. In his account, Sinsheimer explains how some "nascent thoughts" (concerning the complete sequencing of a few microorganisms), "visions" (related to the "potential application of genetic knowledge to the human condition"), "concerns" (involving that same "rational approach" to "genetically based disorders"), and "ambitions" (in connection with his job as Chancellor of UC Santa Cruz) all "coalesced" in the mid-1980s.[32] In marking the origin-event of the Santa Cruz meeting, Sinsheimer deploys an array of terms that signify phenomena whose own origins are at best ambiguous, and at least partially anticipatory.

Sinsheimer then gathers all these rather temporally ambiguous originating forces under one "immediate" sign: money. More precisely, a *future potential* of money:

> The immediate impetus for their coalescence was the potential availability of funds for a Human Genome project. UC had received a large gift – $36 million – toward the construction of the 10-m telescope. Through a complex series of events, UC was at that time obliged to return the funds. A confluence of ideas led to the thought that this money might instead be used to launch an Institute to Sequence the Human Genome at UC Santa Cruz.[33]

The "complex series of events" involving the donors and what eventually became the Keck telescope in Hawaii is elucidated nicely in *The Gene Wars*.[34] But in all other accounts, the complexity of the "confluence" is forgotten or elided, and what persists is the trope of the Santa Cruz meeting as a simple origin, a "seed". The important historico-genetic elements are cloned: the names of Sinsheimer and Santa Cruz, so that credit can accrue in the proper accounts; the "idea" of sequencing, so that the proper elevation is ensured; the grand scale and centralized character of the proposal, so that historical continuity can be projected through the later Human Genome Project. These cloned segments of history are then spliced into subsequent narratives, free of all "junk" elements. The entangled confluence of other causes is washed away, discouraging any aberrant expressions – such as that an "origin" of the Human Genome Project lies in hordes of Americans purchasing Volkswagens and BMWs imported from Germany (since this is how the potential donor had amassed the fortune that became the "foundation" whose funds his widow was then trying to give away). Or that an "origin" of the Human Genome Project could be retrospectively PET-scanned out of the billions of neuronal firings within Sinsheimer's skull that, given our pathetic current understanding, we benightedly call "visions", "concerns", and "ambitions". "Nascent thoughts", indeed.

Still, there is something to be gained by taking Sinsheimer's words at face value: even if there was no originary origin but only a "complex series of events" coupled to a series of "nascent" anticipations, there was an "immediate impetus", and the impetus was a potentiality – that is, a promising opportunity. A project to sequence the human genome – whether or not it was feasible, profitable, or advisable – had the value of being able to leverage funds from a kind of futures market.

The Future on the Line: Santa Fe 1986

When we last saw Charles DeLisi, he was being written into the reliable "Eureka!" narrative, "thinking hard" and "looking up" ("suddenly", in the account most true to the eureka-genre) from a report on the detection of heritable mutations. As a result of that flash from the uplifted future (and here I am myself condensing a complex series of events that I have detailed elsewhere),[35] the workshop at Santa Fe was convened in the first week of March 1986. According to DeLisi's *American Scientist* account, this workshop brought together "scientific leaders from industry, academia, and the national laboratories" to discuss the "technical feasibility of sequencing the human genome by the end of the century ... It was reminiscent of one of those rare moments in the early phase of major new ventures, such as the Manhattan Project at Los Alamos, or explorations of outer space, that capture the collective imagination of a community".[36]

Like so much else in the world of genomics, the meeting was conceived, organized, and held with remarkable speed. And once again, while the historical marker "DeLisi" is usually snipped out of its surrounding elements and bound one-to-one with the markers for the meeting – e.g. "DeLisi convened a similar workshop [to the one at Santa Cruz] on the same subject at Los Alamos"[37] – the work of convening and producing the Santa Fe meeting should in fact be credited to an ensemble of agents operating collectively.

In other words, DeLisi, in the nuclear command center of Washington, DC, could only "originate" by being part of an active network of cells disseminated through the American landscape, especially the mesas of New Mexico. DeLisi charged Mark Bitensky, a senior fellow at Los Alamos and head of its Life Sciences Division, with the organizational tasks of selecting participants, making all the arrangements, running the meeting, and preparing a report. He was assisted in this by Ed Hildebrand and Robert Moyzis. At the time, Moyzis was group leader of the Genetics Division at Los Alamos; I spoke with him five years later, after he had been named director of the genome center there. The part of our conversation excerpted here should be read for the themes of memory, connection, and, indirectly as always, the future:

> So December of '85. I remember this very distinctly. I had never met Charles DeLisi ... So I got on the phone with this guy – by now it was probably January when we connected. And then it was like, I want to organize this meeting in Santa Fe next month – so this is one month lag time on a meeting – and why don't you guys just split up and call all the relevant people you can think of, and essentially just do this, just get as many people into Santa Fe with a month's notice as you can get. And the reason this sticks in my mind is because not only did we make so many phone calls, the response was almost uniformly about the same. I'd get on the phone and say, "I'm calling to invite you to a meeting we're going to have next month on the possibility of sequencing the human genome", and usually there was this kind of dead silence on the other end of the phone.

> But almost everybody uniformly said, "Oh, sounds great". So while there was this initial reaction all the time of almost real surprise, I was surprised by a) first of all, almost everybody we asked came, which is a real rarity, especially with that kind of lag time; and b) that almost everybody was uniformly positive, too, at that point in time. That, I guess, somewhat surprised me, that everybody said, "Oh, okay, sounds pretty ambitious, but sounds like a good meeting".[38]

Historians will continually establish new connections, and events and words are bound to take on new tones. I had initially placed that interview excerpt in the context of the debates that were preoccupying both my interviewees and my historical sensibilities at the time: could and should a Human Genome Project emphasize sequencing over mapping and other approaches? Now I couldn't help but be struck by what I previously glided over: the image of a young molecular biologist working for days with a telephone pressed to his ear, catheterized to that most quotidian of technological systems, simultaneously announcing and listening for the (possibly) impending arrival of an even more intricately webbed system of connectivity. And hearing "dead silence" – for a fleeting moment.

Maybe that's why I was haunted by what I had previously regarded as an interesting but innocuous transcript excerpt, as I re-read it over and over while writing this present essay. The two words "dead silence" practically shouted at me, but I couldn't quite make out the meaning. I tried many times to re-establish a connection with this voice from the past, who was in turn re-connecting to his own past, a past which spoke "very distinctly", without the static or hum of a long distance link-up.

Perhaps the only way to read "dead silence" is through the trope of catachresis: deliberately *forcing* a name which does not fit (e.g. "the foot of a mountain"). In my book, then, "dead silence" is another name for "the future". The "dead silence" that Moyzis initally reports hearing, over and over, was not "on the other end" so much as it was *on the line* between them.[39] When the connections were first established in this genomics conference call that was rapidly accruing more and more parties, the future of a fully sequenced human genome spoke in the form of a silence. The surprise, the misrecognition, the silence was momentary, however; the future's voice was already distinct enough to be recognized in early 1986. The future's unthinkable dead silence becomes the recognizable voice of the future anterior: "Oh, it's you ..."

The Santa Fe meeting, then, is better marked by a video-loop image of Robert Moyzis on the phone, dialing and re-dialing and being overtaken by surprise as dead silence sprang quickly to life, than it is by the single frame of Charles DeLisi "thinking hard" and "looking up" from a text.

In the early months of 1986, a long-distance genomics network was emerging along with a future anterior of a fully sequenced human genome, with less and less "lag time" between them.

And just a few months later, there was no silence and no lag time at all but everyone speaking at once.

Finished at the Start: Cold Spring Harbor 1986

The last of the conventional origin-events of the Human Genome Project to be re-read here is the animated discussion that took place in June 1986 at the annual Cold Spring Harbor symposium, that year entitled "The Molecular Biology of *Homo sapiens*".

The many scientific achievements and promises that are reported at these important symposia are always covered in the scientific press. But that year, the first news out of Cold Spring Harbor reported by *Science* concerned not the usual spectrum of noteworthy developments, but this one discussion; *Science* subscribers had to wait another two weeks for a report on the rest of the meeting.[40] Already, the very prospect of a Human Genome Project was pre-empting other biological events.

Consciously or unconsciously, the account in *The Gene Wars* of this hastily-scheduled discussion is quite telling in its language. It gestures toward the uncontrollable flows that are so often overshadowed in the more tightly controlled, rationally planned discourses of origin stories. Paul Berg is shown trying to "channel" the discussion, but "the fusillades became too intense to contain". Remarks by David Botstein, who could "no longer contain his volcanic energy", "broke the dam" and prompted diverse commentaries from the floor. Cautious statements of a DOE representative were again "largely swept aside as the dam broke", and Berg "struggled intermittently and unsuccessfully to contain the flood", but "the emotional torrent was simply too strong".[41] Almost everything was thrown into the turbulent mix: scientific arguments, strategic concerns, speculations on what congressional representatives would find appealing, economic arguments, cultural concerns regarding scientific boredom, baleful analogies to the space program, and nationalist worries about Japanese superiority in biotechnology.

One of the symposium attendees who entered the discussion at this point was Maxine Singer, an eminent scientist perhaps most known outside the scientific community for her role in calling attention in 1973 to the potential hazards of recombinant-DNA research.[42] *The Gene Wars* characterizes Singer's comments as "focusing on the notorious failure of science to predict its future".[43] While not inaccurate, this gloss requires a closer and different reading. Here are Singer's comments to which *The Gene Wars* seems to be referring:

> The other night, Dan Koshland put up a slide of predictions about science from a group of very famous scientists of the past. And we all had a marvelous time laughing at their wisdoms. There are many other examples of that, and therefore I come to the conclusion that we won't know whether it was worth doing a project like this, as a project *per se*, until we're more or less finished with it, and see what we've learned from it. But what we do know is that the history of the last thirty years has told us, that when we do biochemistry and when we do DNA sequencing, in conjunction with genetics, then we learn things, and we learn things at a very good rate. There are people who say, for example, that the sequencing of the SV40 genome

> really led us forward, but it led us forward because ... it went hand in hand with the genetics. And that's why I feel that an approach which includes mapping and genetics and cDNAs, and things that we have some sense of the function of, and taking this sequencing as we go with that, makes a lot more sense than devoting a large amount of resources, both talent and money, to something which we can all agree is ultimately worth knowing, probably, but along the way I think we'll move faster in the kinds of things that we like to know and that will be useful for science, for medicine, for agriculture, for whatever.[44]

Time has been kind to Singer's arguments, which amount to a rough synopsis of how the Human Genome Project actually came to be defined over the next few years of negotiations: the development of genetic maps, the use of cDNAs (complementary-DNA, produced from the messenger-RNA of expressed genes) as important tools, and other lines of scientific work were emphasized over the mass production of sequence information. But if the "take-home message" was easy for *The Gene Wars* to retrospectively decode, other participants at the time were less sure of the proper reading frame. David Botstein, who had already spoken passionately and at some length against a large-scale, sequencing-intensive project, experimented with a reinterpretive response:

> Maybe this isn't what Maxine meant to say, but this is what it meant to me, and I think it's a fundamental point. We don't know what – we don't want to know just the sequence of nucleotides or the sequence of genes or the sequence of promoters. We want to know what it all means for biological function, OK? That's the big problem. And my claim is that, for better or for worse, knowing the nucleotide sequence is just the very bare beginning. I think that's what Maxine was trying to get at.

Keep both Singer's and Botstein's remarks in the back of your mind for a moment, and consider the subsequent remarks injected into this discussion by Walter Gilbert. Gilbert had by then become the most prominent scientist advocating a sequencing-intensive project; he had been a powerful voice at both the Santa Cruz and Santa Fe meetings, and had opened the discussion at Cold Spring Harbor by presenting his vision of such a large-scale endeavor. Gilbert responded to Singer and Botstein with what would become for him (and a few others) a much-used set of phrases over the next few years. This project was not "all of biology" and should not be considered as definitive or all-encompassing; it was a "tool" with which one could do biology more efficiently, and differently:

> [T]he comment you're making, Maxine, is really, let us do all of biology. That's true, that the problem of, let's say, human biology, is ultimately to know all of the cDNAs, all the connections between genes and function, and that is the measure, that is the core of the biological effort. If you can dress

> that up and say, this [human genome project] is a shorthand for that project – I don't think this is a shorthand for the other. It shouldn't be. At best, this sort of project is a tool to use toward the other.

With these three linked commentaries from the Cold Spring Harbor discussion in place now, perhaps we can better triangulate on the future anterior and how it might be seen to operate, through some half-present cybernetic logic loop in time, on the Human Genome Project and on biology itself.

Singer's initial comments didn't so much question science's ability to predict its own future as they pointed to how science necessarily *relies on its own future.* Its present "worth" will only have been established in the future. In my reading, the "fundamental point" that both Singer and Botstein seem to have been "trying to get at" – the oblique paths of the future anterior are always difficult to traverse – is that when it comes to biological research, and perhaps to biological entities themselves, there is no fundamental point. Neither organisms nor biology have a rock-bottom "beginning". "Trying to get at it" is indeed possible, and even necessary, but one will never arrive, in part because the "very bare" point of origin only exists or can only be approached because of the other term here that defines this discursive space: "what it all means". DNA sequence information (and its accelerating pursuit) makes sense only within the context of "all of biology" – a context which technically exists only in the future. A future, moreover, that *will have been produced* out of the pursuit of the DNA sequence information whose meaning is underwritten by this yet-to-be-produced future biology.

I have selected these few excerpts from the extremely rich and wide-ranging Cold Spring Harbor symposium because they mark out the oppositions of the discursive field: The Human Genome Project, and biology more generally, occurs *between* DNA and organism, *between* "very bare beginning" and "what it all means", *between* an origin and an "ultimate", neither of which is either fully present or absent, because they trace the limits of the conceptual system that animates the entities within it. The two terms underwrite each other and require each other, but neither one is more "fundamental" than the other. The irresolvable tension between these terms is a productive one, the tension of the force-field that they establish. One does not know what "all of biology" is, yet one presumes, in advance, that DNA sequences are "fundamental" to it – or rather, that DNA sequences will have been fundamental to it. It's this will-have-been that is the odd logic of the future anterior.

Or written somewhat differently: one promises that DNA sequences are simultaneously the fundamental origin of the future of biology, and of the biology of the future. In his own way, Walter Gilbert understood this promissory structure of genomics and the Human Genome Project better than almost anyone else. Gilbert's main reason for advocating a full-throttle, high-powered pursuit of genomics was not that it would complete our understanding of "what it means to be human", let alone "all of biology", but that concerted efforts to map and sequence genomes, and to develop the new machines, databases, and techniques for such an effort, was the way to leverage an altogether different future for biology – his much-vaunted "paradigm shift in biology".[45] Such an anticipated but unknown transformation couldn't be guaranteed, but it could be promised – that is, affirmed through one's words and deeds.

The promissory character of genomics means that it is also a kind of bet or gamble – a quality which would soon become even more evident to Gilbert, as he attempted over the next few years to launch a corporation that would produce, copyright, and sell DNA-sequence information. Gilbert promised venture capitalists and major pharmaceutical corporations that such sequence information would be a valuable commodity. But no one was buying it, in large part because in 1987 such a future was too uncertain and too distant. Investors could hear only dead silence from the future, and Gilbert's corporation went unrealized. Only five years later, however, commercial genomics companies like Incyte Pharmaceuticals, Human Genome Sciences, Millenium Pharmaceuticals, Celera, and others had incorporated and were thriving, able to bank on Gilbert's future which in the meantime had arrived, or at least became much easier to hear.

But we're somewhat ahead of ourselves again. Perhaps we can now better understand Gilbert's resistance to "shorthand", his suspicion of a writing that telegraphically collapses the profusion of words and things that constitute an organism, a discipline, or a social project, into a few hastily inked marks. My re-articulation of his argument: Don't write "all of biology" in the shorthand form of "DNA sequences" (in the scientific field) or "the Human Genome Project" (in the social and political field), produce instead, through the continual re-affirmation of and re-dedication to the promise of genomics, the new writing instruments that will re-write a future biology in all its profuse detail. The object of the Human Genome Project will have been to re-write biology into a biology of continual re-writing, to produce a different set of discursive (i.e. material, social, textual) practices; the object will not have been to condense or reduce biology into any kind of "shorthand", that writing technique that partakes of the rhetorical trope of synecdoche.

The Once and Future Organism, and Other Funny Things

Gilbert was also the genomics advocate most closely associated with the phrase, "the holy grail of genetics". In some ways, it's sad that this rhetoric, so popular in early discussions and media coverage of the Human Genome Project, fell out of favor. That it was largely replaced by "infrastructure building" as a legitimating trope is also quite understandable.[46] Infrastructure is all origin – mundane, grounded, concrete – with hardly a whiff of millennial dreams or other futural fantasms. Despite its obvious faults, the invocation of Christianized Celtic desires for "the once and future king" at least sheltered an important truth about the conceptual and social structure of the Human Genome Project: it awaits the arrival of the unifying sovereign from the past/future.

But so, in a way, do the multiple organisms – fruit flies, yeast, mice, humans, and so many others – that are the object of genomics today.

In *On Beyond Living*, Richard Doyle analyzes the rhetorics of molecular biology and the life sciences more generally in the second half of the twentieth century. A scientific field so dependent on the widely disseminated concept-metaphor of "language" demands a precise and relentless pursuit of how "language itself" works to simultaneously constitute a set of methods and questions, a conceptual field, and an object of study. Doyle analyzes the "rhetorical software" that is so absolutely vital to both the life sciences today, and to the organisms that populate them.

"The future" is a recurrent trope in Doyle's analysis, and is especially evident in his treatment of François Jacob and Jacques Monod's efforts to install control and "genetic regulation" in the genetic discourse in the early 1960s. Reading Jacob and Monod's scientific and autobiographical texts, Doyle shows how the concept of the origin – in this case, the origin that lies in the "genetic program" that is then "expressed" to become an organism – is both impossibly and inextricably bound to its future:

> The simple and brutal possibility condition for expression is an organism, an organism that must in some sense exist – live – "before" the construction of itself through a strand of DNA and its messenger. By "containing" the means of protein synthesis within the genome, Jacob and Monod also contain the time necessary for that execution. The claim that the genome, if not the structural genes alone, is necessary and sufficient for the "definition" of the structure of proteins depends upon the erasure of the nucleic acids' dependence on the cytoplasm and the organism, an erasure that is made possible by the slippage between "instruction" and "construction" in Jacob and Monod's rhetoric of definition. This slippage effaces the process of protein synthesis, a process that requires time and an organism, not just a genetic fiat.[47]

The "shorthand" which Walter Gilbert warned against at Cold Spring Harbor is in fact the very shorthand that molecular biology (and, less intensely, genetics before that) had employed so productively for decades. It is a shorthand that collapses the difference between an organism and its genetic program – a difference bound up with the question of time and development. With Monod and Jacob's operon model of the gene and regulation, the genetic program becomes a synecdoche for the organism. The part stands in for the whole. But "the impossible anteriority of a gene abstracted from a developed organism", Doyle writes, "returns in the form of the problem of embryology, the discipline that deals with the development of organisms over time, the very time deleted here by the synecdoche of the genetic program".[48]

Doyle's intent is not to expose the "fallacy" of this synecdoche, or of the related metonyms, metaphors, chiasmas, catachreses, and other unavoidable tropes of any discourse such as the life sciences. "My aim here is not to point out narratival holes ... Rather, I am concerned with the effect of these holes on the narrative ... What is produced through this impossible reliance on the future is not impossibility but tension, a suspension or oscillation between temporalities that allows the complexity of a dynamic system to be described".[49]

This may seem like a rather funny way of talking about the life sciences, but that may be exactly why it helps us understand what Evelyn Fox Keller has called the "funny thing [that] happened on the way to the holy grail". Keller traces what she calls the "discourse of gene action" – in which the gene is synecdochally abstracted from the organism and granted legislative authority – back to the split between embryology and genetics effected by T.H. Morgan. "[E]ven in the early days of genetics", she writes, "when the gene was still merely an abstract concept and the necessity of nuclear-cytoplasmic interactions was clearly understood, geneticists of Morgan's school tended to assume that these hypothetical particles, the genes, must somehow lie at the root of development".[50] She too emphasizes the productivity of such a move:

> But in introducing this particular way of talking, the first generation of American geneticists provided a conceptual framework that was critically important for the future course of biological research ... It enabled geneticists to get on with their work without worrying about the lack of information about the nature of such action – to a considerable degree, it even obscured the need for such information.[51]

The discourse of gene action has allowed geneticists to leverage a future without a full guarantee from a grounding past. Even in its synecdochal incompleteness and temporally impossible shorthand, it provided "powerful rationales and incentives for mobilizing resources, for identifying particular research agendas, for focusing our scientific energies and attention in particular directions ... And it would be foolhardy to pretend it has not worked well. The history of twentieth-century biology is a history of extraordinary success; genetics – first classical, then molecular – has yielded some of the greatest triumphs of modern science". But particularly with the intensification of this discourse through the infrastructure provided through the Human Genome Project, something "funny" has begun to happen within this discourse, argues Keller. The "extraordinary progress" we've made with our infrastructure for manipulating genes "has become less and less describable within the discourse that fostered it. The dogmatic focus on gene action called forth a dazzling armamentarium of new techniques for analyzing the behavior of distinct gene segments, and the information yielded by those techniques is now radically subverting the doctrine of the gene as sole (or even primary) agent".[52] The future, repressed, returns.

I know that it's "funny" to write about the life sciences, and about the history of the Human Genome Project, with phrases such as those employed in the present essay concerning the future (anterior) and its oblique effects. But the truth is sometimes funny, as the scientists at the Human Genome I conference recognized when they heard Watson's declaration of the official (futural) beginning of the HGP. As something that overtakes a person, laughter is a sign or effect of that more general overtaking of the subject that psychoanalytic theory has analyzed. Slavoj Zizek asks: "What is a 'journey' into the future if not this 'overtaking' by means of which we supposed in advance the presence in the other of certain knowledge ... This knowledge is an illusion, it does not exist in the other, the other does not really possess it, it is constituted afterwards through our – the subject's – signifier's working: but it is at the same time a necessary illusion because we can paradoxically elaborate this knowledge only by means of an illusion".[53] The other in its various instantiations – a foundational DNA, a causality grounded exclusively in the past, or the creators of the origins of the Human Genome Project – is a necessary and generative illusion retroactively summoned, anteriorized from the future.

My pursuit of the future anterior is akin to what Paul Forman describes as the historian's responsibility "to make evident the overdetermination of the emergent discovery". Like other overtakings such as laughter, the "source" of such historical overdeterminations as we see in the Human Genome Project lies outside the systems of the sciences and their history. Forman paradoxically contrasts overdetermination to inevitability:

> Overdetermination ... implies both less and more than inevitability. Less, because it assumes no transcendent reality ... whose activity upon our world is merely and literally metaphorical. More, because holding to the concept of causality while disbelieving in a unique chain of causation, historians and their readers are required to recognize and wrestle with the very large number of characteristically very different material, social, and personal factors that act not metaphorically, but actually upon the discoverer.[54]

My funny way of re-reading these historical events is not intended to imply that the Human Genome Project was inevitable, or pre-programmed in the preformationist sense. It's not intended to imply that the actions of individuals (Watson, DeLisi, White, Moyzis, and so many others) were not important, or to say that the Human Genome Project was not indeed to some extent a "personality-driven thing", in the words of Ray White. It's not intended to imply that the various expert panels that gathered around tables in Santa Cruz, Santa Fe, Washington DC, or elsewhere didn't exert rather powerful effects. Obviously, these have been very popular, compelling, and productive ways to account for the origins of the Human Genome Project, and tell us a great deal about its formation.

But they are "shorthand" accounts, allegorical companions to the discourse of gene action whose ongoing history they recount. Such shorthand devices are an unavoidable feature of language and thought, a conceptual structure and rhetoric in which what will count as an origin or impetus will have been circumscribed in specific ways, with a number of specific effects. As we know from literary theory, every insight is inextricably bound to a blindness. This is a necessary effect of the limits of any system, that "necessary illusion" that makes both historiography and molecular biology possible and productive, as analyzed in my re-readings of Watson's declaration, the Santa Cruz meeting, and the Cold Spring Harbor discussion. Another effect is the bracketing or marginalization of "non-genetic" causes, as discussed in the sections on the Alta summit and the Santa Fe conference, where complex interactions in a disseminated system are elided or neglected.

The experimental figures of the future anterior and the promise of another way of writing the history of genomics. That future historiography – to which the present essay can only gesture – would also exhibit certain homologies with the practices and theories of the life sciences. Rather than the tropes of codes, programs, foundations, and origins, however, perhaps the future writing of histories and organisms will have produced, through persistent tinkering, rhetorical and other machines better suited to those phenomena grouped under the heading of *emergence* that so interest historians and biologists alike, yet prove so elusive. *That* history and that biology will have learned how to think and write of the subtle and sudden forcefulness, the untimeliness, of the promise and the future anterior that have somehow, now, always and already overtaken us – not as dead silence, but as thoroughly *quick* one.

Rensselaer Polytechnic Institute

NOTES

[1] Elizabeth Pennisi, "Academic Sequencers Challenge Celera in a Sprint to the Finish", *Science* 283 (19 March 1990), pp. 1822.

[2] This is not to suggest that these accounts omit any suggestion of disorder, disagreement, conflict, or chance occurrence. It is more a question of dominant effect [see Roman Jakobson, "The Dominant", pp. 41–46 in *Language and Literature,* ed. K. Pomorska and S. Rudy (Cambridge: Harvard University Press, 1987)], or of a text and its margins [see Jacques Derrida, "Outwork, Prefacing", in Barbara Johnson (trans.) *Dissemination* (Chicago: University of Chicago Press, 1981)], pp. 2–64.

[3] George E. Marcus, "Critical Cultural Studies as One Power/Knowledge Like, Among, and in Engagement With Others", in *Ethnography Through Thick and Thin* (Princeton: Princeton University Press, 1998), pp. 203–230, on p. 208.

[4] James D. Watson, "The Human Genome Project: Past, Present, and Future", *Science* 248 (1990), pp. 44-49.

[5] Michael A. Fortun, *Making and Mapping Genes and Histories: The Genomics Project in the United States 1980–1990* (unpublished Ph.D. dissertation, Harvard University, 1993).

[6] Michael Fortun, "Projecting Speed Genomics", *The Practices of Human Genetics*, Michael Fortun and Everett Mendelsohn, eds. (Dordrecht: Kluwer, 1999), pp. 25–48.

[7] See Jacques Derrida, "At this very moment in this work here I am", in *Re-Reading Levinas*, Robert Bernasconi and Simon Critchley, eds. (Bloomington: Indiana University Press, 1991).

[8] Werner Hamacher, "LECTIO: De Man's Imperative", in *Reading De Man Reading*, ed. Lindsay Waters and Wlad Godzich (Minneapolis: University of Minnesota Press, 1989), pp. 197–198.

[9] J.L. Austin, *How to Do Things with Words* (Oxford University Press, 1976). An essential supplement to Austin's work is Jacques Derrida, *Limited Inc* (Evanston: Northwestern University Press, 1988).

[10] James D. Watson, "Organization: NIH", address delivered at Human Genome I meeting, 2–4 October 1989, San Diego, author's transcript.

[11] Daniel J. Kevles, "Out of Eugenics", in *The Code of Codes: Scientific and Social Issues in the Human Genome Project*, Daniel J. Kevles and Leroy Hood, eds. (Cambridge, Mass.: Harvard University Press, 1992), p. 3. Just as the phrase "the King of Kings" is intended to trump any revisionist project in the religious realm, the very title of this book can be read as an attempt to foreclose any recession of the origin in the domain of the life sciences and their history. The

foreclosure can never in fact succeed, however, since one can (and should) always ask, "What codes 'the code of codes'?", "What codes the 'code of "the code of codes" '?", and so on.

[12] *Ibid.*, p. 18.

[13] Robert Cook-Deegan, *The Gene Wars: Science, Politics, and the Human Genome* (New York: W.W. Norton, 1994). The many materials he gathered are deposited at the Human Genome Archive, National Reference Center for Bioethics Literature, Georgetown University, Washington, DC.

[14] Cook-Deegan, *The Gene Wars* (cit. n. 13), p. 96.

[15] Charles DeLisi, "The Human Genome Project", *American Scientist* 76 (Sept.-Oct. 1988), pp. 488–493; Robert M. Cook-Deegan, "The Alta Summit, December 1984", *Genomics* 5 (1989), pp. 661–663. This meeting is also highlighted in Cook-Deegan's later writings: Robert Cook-Deegan, "The Human Genome Project: The Formation of Federal Policies in the United States, 1986–1990", *Biomedical Politics*, Kathi E. Hanna, ed. (Washington, DC: National Academy of Sciences Press, 1991), pp. 99–168; Cook-Deegan, *The Gene Wars*, pp. 95–96.

[16] Charles R. Cantor, "Orchestrating the Human Genome Project", *Science* 248 (6 April 1990), pp. 49–51.

[17] Robert Shapiro, *The Human Blueprint: The Race to Unlock the Secrets of Our Genetic Code* (New York: St. Martin's Press, 1991), pp. 220–221.

[18] U.S. Department of Energy, *Human Genome: 1989–90 Program Report*, Office of Energy Research, Office of Health and Environmental Research; see p. 2 and Appendix B.

[19] Cook-Deegan, "The Human Genome Project: The Formation of Federal Policies in the United States, 1986–1990" (cit. n. 15), p. 662.

[20] DeLisi, "The Human Genome Project" (cit. n. 15), p. 489.

[21] Cook-Deegan, "The Human Genome Project: The Formation of Federal Policies in the United States, 1986–1990" (cit. n. 15), p. 622.

[22] Shapiro, *The Human Blueprint* (cit. n. 15), p. 221.

[23] Robert Kanigel, "The Genome Project", *The New York Times Magazine*, 13 December 1987, pp. 44ff; on p. 98.

[24] Kevles, "Out of Eugenics" (cit. n. 11), p. 18.

[25] Ray White was one of the co-authors of the historical paper that first described how restriction fragment-length polymorphisms (RFLPs) could be used to construct dense, full-coverage genetic linkage maps for humans; see David Botstein, Raymond L. White, Mark Skolnick, and Ronald W. Davis, "Construction of a genetic linkage map in man using restriction fragment length polymorphisms", *American Journal of Human Genetics* 32 (1980), pp. 314–331.

[26] Author's interview with Ray White, October 16, 1991; see Fortun, *Making and Mapping Genes and Histories* (cit. n. 5), Chapter 3, for a fuller account of the development of genetic linkage maps with restriction fragment length polymorphisms (RFLPs) in the 1980s, and White's important role in this work.

[27] Watson, "The Human Genome Project: Past, Present, and Future" (cit. n. 4), p. 45.

[28] Robert Sinsheimer, "Historical Sketch: The Santa Cruz Workshop – May 1985", *Genomics* 5 (1989), pp. 954–956. In addition to this more immediate historical origin story, Sinsheimer has also offered a tale of more epic proportions based on evolutionary, if not cosmic, significance. Sinsheimer has argued that "the Human Genome Initiative is a hinge point in biological evolution" and "a turning point in human history" in which both the past ("we are the first to comprehend our origins and now to reveal the very genesis of our being") and the future (in which this "epic venture of discovery" will "hopefully alleviate human genetic flaws that produce so much misery and suffering") come into focus; Sinsheimer, "The Human Genome Initiative", *FASEB Journal* 5 (1991), p. 2885. Another article opens in a similar vein: "The Human Genome Project represents the convergence of three billion years of biological evolution and ten thousand or more years of cultural evolution – and their interaction is bound to change both, profoundly. From now on, their futures will be indissolubly linked". Sinsheimer, "Whither the Genome Project?", *Hastings Center Report* (July/August 1990), p. 5.

[29] Produced WGBH in Boston, this "Nova" program was titled "The Book of Life", and originally aired on public broadcasting stations on Halloween, October 31, 1989.

[30] Cook-Deegan, *The Gene Wars* (cit. n. 13), pp. 79, 84.

[31] Sinsheimer, "Historical Sketch" (cit. n. 28), pp. 955–956.

[32] *Ibid.*, p. 955.

[33] *Ibid.*

[34] Cook-Deegan, *The Gene Wars* (cit. n. 13), pp. 80–82.

[35] See Fortun, *Making and Mapping Genes and Histories* (cit. n. 5), Chapter 2.

[36] DeLisi, "The Human Genome Project" (cit. n. 15), p. 489. The question of what time-zone a "collective imagination" resides in is one I will have to defer. But I would like to add a note of appreciation and even praise for the BOGSATiated histories that I critique. The Santa Fe meeting, as DeLisi's description here suggests, is noteworthy not only for having enlarged the community that was part of the discussion, but for changing the sociality of the discussion as well. Reading down the long list of participants, one sees that there were not only far more people at the Santa Fe meeting than at either the Alta or Santa Cruz meetings, but that there was a good mix of scientists famous and obscure, representing diverse disciplines and practices, from universities, corporations, and government laboratories – and there was a better percentage of women present (4 women and 39 men) than there were at Alta (19 men; see Cook-Deegan, "The Alta Summit") or at Santa Cruz (where Helen Donis-Keller was the sole woman among 18 participants; see Sinsheimer, "Historical Sketch"). Whether or not this tells us anything about various cultures of research within DOE or other communities, or about gender and molecular genetics more broadly, are questions for further interviewing. I only note the fact that such suggestive starting points for inquiry can be extracted from BOGSAT accounts, and that the organizers of the Santa Fe workshop seem to deserve some credit for achieving a degree of diversity, for whatever reasons, that other "origin" events did not. For the list of Santa Fe participants, see Mark Bitensky, *Sequencing the Human Genome: Summary Report of the Santa Fe Workshop, March 3-4, 1986*, Office of Health and Environmental Research, U.S. Department of Energy (Los Alamos National Laboratory, 1986).

[37] Kevles, "Out of Eugenics" (cit. n. 11), p. 19.

[38] Author's interview with Robert Moyzis, October 22, 1991.

[39] My reading here is no doubt influenced by having been on the line too long with Avital Ronell, who suggests that the historian's job is akin to that of the switchboard operator: "When I'm on the job, I shall try to make a connection on a somewhat complicated switchboard that always threatens to jam up. This will be no reading of the Purities, but an attempt to move back and forth between what lights up before us. This includes the call from the past which tries to disguise its voice while at the same time telling us that the future is on the line. Nothing happens on this switchboard that does not announce itself as coming from the future". Avital Ronell, *Finitude's Score: Essays for the End of the Millenium* (Lincoln: University of Nebraska Press, 1994), p. 221. See also Ronell's extended reading of the telephone's nervous systems, *The Telephone Book: Technology, Schizophrenia, Electric Speech* (Lincoln: University of Nebraska Press, 1989).

[40] See Roger Lewin, "Proposal to Sequence the Human Genome Stirs Debate", *Science* 232 (1986), pp. 1598–1600; and *idem*, "Molecular Biology of *Homo sapiens*", *Science* 233 (1986) pp. 157–8.

[41] Cook-Deegan, *The Gene Wars* (cit. n. 13), pp. 110–13.

[42] See Sheldon Krimsky, *Genetic Alchemy: The Social History of the Recombinant DNA Controversy* (Cambridge, Mass.: MIT Press, 1982).

[43] Cook-Deegan, *The Gene Wars* (cit. n. 13), p. 113.

[44] Author's transcript of taped discussion at Cold Spring Harbor meeting, "The Molecular Biology of *Homo sapiens*", June 1986; tape recording made by C. Thomas Caskey and deposited by Robert Cook-Deegan at the Human Genome Archive, National Reference Center for Bioethics Literature, Georgetown University, Washington, DC.

[45] Walter Gilbert, "Towards a Paradigm Shift in Biology", *Nature* 349 (January 10, 1991), p. 99. The article also serves to mark the long distance between the days of Thomas Kuhn, when paradigm shifts were accounted for centuries after their occurrence, and today's anticipation of them.

[46] For an analysis of this shift in legitimating arguments for the Human Genome Project from the "Holy Grail" to the "Route One of Genetics", see Fortun, *Making and Mapping Genes and Histories* (cit. n. 5), Chapter 5; see also Fortun and Herbert J. Bernstein, *Muddling Through: Pursuing Science and Truths in the 21st Century* (Washington, DC: Counterpoint Press, 1998), Chapter 7, "Producing Multiplicities: Inquiry Infrastructures for Molecular Genetics".

[47] Richard Doyle, *On Beyond Living: Rhetorical Transformations of the Life Sciences* (Stanford: Stanford University Press, 1997), p. 75.

[48] *Ibid.*, p. 76.

[49] *Ibid.*, p. 78.

[50] Evelyn Fox Keller, *Refiguring Life: Metaphors of Twentieth-Century Biology* (New York: Columbia University Press, 1995), pp. 6–7.

[51] *Ibid.*, pp. 10–11.

[52] *Ibid.*, pp. 21–22.

[53] Cited in Doyle, *On Beyond Living* (cit. n. 47), p. 77.

[54] Paul Forman, "Inventing the Maser in Postwar America", *Osiris* 7 (2nd series, 1992), pp. 105–134, on p. 129.

IV

POSTSCRIPT:
EVERETT MENDELSOHN AND THE MIDDLE EAST

YARON EZRAHI

"SPEAKING JUSTICE TO POWER": EVERETT MENDELSOHN AS AN INTERNATIONAL PUBLIC INTELLECTUAL

ABSTRACT

The international public intellectual functions as an outsider deriving authority in given local situations from faith in universal moral principles that are expected to impose some limits on the actions of states, nations and individuals. As such an international public intellectual, Everett Mendelsohn has negotiated between conflicting forces and ideologies both in the United States (in protests against the Vietnam War) and in the Middle East (in the protracted Israeli-Palestinian conflict). A "moral realist", Everett has been particularly effective in buttressing moral arguments with supportive facts. His faith and hope in human beings became a major contribution to discussions between opposing sides in the international public sphere.

Unlike national public intellectuals, international intellectuals are basically outsiders who derive their authority from faith in the existence of some universal, moral principles which are expected to impose limits on the actions of states, nations and individuals everywhere. Since this belief in universal moral principles and their superior authority over particular values such as national greatness, political hegemony, economic interests and the like, is not universal, those who claim to represent such principles and insist on their application in any particular context, are likely to be exposed, resisted, abused, insulted or ignored. In addition to their other skills, therefore, such international intellectuals must have a kind of stubbornness and toughness which allows them to persist in pursuing their goals under fire.

Unlike national intellectuals, international intellectuals do not usually try to operate directly as actors in the domestic political arena of the countries they scrutinize or criticize. Although they may try to alert citizens to the consequences of violations of universal, moral principles by their own governments, they usually try to deploy their influence in the international arena where nations and governments struggle to win the support of world public opinion. Characteristically, these international public intellectuals attempt to ally themselves

G.E. Allen and R.M. MacLeod (eds.), Science, History and Social Activism: A Tribute to Everett Mendelsohn, 365–368.

with local cosmopolitans who can translate their fear of international censorship into domestic political pressures on politicians, generals and other national leaders involved in actual or potential violations of international norms. Often, of course, the court of international public opinion is merely an idealized projection of a desirable frame whose actual institutional and behavioral embodiments are fragmentary, elusive and unstable. The prospects for the development of such an internationally shared civic culture of moral discipline are, of course, a matter of wide disagreements. In my opinion, they are neither very good nor negligible. In any case, in the absence of a sufficiently stable support system, international public intellectuals depend even more upon their unique skills in promoting the application of these principles via a host of strategies, programs or roles, and on their ability to navigate among conflicting interests and perspectives in order to decide judiciously when to keep a low profile and when to approach the mass media.

Everett Mendelsohn belongs to this very special class of often lonely, self-charged, international intellectuals working as mediators or facilitators in foreign lands. Although many years ago Everett worked on the Middle East conflict within the framework of the Quakers' Friends Committee, he has for decades since operated as a lone, single, peace entrepreneur, travelling extensively between regional adversaries and sometimes talking to involved American officials in the pursuit of viable alternatives to the use of violence, the frequency of which has been a characteristic of the Middle East.

Any attempt to briefly characterize Everett's extensive, long-standing involvement as a mediator, facilitator or "mitigator" in the Arab-Israeli conflict is likely to distort an enormously rich and complex network of activities. But, if I were pressed to give such a characterization, at the risk of simplification, I would say that common to his activities is the desire to empower lay publics and leaders "to speak justice to power", to advance and reinforce moral considerations in an arena usually dominated by national, if not nationalistic, strategic, military and political interests.

The moral-intellectual critique of power has, of course, marked Everett's work not only in the Middle East but also as a professor of the history of science and a scholar working on the relations between science, politics and society. Much of his scholarly work and his record as a teacher has been devoted to unearthing and exposing to intellectual and normative scrutiny, the often-hidden links between knowledge, political and military powers and economic interests. Very early, perhaps earlier than most scholars and observers in this field, Everett discerned and called attention to the abuse of the enlightenment vision of the role of science and its mission in the service of democratization. Instead of augmenting the powers of laymen to hold their governments accountable and influence public policy-making, science and technology, according to Everett, were all too often enlisted in the service of the national state and its clerks. Only rarely supporting lay critics of arbitrary power and failed government actions, the knowledge and authority of experts were used most frequently to rationalize government actions and win the trust and sacrifices of credulous citizens.

Everett Mendelsohn's work in the Middle East is related to his long-standing critique of the modern nation-state, its cynical use of science and technology to disempower lay opposition or criticism, to enhance its military power and consequently also destabilize the international order. In the particular context of the Middle East, Everett's concern was, for a long time, largely directed to the role of the big powers in seeking to augment their influence by arming the adversaries with increasingly more advanced weapons. Eventually he directed more attention to the direct relations between the adversaries themselves.

Everett's moral authority in the Arab-Israeli context has gained from his record during the 1960s as a critic of his own government's policies in Vietnam. In both the domestic American and the Middle East contexts, Everett was particularly effective in exposing the fragile grounds on the basis of which experts were used by clerks to exclude laymen as participants or critics. Everett often noted in these contexts that technical training is not a substitute for moral education and that instrumental criteria conceal rather than discard the relevance of ethical choices. He characteristically targeted the special affinities between the arrogance of power and the arrogance of expert knowledge, including in the Middle East context of professional "orientalists". He could bring many examples of how the arrogance of experts led them to believe for decades that intellectually certified knowledge almost automatically embodies public values. Everett, moreover, brought with him to the profoundly fatalistic cultures of the Middle East, the insight that what is regarded as inevitable or inexorable is not independent of shared beliefs and that in politics Darwinian notions of human conflicts draw on vulgar and usually untenable notions of scientific theories.

Considering the scope and diversity of his skills and insights, Everett's attention to the Middle East has been a very special gift to both Arabs and Jews. In numerous meetings (some of which I had attended) held especially during the 1980s and 1990s, Everett played a major role in facilitating contacts and communications across deeply entrenched suspicions and pyramids of sacrifice.

As a former student of Everett's, I could not fail to notice how his extraordinary skills as an inspiring teacher and his qualities as a listener in the classroom were converted in these charged political and emotional situations into assets of mediation and moderation. Beyond these skills, Everett introduced into this conflict situation the particular approach of the moral realist. By moral realism, I do not mean the approach which tempers moral idealism by pragmatic considerations. In the case of Everett Mendelsohn, moral realism means mostly the desire and capacity to buttress moral arguments by supportive facts and by realistic analysis of what is politically feasible. On many occasions I witnessed Everett carefully and painstakingly constructing a diagnosis of which forces and trends could be enlisted, or counted on, in advancing a more just solution to the Arab-Israeli conflict.

At an international Arab-Israeli conference held in Europe following the Gulf War, in the course of which Palestinian support for Sadaam Hussein produced a deep crisis in the Israeli peace movement, Everett characteristically defended Palestinian self-determination as both an historically unconditional moral imperative and a politically expedient move to defuse the kind of frustrations and humiliations which make Palestinians vulnerable to the most extreme fundamentalist or nationalist influences. In the same speech, Everett presented a realistic analysis of domestic Israeli political forces discussing the elements which already before the rise of Rabin to power, and the breakthrough achieved in the secret Oslo talks, indicated to him the trend which later, in fact, would soften rigid Israeli attitudes and policies cultivated during Shamir's Likud government. "By contrast to Israel's governmental reluctance" he observed, "public opinion polls show significant support in Israel for joining a peace process and ultimately for making some territorial compromises in return for a durable and secure peace". This, of course, was the trend in Israeli public opinion which later allowed Rabin and Peres to embark on difficult and risky peace negotiations with the Arabs.

Following this observation, Everett moved on to insist on the importance of "delineating a clear role for Israel within the Middle East [breaking] the stereotypical views that have long been held that the Arabs are reluctant partners to agreements, and that Israel can ultimately rely on American backing independent of what position it adopts in the peace process. The current situation", he concluded, "is fluid, and with some insightful help and rigorous action, solutions may be achieved". In his concluding remarks, Everett again took the high ground of a concerned international participant observer discerning a harmony between realism and ethical concerns, between the morality and the political necessity of Palestinian self-determination. In the above, as in many other speeches and interventions made by Everett while addressing groups and audiences of all shapes and sizes, the common drive has been to explore, examine and where possible exploit the diverse, mostly latent, potential resources for the politics of peace. The groups consisting of Arabs and Jews who met in Europe, the Middle East and the U.S., in order to discuss the conflict informally, often came to depend on Everett's remarkable ability to show sympathy and to simultaneously respond to the fears, anxieties and concerns of both sides.

Sometimes, no doubt, Everett's positions appeared overly naive or rationalistic to one side or the other. Among the deeply fatalistic and pessimistic Middle Easterners, his optimism and respect for good moral and political arguments was bound to sometimes appear "too American", floating above the realities of suffering, displacements, humiliations and existential fears experienced by both Arabs and Jews. Yet it is precisely such experiences which produce "realism" as a self-paralyzing frame of mind, and it is precisely such "innocent" optimism which, by contrast, produces the very expectations which can lead to change. Everett has been able to function as an agent of the best that the American democratic spirit could offer to us in the Middle East, a tradition of faith and hope in human beings which, at certain moments of historical grace, can turn into a self-fulfilling prophecy.

The Israel Democracy Institute, Jerusalem

BIBLIOGRAPHY: THE PUBLISHED WRITINGS OF EVERETT MENDELSOHN

The bibliography below represents the complete published work of Everett Mendelsohn to date. We are particularly grateful to Yuri Hospadar of the Department of History of Science at Harvard for his assistance in preparing and checking the details of this listing [Editors].

Books

A Treasury of Scientific Prose. A Nineteenth-Century Anthology, I. Bernard Cohen, Howard Mumford Jones, and Everett Mendelsohn, eds. (Boston: Little Brown, 1963).

Heat and Life: The History of the Theory of Animal Heat (Cambridge: Harvard University Press, 1964).

Human Aspects of Biomedical Innovation, E. Mendelsohn, J.P. Swazey and I. Taviss, eds. (Cambridge: Harvard University Press, 1971).

Science and Values, E. Mendelsohn, Arnold Thackray, eds. (New York: Humanities Press, 1974).

Topics in the Philosophy of Biology, Everett Mendelsohn and Marjorie Grene, eds. (Dordrecht: Reidel, 1976).

The Social Production of Scientific Knowledge. Sociology of the Sciences Yearbook, Vol. 1, Everett Mendelsohn, Peter Weingart and R. Whitley, eds. (Dordrecht/Boston: Reidel, 1977).

The Social Assessment of Science. Report 13, Everett Mendelsohn, Dorothy Nelkin and Peter Weingart, eds. (Bielefeld: Wissenschaftsforschung, 13 Report, Universität, Bielefeld, 1978).

Sciences and Cultures, Anthropological and Historical Studies of the Sciences. Sociology of the Sciences Yearbook, Vol. V., Everett Mendelsohn, Yehuda Elkana, eds. (Dordrecht/Boston: Reidel, 1991).

Principal Author *A Compassionate Peace, A Future for the Middle East* (New York: Hill and Wang; London: Penguin Books, 1982); revised edition, *A Compassionate Peace, A Future for Israel, Palestine, and the Middle East* (New York: Hill and Wang, 1989).

Transformation and Tradition in the Sciences, Everett Mendelsohn, ed. (Cambridge: Cambridge University Press, 1984).

Nineteen Eighty-Four: Science Between Utopia and Dystopia. Sociology of the Sciences Yearbook, Vol. 8, E. Mendelsohn and Helga Nowotny, eds. (Dordrecht/Boston: Reidel, 1984).

Science, Technology and the Military. Sociology of the Sciences Yearbook, Vol. 12, E. Mendelsohn, Merritt Roe Smith and Peter Weingart, eds. (Dordrecht/Boston: Kluwer Academic Publishers, 1988).

G.E. Allen and R.M. MacLeod (eds.), Science, History and Social Activism: A Tribute to Everett Mendelsohn, 369–373.

Technology, Pessimism, and Postmodernism. Sociology of the Sciences Yearbook, Vol. 17, E. Mendelsohn, Yaron Ezrahi and Howard Segal, eds. (Dordrecht/Boston: Kluwer Academic Publishers, 1994); republished as paperback, Amherst: University of Massachusetts Press, 1995.

Biology as Society, Society as Biology: Metaphors. Sociology of the Sciences Yearbook, Vol. 18, Everett Mendelsohn, Sabine Maasen and Peter Weingart, eds. (Dordrecht/Boston: Kluwer Academic Publishers, 1995).

E. Mendelsohn, Jeffrey Boutwell, *Israeli-Palestinian Security: Issues in the Permanent Status Negotiations* (Cambridge: American Academy of Arts and Sciences, 1995)

The Practices of Human Genetics. Sociology of the Sciences Yearbook Vol. 19, Everett Mendelsohn and Michael Fortun, eds. (Dordrecht/Boston: Kluwer Academic Publishers, 1999).

Science in Culture, Everett Mendelsohn, Peter Galison and Stephen R. Graubard, eds. (New Brunswick, N.J.: Transaction Publishers, 2001). Originally published as a special issue of *Daedalus*, Winter 1998.

Life Sciences in the Twentieth Century, Biographical Portraits, Everett Mendelsohn, ed. (New York, Charles Scribner's Sons, 2001).

Life Sciences Before the Twentieth Century, Biographical Portraits, Everett Mendelsohn, ed. (New York: Charles Scribner's Sons) [In Press 2002].

Articles

"From Innate Heat to Animal Heat Metabolism: A Critical Survey", *Actes IX*° Congrés International d'Histoire des Sciences, Barcelona-Madrid (Paris: Hermann, 1959), pp. 388–398.

"Animal Heat in the Eighteenth Century: The Early Chemical Phase", *Colloque International de l'Histoire de Chimie aux XVIIIè Siécle* (Paris: Hermann, 1959).

"John Lining and His Contribution to Early American Science", *Isis* 51 (1960), pp. 278–292.

"Schwann's Mistake", *Actes X*° Congrés International d'Histoire des Sciences, Ithaca: Paris: Hermann, 1962 (1964), pp. 967–970.

"Controversy Over the Site of Heat Production in Animal Body", *Proceedings of the American Philosophical Society* 105 (1962), pp. 412–420.

"Scientists and the Development of Atomic Weapons", *Society for Social Responsibility in the Sciences Newsletter*, April, 1962, pp. 1–3.

"Science in America: The Twentieth Century", *The Evolution of American Thought,* A.M. Schlesinger, Jr. and M. White, eds. (Boston: Houghton Mifflin, 1963), pp. 432–445.

"The Biological Sciences in the Nineteenth Century: Some Problems and Sources", *History of Science* 3 (1964), pp. 39–59. Reprinted in Bobbs-Merrill Reprint Series in History of Science, 1966.

"Cell Theory and the Development of General Physiology", *Archives Internationales d'Histoire des Sciences* 65 (1964), pp 419–429. Reprinted in *Voice of America Forum History of Science Series Lectures*, 8 (Washington, D.C.: US Information Agency, 1964), pp. 1–8.

"The Changing Nature of Physiological Explanation in the Seventeenth Century", *Melanges Alexandre Koyré: L'Avenure de la Science* (Paris: Hermann, 1964), pp. 367–386.

"La vie scientifique aux Etats-Unis au XX siècle", *Histoire générale des sciences. Tome III, pt. 2, La Science Contemporaire* (Paris: Presse Univsité France, 1964), pp. 914–923.

"The Emergence of Science as a Profession in Nineteenth-Century Europe", *The Management of Scientists* (Boston: Beacon Press, 1964), pp. 3–48.

"Science Has a Social Context: Comments upon Papers by Dr. Hans Bahrdt and Dr. Jacob Schmookler", in *Economic and Social Factors in Technological Research and Development*, Richard A. Tybout, ed. (Columbus: Ohio State University Press, 1965), pp. 51–58.

"Physical Models and Physiological Concepts: Explanation in Nineteenth-Century Science Biology", *Boston Studies in the Philosophy of Science* 2 (1965), pp. 201–219.

"The Context of Nineteenth-Century Science", *Golden Age of Science.* B.Z. Jones, ed. (New York/Washington: Simon & Schuster/Smithsonian Institution, 1966), pp. xiii–xxvii.

"The History of Science" *New Book of Knowledge* (Danbury, CT, Grolier Press,1966), pp. 60–76.

"The Confrontation of Science, Technology and Modern Society", *Contemporary Civilization* 4, James Findlay, ed. (Glenview, IL: Scott, Foresman, 1967), pp. 182–196.

"A Definition of the Irreversible Coma, Report of the Ad Hoc Committee of the Harvard Medical School to Examine the Definition of Brain Death", with Henry K. Beecher *et al.*, *Journal of the American Medical Association*, 205 (1968), pp. 337–340.

"Philosophy and Overview: The Pre-History of Biomedical Engineering" *Dimensions of Biomedical Engineering, University of Pittsburgh, School of Engineering Publication Series*, E. Salkovitz, L. Gerende, and L. Wingard, eds. (San Francisco: San Francisco Press, 1968), pp. 7–48.

"Three Scientific Revolutions" in *Science and Policy Issues,* Paul J. Piccard, ed. (Itasca, Illinois: F.E. Peacock Publishers, 1969), pp. 19–35.

"Philosophical Biology vs. Experimental Biology: Spontaneous Generation in the Seventeenth Century", *Acts 12th International Congress of History of Science*, Vol. Ib, 1968 (Paris: Blanchard, 1971), pp. 201–229.

"A Human Reconstruction of Science", *Boston University Journal* (Spring, 1973), pp. 45–52.

"Should Science Survive its Success?" *Tech Engineering News* 55 (October, 1973), pp. 16–24; expanded version published in *For Dirk Struik,* R.S. Cohen *et al.*, eds. (Dordrecht: D. Reidel, 1974), pp. 373–389.

"Revolution and Reduction: The Sociology of Methodological and Philosophical Concerns in Nineteenth-Century Biology", in *Interaction Between Science and Philosophy*, Yehuda Elkana, ed. (Jerusalem/New York: Humanities Press, 1974), pp. 407–427.

"Comparative Studies in Science and Medicine: Problems and Perspectives", in Arthur Kleinman *et al.*, eds. *Medicine in Chinese Cultures: Comparative Studies of Health Care in Chinese and Other Societies* (Washington, D.C: U.S. Dept. of Health, Education, and Welfare, Public Health Service, National Institutes of Health, U.S. Govt. Print. Office, 1975), pp. 659– 667.

"Value Choices Involved in Nuclear Development", *Proceedings of the Symposium on Nuclear Energy and Public Health and Welfare in Colorado* (Boulder, CO: American Friends Service Committee, November, 1975), pp. 41–50.

"Values and Science: A Critical Reassessment", *The Science Teacher* 43 (1976), pp. 21–23 (Washington: U.S. Department of Health, Education, and Welfare, 1975).

"The Crisis of Values in Modern Science: Prognosis for the Future", *Science and Society: A Symposium* (Rhode Island College, Providence, RI: Rhode Island College Press, 1976), pp. 20–23.

With James Sorenson and Judith Swazey "Recombinant DNA–Science as a Social Problem, Appendix", *Special Study* (Washington: National Commission for the Protection of Human Subjects of Biomedical and Behavioral Research, December 1976), pp. 317–335.

"The Internationalization of Science", *Social Implications of the Scientific and Technological Revolutions* (New York: UNESCO, 1981 [1977]), pp. 4–17.

"The Social Construction of Scientific Knowledge", in *Social Production of Scientific Knowledge*, E. Mendelsohn, P. Weingart, and J. Whitley, eds. (Dordrecht: Reidel, 1977), pp. 3–26.

"The Social Assessment of Science: Issues and Perspectives", in *The Social Assessment of Science*, E. Mendelsohn, D. Nelkin and P. Weingart, eds. (Bielefeld: Wissenschaftsforschung 13, Report, 1978), pp. 321.

"Historia da Ciencia e Estudos em Politica Cientifica", *Anais do Seminario Internacional de Estudos Sobre Politica Cientifica* (Rio de Janeiro: Conselho Nacional de Desenvovimento Cientifico e Technologico, 1978), pp. 1–17.

"Current Challenges to Contemporary Values in Science and Technology", Counterpoint Forum: *Science and Ethics* (Perth, Australia: Murdoch University Press, 1979), pp. 3–21.

"Science as History", *Harvard Magazine* 82 (Nov–Dec 1979). pp. 71–75.

"Die Notwendigkeit einen Humanen Rekonstruktion der Wissenschaft", in *Brauchen Wir Eine Andere Wissenschaft?, X* Salzburger Humanismusgespräch, Oskar Schatz, ed. (Salzburg: Verlag Styria, 1980).

"The Continuous and the Discrete in the History of Science", in *Constancy and Change in Human Development*, J. Kagan and O. Brim, eds. (Cambridge: Harvard University Press, 1980), pp. 75–112; revised and reprinted in *Journal of Social Reconstruction* 1 (1980), pp. 131.

"New U.S. Missiles in Europe: Challenge to Arms Control and Disarmament", *New Perspectives* 10 (1980), pp. 7–8.

"The Historian Confronts the Bomb", *Proceedings of the Symposium, The Role of the Academy in Addressing the Issues of NuclearWar*, Washington, DC: March 25–26, 1982 (Geneva, NY, Hobart and William Smith College Press, 1982), pp. 44–59.

"Knowledge and Power in the Sciences", in *Science Under Scrutiny*. R.W. Home, ed. (Dordrecht/ Boston: Reidel, 1983), pp. 31–47; revised and reprinted in *The Kaleidoscope of Science* (Dordrecht: Reidel, 1986), pp. 225–240.

"Science, Power and the Reconstruction of Knowledge", in *Knowledge and Higher Education*. Gunnar Bergendal. ed. (Stockholm: National Board of Universities and Colleges, 1983), pp. 47–83.

"Frankenstein at Harvard, The Public Politics of Recombinant DNA Research", in *Transformation and Tradition in the Sciences,* E. Mendelsohn, ed. (Cambridge: Cambridge University Press, 1984), pp. 317–335.

"Disinventing Nuclear War", *Antioch Notes* 55(2) (1984), pp. 1–6.

"The Origin of Life and the Materialism Problem", *Revue du Metaphysique et Morale* 90 (1985), pp. 15–25.

"Historians and the Nuclear Temptation", *Jahrbuch, Wissenschaftskollege zu Berlin, 1983/84* (Berlin: Wissenschaftskolleg zu Berlin, 1985), pp. 217–228.

"Knowledge and Power in the Sciences", in *The Kaleidoscope of Science*, E. Ullman-Margalit, ed. (Dordrecht: Reidel, 1986), pp. 225–240.

"Der Kommentar: Die Notwendigkeit einer humanen Rekonstruktion der Wissenschaft: Humanität, Vernunft und Moral in der Wissenschaft", *Dialektik* 14 (1987), pp. 13–26.

"The Political Anatomy of Controversy in the Sciences", in *Scientific Controversies: Case Studies in the Resolution and Closure of Disputes in Science and Technology,* Tristram Engelhardt and Arthur Caplan, eds. (Cambridge: Cambridge University Press, 1987), pp. 93–124.

"Science and Democratic Education for the 21st Century", *Integrating Knowledge: A Report on the CETE Conference on Undergraduate Teacher Education and the Liberal Arts*, Middlebury, Vermont, October, 1988 pp. 21–30.

"Robert K. Merton: The Celebration and Defense of Science", *Science and Context* 3 (1989), pp. 269–289.

"Science, Technology and the Military: Patterns of Interaction", in *Science, War and Peace,* Jean-Jacques Salomon. ed. (Paris: Economica, 1990), pp. 49–70; translated in Portugese, "Cienca, Technlogia e Sistema Militar, Modelos de Interaccao", *Cienca, Tecnologia, Sociedade* 13/14, 1991, pp. 5–21.

"Prophet of Our Discontent: Lewis Mumford Confronts the Bomb", in *Lewis Mumford, Public Intellectual,* Thomas and Agatha Hughes, eds. (Oxford: Oxford University Press, 1990), pp. 343–360.

"The Social Locus of Scientific Instruments", in *Invisible Connections: Instruments, Institutions and Science,* Robert Bud and Susan Cozzens, eds. (Bellingham, WA: SPIE Optical Engineering Press, 1992), pp. 5–22.

"Religious Fundamentalism and the Sciences", in *Fundamentalisms and Society, Reclaiming the Sciences, the Family, and Education,* Martin Marty and R. Scott Appleby, eds. (Chicago: University of Chicago Press), 1993, pp. 23–41.

"Grasping the Elusive Peace in the Middle East", *Harvard Middle Eastern and Islamic Review* 1 (1994), pp. 1–16.

"The Politics of Pessimism: Science and Technology circa 1968", in *Technology, Pessimism, and Postmodernism. Sociology of the Sciences Yearbook Vol. 17, 1993*, Yaron Ezrahi, Everett. Mendelsohn, and Howard Segal. eds. (Dordrecht/Boston: Kluwer Academic Publishers, 1994), pp. 151–173. Republished in paperback (Amherst MA: University of Massachusetts Press, 1995), pp. 151–173.

"Thinking Like a Mountain: The Epistemological Puzzle of Environmentalism", in *Grenzüberschreitungen in der Wissenschaft.* Peter Weingart, ed. (Baden-Baden: Nomos Verlagsgesellschaft, 1995), pp. 152–167.

"Science and the Construction of the Idea of Europe", *Vest* 8(25) (1995), pp. 59–64.

"Can Oslo Survive?" *Harvard Middle Eastern and Islamic Review* 3 (1996), pp. 1–19.

"Science, Scientists and the Military", in *Science in the Twentieth Century*, John Krige and Dominique Pestre, eds. (Reading, UK: Harwood Academic Publishers), 1997, pp. 175–202.

"Is Public Policy Lagging Behind the Sciences?" in *Genomics, Healthcare and Public Policy*, Paul Williams and Sarah Clow, eds. (London: Office of Health Economics, 1999), pp. 64–93.

"The Eugenic Temptation: When Ethics Lag Behind Technology", *Harvard Magazine* (March-April, 2000), pp. 39–41, 105–106.

NOTES ON CONTRIBUTORS

Mark B. Adams was a student of Everett Mendelsohn at Harvard 1962–1970, where he earned his B.A. (1966), M.A. (1968), and Ph.D. (1973). In 1970 he moved to the Department of History and Sociology of Science at the University of Pennsylvania, where he helped to found the program and has served variously as its Undergraduate and Graduate chair. His teaching and research interests have focused on the history of Russian and Soviet science, the history of biology (especially evolution, genetics, population genetics, and eugenics), and science and literature (especially science fiction). His eighty publications include *The Wellborn Science: Eugenics in Germany, France, Brazil, and Russia* (Oxford, 1990); *The Evolution of Theodosius Dobzhansky* (Princeton, 1994); and, most recently, "The Visionary Biology of J.B.S. Haldane" [*Journal of History of Biology*, 33 (2000), 257–291].

Garland Allen received his B.A. in Biology and English from the University of Louisville, and an MAT in English from Harvard, teaching high school biology for three years (1958–61) before returning to graduate school at Harvard in the history of science, to work under Everett Mendelsohn and Ernst Mayr. He is the author of *Life Science in the Twentieth Century* (Wiley, 1975; Cambridge University Press, 1978) and *Thomas Hunt Morgan, the Man and his Science* (Princeton University Press, 1978). His interests in the history of science and biology grew from his experience as a Teaching Fellow in George Wald's introductory biology course, Natural Sciences 5. Following two years as Allston-Burr Senior Tutor in Quincy House at Harvard, he took the job he still holds in the Biology Department at Washington University, St. Louis, where he teaches both the history of biology and introductory biology to biology majors. He writes on the history of genetics, evolution and development, and the history of American eugenics, and has co-authored seven biology textbooks. In 1998, he presented the Sarton Award Lecture at the AAAS. Since 1998, he has been co-editor of the *Journal of the History of Biology.*

G.E. Allen and R.M. MacLeod (eds.), Science, History and Social Activism: A Tribute to Everett Mendelsohn, 375–381.

Mark V. Barrow, Jr., is an associate professor in the History Department at Virginia Tech, Blacksburg, Virginia, where he teaches environmental history, the history of science, and American history. He received a B.A. in History from the University of Florida (1986) and a Ph.D. in the history of science from Harvard University (1992), with Everett Mendelsohn as his advisor. His first book, *A Passion for Birds: American Ornithology after Audubon*, was published by Princeton University Press in 1998. His current research focuses on the evolution of Anglo-American attitudes toward anthropogenic (human-induced) extinction from the late 18th century to the Endangered Species Act. Since 1990, he has served as Book Review Editor for the *Journal of the History of Biology*.

Peter Buck is Dean of the Summer School and Senior Lecturer on the History of Science at Harvard. He has also taught at U.C.L.A. and M.I.T. His first book, *American Science and Modern China, 1876–1936* (Cambridge University Press, 1980), examined the transfer of scientific ideas and organizations from the Untied States to China in the late 19th and early 20th centuries. He has written a variety of essays and articles on topics ranging from vital statistics, the history of the social sciences, and child health, to studies of army life during the Second World War. With Barbara Rosenkranz, he is writing a book, *Healthy, Wealthy and Wise*, which discusses community medicine, public health, and applied social research in America between the two world wars.

Fred Churchill is Professor Emeritus at Indiana University, where he taught and served for many years as Chairman of the Department of History and Philosophy of Science. He received his B.A. and Ph.D. from Harvard, as a graduate student of Everett Mendelsohn. He has written on the history of late nineteenth and early twentieth-century embryology, particularly the work of August Weismann. With Helmut Risler, he recently published two volumes of Weismann's edited letters (Universitätsbibliothek, Freiburg, 1999). He has also worked on a variety of issues related to cytology in the late nineteenth century.

Yaron Ezrahi studied political science at the Hebrew University of Jerusalem and at Harvard, where he received his doctorate in 1972. He is an authority on the theory and culture of modern democracies, on which he has written many articles and two books, including *The Descent of Icarus: Science and the Transformation of Contemporary Democracy* (Harvard University Press, 1990), and *Rubber Bullets: Power and Conscience in Modern Israel* (farrar, Straus and Giroux, 1997), which won the Jewish National Book Award for that year. He has served as a consultant to the White House, the OECD, and the Carnegie Commission on Science and Government, and has been a consultant to the Israeli Parliament and the Israeli National Academy of Science. He is well known as a commentator on the Arab-Israeli conflict, Israeli politics, and the peace process in the Middle East.

Raphael Falk is professor Emeritus of genetics at the Hebrew University of Jerusalem. He was educated at the Hebrew University and at the University of Stockholm, and has been on the academic staff of the Hebrew University since 1955. He has been a visiting professor at Columbia University; the University of Oregon; the Institute for Advanced Studies, Berlin; the University of Massachusetts at Amherst; and Harvard. He has done experimental research in population genetics, chromosome organization, and developmental genetics of *Drosophila*. Since 1984, he has been studying the history and philosophy of genetics and evolution, a topic on which he has published a number of papers. He is also co-editor of a collection of a special issue of *Science in Context* (1998), entitled, "Eugenic Thought and Practice: A Reappraisal" (Cambridge University Press).

Mike Fortun is a historian of science whose work has focused on molecular biology, human genetics, and the biotechnology industry. He is a co-founder and Executive Director of the Institute for Science and Interdisciplinary Studies (ISIS) at Hampshire College in Amherst, Massachusetts, a non-profit organization which conducts scientific research; provides technical assistance on environmental issues; and sponsors publications and public events which cultivate the critical yet respectful science literacy crucial to democratic society. He is co-author, with physicist Herbert J. Bernstein, of *Muddling Through: Pursuing Sciences and Truth in the 21st Century* (Counterpoint Press, 1998), and served as co-editor, with Everett Mendelsohn, of *The Practice of Human Genetics* (Yearbook of the Social Sciences,1999).

Anne Harrington is Professor of the History of Science at Harvard, specializing in the history of psychiatry, neuroscience, and the other mind sciences. As an undergraduate, she studied in Harvard's History and Science Program, where she was inspired by Everett Mendelsohn to pursue graduate study. She received her D.Phil. in the History of Science from Oxford in 1985, and has held postdoctoral fellowships at the Wellcome Institute for the History of Medicine, and the University of Freiburg. She is currently Director of the Harvard University Mind, Brain, and Behavior Interfaculty Initiative, and was for six years a member of the MacArthur Foundation Research Network on Mind-Body Interactions. She is the author of two books with Princeton University Press, *Re-enchanted Science: Holism and German Culture from Wilhelm II to Hitler* (Princeton University Press, 1997), and *Medicine, Mind and the Double Brain: A Study in Nineteenth-Century Thought* (Princeton University Press, 1987) and the editor of two collections. The book on which she is currently working, an anthropological and historical investigation of the making and current status of modern mind-body medicine in America, is entitled *Stories under the Skin*.

Lily Kay was, until her untimely death in 2000, a member of the Program in Science, Technology and Society at MIT, where she taught with Everett Mendelsohn, and wrote on the history of molecular biology. She was the author of *The Molecular Vision of Life: Caltech, the Rockefeller Foundation, and the Rise of the New Biology* (Oxford University Press, 1993), *Who Wrote the Book of Life? A History of the Genetic Code* (Stanford University Press, 1999), and numerous articles on the history of molecular biology and its patronage in the 1930s by the Rockefeller Foundation. At the time of her death, she was working on the history of research in neural networks.

Roy MacLeod is Professor of History at the University of Sydney. He studied with Everett Mendelsohn at Harvard, where he took his B.A. in History and Science in 1963. He received his doctorate in History at Cambridge, where he was elected a Research Fellow of Churchill College. In 1970, he established the History and Social Studies of Science Subject Group at the University of Sussex, and was appointed foundation Reader in the subject. In 1978, he was appointed to a foundation chair at the Institute of Education of the University of London, and in 1982, moved to Sydney. He has held visiting appointments in Cambridge, Paris and Amsterdam, at Harvard, the Smithsonian Institution, the University of Indiana, the University of California (Santa Cruz), the University of Umeå, Christ Church, Oxford, and Heidelberg. He has written many essays in social and administrative history, and the social history of science, medicine and technology. His books include *Public Science and Public Policy in Victorian Britain* (Variorum, 1995) and *The Creed of Science* (Variorum, 2000); and many edited volumes, including *Darwin's Laboratory: Evolutionary Theory and Natural History in the Pacific,* with P.F. Rehbock (University of Hawaii Press, 1994); *Technology and the Raj: Technical Transfers and Technological Change in British India, 1780–1945* , with Deepak Kumar (Sage, 1995); and *Science in the Pacific War, 1939–1945* (Kluwer, 2000). He was the founding co-editor of the journal *Social Studies of Science* in 1970, and is now the Editor of *Minerva.*

Jane Maienschein is Professor of Philosophy and of Biology and co-director of the Biology and Society Program at Arizona State University, where she studies the history and philosophy of developmental and cell biology and genetics as a means of exploring the nature of science and scientific change. She is the author of two books, *Transforming Traditions in American Biology 1880–1915* (Johns Hopkins University Press, 1991), and *100 Years Exploring Life. An Autobiography of the Marine Biological* Laboratory (Jones and Bartlett Press, 1989), and editor/co-editor of several volumes, including *The American Expansion of Biology* (Rutgers University Press, 1991), *The American Development of Biology* (University of Pennsylvania Press, 1988). She has been co-editor of the *Journal of the History of Biology* since 1998.

Helga Nowotny was born in Vienna, and holds a doctorate in law from the University of Vienna and a Ph.D. in sociology from Columbia University. She is currently professor in Philosophy and Social Studies of Science at the Swiss Federal Institute of Technology (ETH), Zurich, and Director of the Collegium Helveticum. She has held senior visiting positions at the University of Vienna;

the Wissenschaftskolleg zu Berlin; King's College, Cambridge; the University of Bielefeld; the École des Hautes Études en Sciences Sociales, Paris; and the Center for Social Sciences in Berlin. For many years, she has been a Permanent Fellow of the Collegium Budapest/Institute of Advanced Study, where she is now President of the Scientific Advisory Board. She is a member of the Academia Europaea and serves in an advisory capacity with the European Commission. Her principal research lies in the social studies of science and technology.

Robert Proctor is Distinguished Professor of the History of Science at Pennsylvania State University. He has recently been a Senior Scholar in Residence at the U.S. Holocaust Memorial Museum in Washington, D.C. (1994); a Visiting Scholar at the Hamburger Institut für Sozialforschung (1995); and a visiting scholar at the Max Planck Institute for the History of Science in Berlin (1999–2000). He has been awarded fellowships by the Fulbright Commission, the National Institutes of Health (ELSI), and the National Endowment for the Humanities. He has written many articles and four books: *Racial Hygiene: Medicine under the Nazis* (Harvard University Press, 1988); *Value-Free Science? Purity and Power in Modern Knowledge* (Harvard University Press, 1991); *Cancer Wars: How Politics Shapes What We Know and Don't Know About Cancer* (Basic Books, 1995); and *The Nazi War on Cancer* (Princeton, 1999). His research interests coalesce around the political history and philosophy of science; he has also written on environmental policy, bioethics, molecular anthropology, anthropology and racial theory, and the "social construction of ignorance." He is working on a book on theories of human origins, a book on agates (mineralogy), and a book on the history of evolution. He lives in State College, Pennsylvania, with Londa Schiebinger and their two children.

Shirley A. Roe is Professor of History at the University of Connecticut, having previously taught in the History of Science Department at Harvard. She is author of *Matter, Life and Generation: Eighteenth-Century Embryology and the Haller-Wolff Debate* (Cambridge University Press, 1981), editor of *The Natural Philosophy of Albrecht von Haller* (Arno Press, 1981), and co-editor of *Science Against the Unbelievers: The Correspondence of Bonnet and Needham* (Voltaire Foundation, 1986). She is currently working on a book on the political and social context of biological materialism in the eighteenth century. She was Associate Editor of the *Journal of the History of Biology* from 1981, and (with Everett Mendelsohn) co-editor from 1989 to 1998.

Jean-Jacques Salomon is Honorary Professor of Technology and Society at the Conservatoire National des Arts et Métiers (CNAM) in Paris. He graduated in philosophy and anthropology from the Sorbonne, where he obtained his doctorate in philosophy and history of science. From 1963 to 1983, he created and directed the Science and Technology Policy Division of the Organization for Economic Cooperation and Development (OECD). He founded with Derek de Solla Price and chaired the International Council for Science Policy Studies attached to the ICSU network, and presided over the Standing Committee for Social Sciences of the European Science Foundation (Strasbourg). More recently, he chaired the Collège de la Prévention des Risques Technologiques attached to the Prime Minister's office. He has been an invited professor at MIT, Harvard, Montreal, and at the São Paulo Institute for Advanced Study; and a Visiting Fellow at Clare Hall, Cambridge. His major books include *The Research System* (OECD, 1972–74), *Science and Politics* (MIT Press, 1973), *Mirages of Development: Science and Technology for the Third World* (Lynne Rienner, 1993), *The Uncertain Quest: Science, Technology, Development* (UNU, 1994), *Science, War and Peace* (Economica, Paris 1989), *Le destin technologique* (Balland/Gallimard, 1994), *Survivre à la science: Une certaine idée de l'avenir* (Albin Michel, 2000), *Le scientifique et le guerrier* (Belin, 2001).

Elizabeth Siegel Watkins took both her B.A. in biology (1984) and her Ph.D. in the history of science (1996) from Harvard University. She has worked as a research assistant at the Pasteur Institute in Paris (1982–83), and taught biology at the Middlesex School in Concord, Massachusetts (1985–1989). She is the author of *On the Pill: A Social History of Oral Contraceptives, 1950–1970* (Johns Hopkins University Press, 1998). She lives with her husband and two daughters in Pittsburgh, where she is an adjunct professor of history at Carnegie Mellon University.

Peter Weingart was born in Germany, and studied sociology, economics and political science at Berlin, Freiburg and Princeton. Since 1983, he has taught at the University of Bielefeld, where he has been Professor of Sociology, Director of the Center of Interdisciplinary Research (ZiF), (1989–1994), and Director of the Institute of Science and Technology Studies (1993–present). He has worked closely with Everett Mendelsohn in editing the *Sociology of the Sciences Yearbook*. He has been a Fellow of the Wissenschaftskolleg zu Berlin (1983/84), and is a member of the Berlin-Brandenburg Academy of Science (1997). He has been on the editorial boards of a number of journals, including *Science, Technology and Human Values*, *Minerva*, the *Zeitschrift für Soziologie*, and the *Journal for the History of Biology*.

Gary Werskey entered Harvard as a history graduate student in 1965, and completed his Ph.D. in 1973 under the joint supervision of Stuart Hughes and Everett Mendelsohn. Between 1968 and 1987 he lived in the UK, where he taught science studies and industrial sociology at Edinburgh, Leicester, Bath and London Universities. During this period, he co-founded the *Radical Science Journal* (1972), wrote *The Visible College: A Collective Biography of British Scientists and Socialists of the 1930s* (Holt, Rinehart and Winston,1979), and undertook a major study of Japanese engineers. In 1987, he moved to Australia, creating at the University of New South Wales an undergraduate leadership program for technologists and an open learning postgraduate course in business and technology. Since 1993, he has been a co-Director of

Educational Partnerships Australia, a management consultancy firm specializing in organizational and personal leadership development. He is also a co-founder of Kanimbla View, a Blue Mountains environmental retreat near Sydney.

Gilbert Whittemore took Everett Mendelsohn's famous "Science and Society" course (Soc Sci 119) as an undergraduate at Harvard. Subsequently, he took a law degree at Harvard, with a special interest in law, science, and technology, and a PhD with Mendelsohn in the history of science. Since then, he has combined law with part-time teaching in the history of science at Harvard, MIT, Brown and Lesley College. From 1994 to 1995 he was a senior staff member of the Advisory Committee on Human Radiation Experiments appointed by President Clinton. Currently, he serves with the law firm of Stalter & Kennedy, LLP, in Boston, and is Chair of the Committee on Regulating Research of the Section on Science and Technology of the American Bar Association. He continues his research in the history of radiation hazards and legal standards.

INDEX

Boston Studies in the Philosophy of Science

Editor: Robert S. Cohen, *Boston University*

1. M.W. Wartofsky (ed.): *Proceedings of the Boston Colloquium for the Philosophy of Science, 1961/1962.* [Synthese Library 6] 1963 ISBN 90-277-0021-4
2. R.S. Cohen and M.W. Wartofsky (eds.): *Proceedings of the Boston Colloquium for the Philosophy of Science, 1962/1964.* In Honor of P. Frank. [Synthese Library 10] 1965 ISBN 90-277-9004-0
3. R.S. Cohen and M.W. Wartofsky (eds.): *Proceedings of the Boston Colloquium for the Philosophy of Science, 1964/1966.* In Memory of Norwood Russell Hanson. [Synthese Library 14] 1967 ISBN 90-277-0013-3
4. R.S. Cohen and M.W. Wartofsky (eds.): *Proceedings of the Boston Colloquium for the Philosophy of Science, 1966/1968.* [Synthese Library 18] 1969 ISBN 90-277-0014-1
5. R.S. Cohen and M.W. Wartofsky (eds.): *Proceedings of the Boston Colloquium for the Philosophy of Science, 1966/1968.* [Synthese Library 19] 1969 ISBN 90-277-0015-X
6. R.S. Cohen and R.J. Seeger (eds.): *Ernst Mach, Physicist and Philosopher.* [Synthese Library 27] 1970 ISBN 90-277-0016-8
7. M. Čapek: *Bergson and Modern Physics.* A Reinterpretation and Re-evaluation. [Synthese Library 37] 1971 ISBN 90-277-0186-5
8. R.C. Buck and R.S. Cohen (eds.): *PSA 1970.* Proceedings of the 2nd Biennial Meeting of the Philosophy and Science Association (Boston, Fall 1970). In Memory of Rudolf Carnap. [Synthese Library 39] 1971 ISBN 90-277-0187-3; Pb 90-277-0309-4
9. A.A. Zinov'ev: *Foundations of the Logical Theory of Scientific Knowledge (Complex Logic).* Translated from Russian. Revised and enlarged English Edition, with an Appendix by G.A. Smirnov, E.A. Sidorenko, A.M. Fedina and L.A. Bobrova. [Synthese Library 46] 1973 ISBN 90-277-0193-8; Pb 90-277-0324-8
10. L. Tondl: *Scientific Procedures.* A Contribution Concerning the Methodological Problems of Scientific Concepts and Scientific Explanation.Translated from Czech. [Synthese Library 47] 1973 ISBN 90-277-0147-4; Pb 90-277-0323-X
11. R.J. Seeger and R.S. Cohen (eds.): *Philosophical Foundations of Science.* Proceedings of Section L, 1969, American Association for the Advancement of Science. [Synthese Library 58] 1974 ISBN 90-277-0390-6; Pb 90-277-0376-0
12. A. Grünbaum: *Philosophical Problems of Space and Times.* 2nd enlarged ed. [Synthese Library 55] 1973 ISBN 90-277-0357-4; Pb 90-277-0358-2
13. R.S. Cohen and M.W. Wartofsky (eds.): *Logical and Epistemological Studies in Contemporary Physics.* Proceedings of the Boston Colloquium for the Philosophy of Science, 1969/72, Part I. [Synthese Library 59] 1974 ISBN 90-277-0391-4; Pb 90-277-0377-9
14. R.S. Cohen and M.W. Wartofsky (eds.): *Methodological and Historical Essays in the Natural and Social Sciences.* Proceedings of the Boston Colloquium for the Philosophy of Science, 1969/72, Part II. [Synthese Library 60] 1974 ISBN 90-277-0392-2; Pb 90-277-0378-7
15. R.S. Cohen, J.J. Stachel and M.W. Wartofsky (eds.): *For Dirk Strůik.* Scientific, Historical and Political Essays in Honor of Dirk J. Struik. [Synthese Library 61] 1974 ISBN 90-277-0393-0; Pb 90-277-0379-5
16. N. Geschwind: *Selected Papers on Language and the Brains.* [Synthese Library 68] 1974 ISBN 90-277-0262-4; Pb 90-277-0263-2
17. B.G. Kuznetsov: *Reason and Being.* Translated from Russian. Edited by C.R. Fawcett and R.S. Cohen. 1987 ISBN 90-277-2181-5

Boston Studies in the Philosophy of Science

18. P. Mittelstaedt: *Philosophical Problems of Modern Physics.* Translated from the revised 4th German edition by W. Riemer and edited by R.S. Cohen. [Synthese Library 95] 1976
ISBN 90-277-0285-3; Pb 90-277-0506-2
19. H. Mehlberg: *Time, Causality, and the Quantum Theory.* Studies in the Philosophy of Science. Vol. I: *Essay on the Causal Theory of Time.* Vol. II: *Time in a Quantized Universe.* Translated from French. Edited by R.S. Cohen. 1980 Vol. I: ISBN 90-277-0721-9; Pb 90-277-1074-0
Vol. II: ISBN 90-277-1075-9; Pb 90-277-1076-7
20. K.F. Schaffner and R.S. Cohen (eds.): *PSA 1972.* Proceedings of the 3rd Biennial Meeting of the Philosophy of Science Association (Lansing, Michigan, Fall 1972). [Synthese Library 64] 1974 ISBN 90-277-0408-2; Pb 90-277-0409-0
21. R.S. Cohen and J.J. Stachel (eds.): *Selected Papers of Léon Rosenfeld.* [Synthese Library 100] 1979 ISBN 90-277-0651-4; Pb 90-277-0652-2
22. M. Čapek (ed.): *The Concepts of Space and Time.* Their Structure and Their Development. [Synthese Library 74] 1976 ISBN 90-277-0355-8; Pb 90-277-0375-2
23. M. Grene: *The Understanding of Nature.* Essays in the Philosophy of Biology. [Synthese Library 66] 1974 ISBN 90-277-0462-7; Pb 90-277-0463-5
24. D. Ihde: *Technics and Praxis.* A Philosophy of Technology. [Synthese Library 130] 1979
ISBN 90-277-0953-X; Pb 90-277-0954-8
25. J. Hintikka and U. Remes: *The Method of Analysis.* Its Geometrical Origin and Its General Significance. [Synthese Library 75] 1974 ISBN 90-277-0532-1; Pb 90-277-0543-7
26. J.E. Murdoch and E.D. Sylla (eds.): *The Cultural Context of Medieval Learning.* Proceedings of the First International Colloquium on Philosophy, Science, and Theology in the Middle Ages, 1973. [Synthese Library 76] 1975 ISBN 90-277-0560-7; Pb 90-277-0587-9
27. M. Grene and E. Mendelsohn (eds.): *Topics in the Philosophy of Biology.* [Synthese Library 84] 1976 ISBN 90-277-0595-X; Pb 90-277-0596-8
28. J. Agassi: *Science in Flux.* [Synthese Library 80] 1975
ISBN 90-277-0584-4; Pb 90-277-0612-3
29. J.J. Wiatr (ed.): *Polish Essays in the Methodology of the Social Sciences.* [Synthese Library 131] 1979 ISBN 90-277-0723-5; Pb 90-277-0956-4
30. P. Janich: *Protophysics of Time.* Constructive Foundation and History of Time Measurement. Translated from German. 1985 ISBN 90-277-0724-3
31. R.S. Cohen and M.W. Wartofsky (eds.): *Language, Logic, and Method.* 1983
ISBN 90-277-0725-1
32. R.S. Cohen, C.A. Hooker, A.C. Michalos and J.W. van Evra (eds.): *PSA 1974.* Proceedings of the 4th Biennial Meeting of the Philosophy of Science Association. [Synthese Library 101] 1976 ISBN 90-277-0647-6; Pb 90-277-0648-4
33. G. Holton and W.A. Blanpied (eds.): *Science and Its Public.* The Changing Relationship. [Synthese Library 96] 1976 ISBN 90-277-0657-3; Pb 90-277-0658-1
34. M.D. Grmek, R.S. Cohen and G. Cimino (eds.): *On Scientific Discovery.* The 1977 Erice Lectures. 1981 ISBN 90-277-1122-4; Pb 90-277-1123-2
35. S. Amsterdamski: *Between Experience and Metaphysics.* Philosophical Problems of the Evolution of Science. Translated from Polish. [Synthese Library 77] 1975
ISBN 90-277-0568-2; Pb 90-277-0580-1
36. M. Marković and G. Petrović (eds.): *Praxis.* Yugoslav Essays in the Philosophy and Methodology of the Social Sciences. [Synthese Library 134] 1979
ISBN 90-277-0727-8; Pb 90-277-0968-8

Boston Studies in the Philosophy of Science

37. H. von Helmholtz: *Epistemological Writings.* The Paul Hertz / Moritz Schlick Centenary Edition of 1921. Translated from German by M.F. Lowe. Edited with an Introduction and Bibliography by R.S. Cohen and Y. Elkana. [Synthese Library 79] 1977
ISBN 90-277-0290-X; Pb 90-277-0582-8
38. R.M. Martin: *Pragmatics, Truth and Language.* 1979
ISBN 90-277-0992-0; Pb 90-277-0993-9
39. R.S. Cohen, P.K. Feyerabend and M.W. Wartofsky (eds.): *Essays in Memory of Imre Lakatos.* [Synthese Library 99] 1976 ISBN 90-277-0654-9; Pb 90-277-0655-7
40. Not published.
41. Not published.
42. H.R. Maturana and F.J. Varela: *Autopoiesis and Cognition.* The Realization of the Living. With a Preface to "Autopoiesis' by S. Beer. 1980 ISBN 90-277-1015-5; Pb 90-277-1016-3
43. A. Kasher (ed.): *Language in Focus: Foundations, Methods and Systems.* Essays in Memory of Yehoshua Bar-Hillel. [Synthese Library 89] 1976
ISBN 90-277-0644-1; Pb 90-277-0645-X
44. T.D. Thao: *Investigations into the Origin of Language and Consciousness.* 1984
ISBN 90-277-0827-4
45. F.G.-I. Nagasaka (ed.): *Japanese Studies in the Philosophy of Science.* 1997
ISBN 0-7923-4781-1
46. P.L. Kapitza: *Experiment, Theory, Practice.* Articles and Addresses. Edited by R.S. Cohen. 1980 ISBN 90-277-1061-9; Pb 90-277-1062-7
47. M.L. Dalla Chiara (ed.): *Italian Studies in the Philosophy of Science.* 1981
ISBN 90-277-0735-9; Pb 90-277-1073-2
48. M.W. Wartofsky: *Models.* Representation and the Scientific Understanding. [Synthese Library 129] 1979 ISBN 90-277-0736-7; Pb 90-277-0947-5
49. T.D. Thao: *Phenomenology and Dialectical Materialism.* Edited by R.S. Cohen. 1986
ISBN 90-277-0737-5
50. Y. Fried and J. Agassi: *Paranoia.* A Study in Diagnosis. [Synthese Library 102] 1976
ISBN 90-277-0704-9; Pb 90-277-0705-7
51. K.H. Wolff: *Surrender and Cath.* Experience and Inquiry Today. [Synthese Library 105] 1976
ISBN 90-277-0758-8; Pb 90-277-0765-0
52. K. Kosík: *Dialectics of the Concrete.* A Study on Problems of Man and World. 1976
ISBN 90-277-0761-8; Pb 90-277-0764-2
53. N. Goodman: *The Structure of Appearance.* [Synthese Library 107] 1977
ISBN 90-277-0773-1; Pb 90-277-0774-X
54. H.A. Simon: *Models of Discovery* and Other Topics in the Methods of Science. [Synthese Library 114] 1977 ISBN 90-277-0812-6; Pb 90-277-0858-4
55. M. Lazerowitz: *The Language of Philosophy.* Freud and Wittgenstein. [Synthese Library 117] 1977 ISBN 90-277-0826-6; Pb 90-277-0862-2
56. T. Nickles (ed.): *Scientific Discovery, Logic, and Rationality.* 1980
ISBN 90-277-1069-4; Pb 90-277-1070-8
57. J. Margolis: *Persons and Mind.* The Prospects of Nonreductive Materialism. [Synthese Library 121] 1978 ISBN 90-277-0854-1; Pb 90-277-0863-0
58. G. Radnitzky and G. Andersson (eds.): *Progress and Rationality in Science.* [Synthese Library 125] 1978 ISBN 90-277-0921-1; Pb 90-277-0922-X
59. G. Radnitzky and G. Andersson (eds.): *The Structure and Development of Science.* [Synthese Library 136] 1979 ISBN 90-277-0994-7; Pb 90-277-0995-5

Boston Studies in the Philosophy of Science

60. T. Nickles (ed.): *Scientific Discovery.* Case Studies. 1980
ISBN 90-277-1092-9; Pb 90-277-1093-7
61. M.A. Finocchiaro: *Galileo and the Art of Reasoning.* Rhetorical Foundation of Logic and Scientific Method. 1980 ISBN 90-277-1094-5; Pb 90-277-1095-3
62. W.A. Wallace: *Prelude to Galileo.* Essays on Medieval and 16th-Century Sources of Galileo's Thought. 1981 ISBN 90-277-1215-8; Pb 90-277-1216-6
63. F. Rapp: *Analytical Philosophy of Technology.* Translated from German. 1981
ISBN 90-277-1221-2; Pb 90-277-1222-0
64. R.S. Cohen and M.W. Wartofsky (eds.): *Hegel and the Sciences.* 1984 ISBN 90-277-0726-X
65. J. Agassi: *Science and Society.* Studies in the Sociology of Science. 1981
ISBN 90-277-1244-1; Pb 90-277-1245-X
66. L. Tondl: *Problems of Semantics.* A Contribution to the Analysis of the Language of Science. Translated from Czech. 1981 ISBN 90-277-0148-2; Pb 90-277-0316-7
67. J. Agassi and R.S. Cohen (eds.): *Scientific Philosophy Today.* Essays in Honor of Mario Bunge. 1982 ISBN 90-277-1262-X; Pb 90-277-1263-8
68. W. Krajewski (ed.): *Polish Essays in the Philosophy of the Natural Sciences.* Translated from Polish and edited by R.S. Cohen and C.R. Fawcett. 1982
ISBN 90-277-1286-7; Pb 90-277-1287-5
69. J.H. Fetzer: *Scientific Knowledge.* Causation, Explanation and Corroboration. 1981
ISBN 90-277-1335-9; Pb 90-277-1336-7
70. S. Grossberg: *Studies of Mind and Brain.* Neural Principles of Learning, Perception, Development, Cognition, and Motor Control. 1982 ISBN 90-277-1359-6; Pb 90-277-1360-X
71. R.S. Cohen and M.W. Wartofsky (eds.): *Epistemology, Methodology, and the Social Sciences.* 1983. ISBN 90-277-1454-1
72. K. Berka: *Measurement.* Its Concepts, Theories and Problems. Translated from Czech. 1983
ISBN 90-277-1416-9
73. G.L. Pandit: *The Structure and Growth of Scientific Knowledge.* A Study in the Methodology of Epistemic Appraisal. 1983 ISBN 90-277-1434-7
74. A.A. Zinov'ev: *Logical Physics.* Translated from Russian. Edited by R.S. Cohen. 1983
[*see also* Volume 9] ISBN 90-277-0734-0
75. G-G. Granger: *Formal Thought and the Sciences of Man.* Translated from French. With and Introduction by A. Rosenberg. 1983 ISBN 90-277-1524-6
76. R.S. Cohen and L. Laudan (eds.): *Physics, Philosophy and Psychoanalysis.* Essays in Honor of Adolf Grünbaum. 1983 ISBN 90-277-1533-5
77. G. Böhme, W. van den Daele, R. Hohlfeld, W. Krohn and W. Schäfer: *Finalization in Science.* The Social Orientation of Scientific Progress. Translated from German. Edited by W. Schäfer. 1983 ISBN 90-277-1549-1
78. D. Shapere: *Reason and the Search for Knowledge.* Investigations in the Philosophy of Science. 1984 ISBN 90-277-1551-3; Pb 90-277-1641-2
79. G. Andersson (ed.): *Rationality in Science and Politics.* Translated from German. 1984
ISBN 90-277-1575-0; Pb 90-277-1953-5
80. P.T. Durbin and F. Rapp (eds.): *Philosophy and Technology.* [*Also* Philosophy and Technology Series, Vol. 1] 1983 ISBN 90-277-1576-9
81. M. Marković: *Dialectical Theory of Meaning.* Translated from Serbo-Croat. 1984
ISBN 90-277-1596-3
82. R.S. Cohen and M.W. Wartofsky (eds.): *Physical Sciences and History of Physics.* 1984.
ISBN 90-277-1615-3

Boston Studies in the Philosophy of Science

83. É. Meyerson: *The Relativistic Deduction.* Epistemological Implications of the Theory of Relativity. Translated from French. With a Review by Albert Einstein and an Introduction by Milić Čapek. 1985 ISBN 90-277-1699-4
84. R.S. Cohen and M.W. Wartofsky (eds.): *Methodology, Metaphysics and the History of Science.* In Memory of Benjamin Nelson. 1984 ISBN 90-277-1711-7
85. G. Tamás: *The Logic of Categories.* Translated from Hungarian. Edited by R.S. Cohen. 1986 ISBN 90-277-1742-7
86. S.L. de C. Fernandes: *Foundations of Objective Knowledge.* The Relations of Popper's Theory of Knowledge to That of Kant. 1985 ISBN 90-277-1809-1
87. R.S. Cohen and T. Schnelle (eds.): *Cognition and Fact.* Materials on Ludwik Fleck. 1986 ISBN 90-277-1902-0
88. G. Freudenthal: *Atom and Individual in the Age of Newton.* On the Genesis of the Mechanistic World View. Translated from German. 1986 ISBN 90-277-1905-5
89. A. Donagan, A.N. Perovich Jr and M.V. Wedin (eds.): *Human Nature and Natural Knowledge.* Essays presented to Marjorie Grene on the Occasion of Her 75th Birthday. 1986 ISBN 90-277-1974-8
90. C. Mitcham and A. Hunning (eds.): *Philosophy and Technology II.* Information Technology and Computers in Theory and Practice. [*Also* Philosophy and Technology Series, Vol. 2] 1986 ISBN 90-277-1975-6
91. M. Grene and D. Nails (eds.): *Spinoza and the Sciences.* 1986 ISBN 90-277-1976-4
92. S.P. Turner: *The Search for a Methodology of Social Science.* Durkheim, Weber, and the 19th-Century Problem of Cause, Probability, and Action. 1986. ISBN 90-277-2067-3
93. I.C. Jarvie: *Thinking about Society.* Theory and Practice. 1986 ISBN 90-277-2068-1
94. E. Ullmann-Margalit (ed.): *The Kaleidoscope of Science.* The Israel Colloquium: Studies in History, Philosophy, and Sociology of Science, Vol. 1. 1986 ISBN 90-277-2158-0; Pb 90-277-2159-9
95. E. Ullmann-Margalit (ed.): *The Prism of Science.* The Israel Colloquium: Studies in History, Philosophy, and Sociology of Science, Vol. 2. 1986 ISBN 90-277-2160-2; Pb 90-277-2161-0
96. G. Márkus: *Language and Production.* A Critique of the Paradigms. Translated from French. 1986 ISBN 90-277-2169-6
97. F. Amrine, F.J. Zucker and H. Wheeler (eds.): *Goethe and the Sciences: A Reappraisal.* 1987 ISBN 90-277-2265-X; Pb 90-277-2400-8
98. J.C. Pitt and M. Pera (eds.): *Rational Changes in Science.* Essays on Scientific Reasoning. Translated from Italian. 1987 ISBN 90-277-2417-2
99. O. Costa de Beauregard: *Time, the Physical Magnitude.* 1987 ISBN 90-277-2444-X
100. A. Shimony and D. Nails (eds.): *Naturalistic Epistemology.* A Symposium of Two Decades. 1987 ISBN 90-277-2337-0
101. N. Rotenstreich: *Time and Meaning in History.* 1987 ISBN 90-277-2467-9
102. D.B. Zilberman: *The Birth of Meaning in Hindu Thought.* Edited by R.S. Cohen. 1988 ISBN 90-277-2497-0
103. T.F. Glick (ed.): *The Comparative Reception of Relativity.* 1987 ISBN 90-277-2498-9
104. Z. Harris, M. Gottfried, T. Ryckman, P. Mattick Jr, A. Daladier, T.N. Harris and S. Harris: *The Form of Information in Science.* Analysis of an Immunology Sublanguage. With a Preface by Hilary Putnam. 1989 ISBN 90-277-2516-0
105. F. Burwick (ed.): *Approaches to Organic Form.* Permutations in Science and Culture. 1987 ISBN 90-277-2541-1

Boston Studies in the Philosophy of Science

106. M. Almási: *The Philosophy of Appearances.* Translated from Hungarian. 1989 ISBN 90-277-2150-5
107. S. Hook, W.L. O'Neill and R. O'Toole (eds.): *Philosophy, History and Social Action.* Essays in Honor of Lewis Feuer. With an Autobiographical Essay by L. Feuer. 1988 ISBN 90-277-2644-2
108. I. Hronszky, M. Fehér and B. Dajka: *Scientific Knowledge Socialized.* Selected Proceedings of the 5th Joint International Conference on the History and Philosophy of Science organized by the IUHPS (Veszprém, Hungary, 1984). 1988 ISBN 90-277-2284-6
109. P. Tillers and E.D. Green (eds.): *Probability and Inference in the Law of Evidence.* The Uses and Limits of Bayesianism. 1988 ISBN 90-277-2689-2
110. E. Ullmann-Margalit (ed.): *Science in Reflection.* The Israel Colloquium: Studies in History, Philosophy, and Sociology of Science, Vol. 3. 1988 ISBN 90-277-2712-0; Pb 90-277-2713-9
111. K. Gavroglu, Y. Goudaroulis and P. Nicolacopoulos (eds.): *Imre Lakatos and Theories of Scientific Change.* 1989 ISBN 90-277-2766-X
112. B. Glassner and J.D. Moreno (eds.): *The Qualitative-Quantitative Distinction in the Social Sciences.* 1989 ISBN 90-277-2829-1
113. K. Arens: *Structures of Knowing.* Psychologies of the 19th Century. 1989 ISBN 0-7923-0009-2
114. A. Janik: *Style, Politics and the Future of Philosophy.* 1989 ISBN 0-7923-0056-4
115. F. Amrine (ed.): *Literature and Science as Modes of Expression.* With an Introduction by S. Weininger. 1989 ISBN 0-7923-0133-1
116. J.R. Brown and J. Mittelstrass (eds.): *An Intimate Relation.* Studies in the History and Philosophy of Science. Presented to Robert E. Butts on His 60th Birthday. 1989 ISBN 0-7923-0169-2
117. F. D'Agostino and I.C. Jarvie (eds.): *Freedom and Rationality.* Essays in Honor of John Watkins. 1989 ISBN 0-7923-0264-8
118. D. Zolo: *Reflexive Epistemology.* The Philosophical Legacy of Otto Neurath. 1989 ISBN 0-7923-0320-2
119. M. Kearn, B.S. Philips and R.S. Cohen (eds.): *Georg Simmel and Contemporary Sociology.* 1989 ISBN 0-7923-0407-1
120. T.H. Levere and W.R. Shea (eds.): *Nature, Experiment and the Science.* Essays on Galileo and the Nature of Science. In Honour of Stillman Drake. 1989 ISBN 0-7923-0420-9
121. P. Nicolacopoulos (ed.): *Greek Studies in the Philosophy and History of Science.* 1990 ISBN 0-7923-0717-8
122. R. Cooke and D. Costantini (eds.): *Statistics in Science.* The Foundations of Statistical Methods in Biology, Physics and Economics. 1990 ISBN 0-7923-0797-6
123. P. Duhem: *The Origins of Statics.* Translated from French by G.F. Leneaux, V.N. Vagliente and G.H. Wagner. With an Introduction by S.L. Jaki. 1991 ISBN 0-7923-0898-0
124. H. Kamerlingh Onnes: *Through Measurement to Knowledge.* The Selected Papers, 1853-1926. Edited and with an Introduction by K. Gavroglu and Y. Goudaroulis. 1991 ISBN 0-7923-0825-5
125. M. Čapek: *The New Aspects of Time: Its Continuity and Novelties.* Selected Papers in the Philosophy of Science. 1991 ISBN 0-7923-0911-1
126. S. Unguru (ed.): *Physics, Cosmology and Astronomy, 1300–1700.* Tension and Accommodation. 1991 ISBN 0-7923-1022-5

127. Z. Bechler: *Newton's Physics on the Conceptual Structure of the Scientific Revolution.* 1991 ISBN 0-7923-1054-3
128. É. Meyerson: *Explanation in the Sciences.* Translated from French by M-A. Siple and D.A. Siple. 1991 ISBN 0-7923-1129-9
129. A.I. Tauber (ed.): *Organism and the Origins of Self.* 1991 ISBN 0-7923-1185-X
130. F.J. Varela and J-P. Dupuy (eds.): *Understanding Origins.* Contemporary Views on the Origin of Life, Mind and Society. 1992 ISBN 0-7923-1251-1
131. G.L. Pandit: *Methodological Variance.* Essays in Epistemological Ontology and the Methodology of Science. 1991 ISBN 0-7923-1263-5
132. G. Munévar (ed.): *Beyond Reason.* Essays on the Philosophy of Paul Feyerabend. 1991 ISBN 0-7923-1272-4
133. T.E. Uebel (ed.): *Rediscovering the Forgotten Vienna Circle.* Austrian Studies on Otto Neurath and the Vienna Circle. Partly translated from German. 1991 ISBN 0-7923-1276-7
134. W.R. Woodward and R.S. Cohen (eds.): *World Views and Scientific Discipline Formation.* Science Studies in the [former] German Democratic Republic. Partly translated from German by W.R. Woodward. 1991 ISBN 0-7923-1286-4
135. P. Zambelli: *The Speculum Astronomiae and Its Enigma.* Astrology, Theology and Science in Albertus Magnus and His Contemporaries. 1992 ISBN 0-7923-1380-1
136. P. Petitjean, C. Jami and A.M. Moulin (eds.): *Science and Empires.* Historical Studies about Scientific Development and European Expansion. ISBN 0-7923-1518-9
137. W.A. Wallace: *Galileo's Logic of Discovery and Proof.* The Background, Content, and Use of His Appropriated Treatises on Aristotle's *Posterior Analytics.* 1992 ISBN 0-7923-1577-4
138. W.A. Wallace: *Galileo's Logical Treatises.* A Translation, with Notes and Commentary, of His Appropriated Latin Questions on Aristotle's *Posterior Analytics.* 1992 ISBN 0-7923-1578-2
Set (137 + 138) ISBN 0-7923-1579-0
139. M.J. Nye, J.L. Richards and R.H. Stuewer (eds.): *The Invention of Physical Science.* Intersections of Mathematics, Theology and Natural Philosophy since the Seventeenth Century. Essays in Honor of Erwin N. Hiebert. 1992 ISBN 0-7923-1753-X
140. G. Corsi, M.L. dalla Chiara and G.C. Ghirardi (eds.): *Bridging the Gap: Philosophy, Mathematics and Physics.* Lectures on the Foundations of Science. 1992 ISBN 0-7923-1761-0
141. C.-H. Lin and D. Fu (eds.): *Philosophy and Conceptual History of Science in Taiwan.* 1992 ISBN 0-7923-1766-1
142. S. Sarkar (ed.): *The Founders of Evolutionary Genetics.* A Centenary Reappraisal. 1992 ISBN 0-7923-1777-7
143. J. Blackmore (ed.): *Ernst Mach – A Deeper Look.* Documents and New Perspectives. 1992 ISBN 0-7923-1853-6
144. P. Kroes and M. Bakker (eds.): *Technological Development and Science in the Industrial Age.* New Perspectives on the Science–Technology Relationship. 1992 ISBN 0-7923-1898-6
145. S. Amsterdamski: *Between History and Method.* Disputes about the Rationality of Science. 1992 ISBN 0-7923-1941-9
146. E. Ullmann-Margalit (ed.): *The Scientific Enterprise.* The Bar-Hillel Colloquium: Studies in History, Philosophy, and Sociology of Science, Volume 4. 1992 ISBN 0-7923-1992-3
147. L. Embree (ed.): *Metaarchaeology.* Reflections by Archaeologists and Philosophers. 1992 ISBN 0-7923-2023-9
148. S. French and H. Kamminga (eds.): *Correspondence, Invariance and Heuristics.* Essays in Honour of Heinz Post. 1993 ISBN 0-7923-2085-9
149. M. Bunzl: *The Context of Explanation.* 1993 ISBN 0-7923-2153-7

Boston Studies in the Philosophy of Science

150. I.B. Cohen (ed.): *The Natural Sciences and the Social Sciences.* Some Critical and Historical Perspectives. 1994 ISBN 0-7923-2223-1
151. K. Gavroglu, Y. Christianidis and E. Nicolaidis (eds.): *Trends in the Historiography of Science.* 1994 ISBN 0-7923-2255-X
152. S. Poggi and M. Bossi (eds.): *Romanticism in Science.* Science in Europe, 1790–1840. 1994 ISBN 0-7923-2336-X
153. J. Faye and H.J. Folse (eds.): *Niels Bohr and Contemporary Philosophy.* 1994 ISBN 0-7923-2378-5
154. C.C. Gould and R.S. Cohen (eds.): *Artifacts, Representations, and Social Practice.* Essays for Marx W. Wartofsky. 1994 ISBN 0-7923-2481-1
155. R.E. Butts: *Historical Pragmatics.* Philosophical Essays. 1993 ISBN 0-7923-2498-6
156. R. Rashed: *The Development of Arabic Mathematics: Between Arithmetic and Algebra.* Translated from French by A.F.W. Armstrong. 1994 ISBN 0-7923-2565-6
157. I. Szumilewicz-Lachman (ed.): *Zygmunt Zawirski: His Life and Work.* With Selected Writings on Time, Logic and the Methodology of Science. Translations by Feliks Lachman. Ed. by R.S. Cohen, with the assistance of B. Bergo. 1994 ISBN 0-7923-2566-4
158. S.N. Haq: *Names, Natures and Things.* The Alchemist Jābir ibn Ḥayyān and His *Kitāb al-Aḥjār* (Book of Stones). 1994 ISBN 0-7923-2587-7
159. P. Plaass: *Kant's Theory of Natural Science.* Translation, Analytic Introduction and Commentary by Alfred E. and Maria G. Miller. 1994 ISBN 0-7923-2750-0
160. J. Misiek (ed.): *The Problem of Rationality in Science and its Philosophy.* On Popper vs. Polanyi. The Polish Conferences 1988–89. 1995 ISBN 0-7923-2925-2
161. I.C. Jarvie and N. Laor (eds.): *Critical Rationalism, Metaphysics and Science.* Essays for Joseph Agassi, Volume I. 1995 ISBN 0-7923-2960-0
162. I.C. Jarvie and N. Laor (eds.): *Critical Rationalism, the Social Sciences and the Humanities.* Essays for Joseph Agassi, Volume II. 1995 ISBN 0-7923-2961-9
Set (161–162) ISBN 0-7923-2962-7
163. K. Gavroglu, J. Stachel and M.W. Wartofsky (eds.): *Physics, Philosophy, and the Scientific Community.* Essays in the Philosophy and History of the Natural Sciences and Mathematics. In Honor of Robert S. Cohen. 1995 ISBN 0-7923-2988-0
164. K. Gavroglu, J. Stachel and M.W. Wartofsky (eds.): *Science, Politics and Social Practice.* Essays on Marxism and Science, Philosophy of Culture and the Social Sciences. In Honor of Robert S. Cohen. 1995 ISBN 0-7923-2989-9
165. K. Gavroglu, J. Stachel and M.W. Wartofsky (eds.): *Science, Mind and Art.* Essays on Science and the Humanistic Understanding in Art, Epistemology, Religion and Ethics. Essays in Honor of Robert S. Cohen. 1995 ISBN 0-7923-2990-2
Set (163–165) ISBN 0-7923-2991-0
166. K.H. Wolff: *Transformation in the Writing.* A Case of Surrender-and-Catch. 1995 ISBN 0-7923-3178-8
167. A.J. Kox and D.M. Siegel (eds.): *No Truth Except in the Details.* Essays in Honor of Martin J. Klein. 1995 ISBN 0-7923-3195-8
168. J. Blackmore: *Ludwig Boltzmann, His Later Life and Philosophy, 1900–1906.* Book One: A Documentary History. 1995 ISBN 0-7923-3231-8
169. R.S. Cohen, R. Hilpinen and R. Qiu (eds.): *Realism and Anti-Realism in the Philosophy of Science.* Beijing International Conference, 1992. 1996 ISBN 0-7923-3233-4
170. I. Kuçuradi and R.S. Cohen (eds.): *The Concept of Knowledge.* The Ankara Seminar. 1995 ISBN 0-7923-3241-5

Boston Studies in the Philosophy of Science

171. M.A. Grodin (ed.): *Meta Medical Ethics*: The Philosophical Foundations of Bioethics. 1995 ISBN 0-7923-3344-6
172. S. Ramirez and R.S. Cohen (eds.): *Mexican Studies in the History and Philosophy of Science.* 1995 ISBN 0-7923-3462-0
173. C. Dilworth: *The Metaphysics of Science.* An Account of Modern Science in Terms of Principles, Laws and Theories. 1995 ISBN 0-7923-3693-3
174. J. Blackmore: *Ludwig Boltzmann, His Later Life and Philosophy, 1900–1906* Book Two: The Philosopher. 1995 ISBN 0-7923-3464-7
175. P. Damerow: *Abstraction and Representation.* Essays on the Cultural Evolution of Thinking. 1996 ISBN 0-7923-3816-2
176. M.S. Macrakis: *Scarcity's Ways: The Origins of Capital.* A Critical Essay on Thermodynamics, Statistical Mechanics and Economics. 1997 ISBN 0-7923-4760-9
177. M. Marion and R.S. Cohen (eds.): *Québec Studies in the Philosophy of Science.* Part I: Logic, Mathematics, Physics and History of Science. Essays in Honor of Hugues Leblanc. 1995 ISBN 0-7923-3559-7
178. M. Marion and R.S. Cohen (eds.): *Québec Studies in the Philosophy of Science.* Part II: Biology, Psychology, Cognitive Science and Economics. Essays in Honor of Hugues Leblanc. 1996 ISBN 0-7923-3560-0
     Set (177–178) ISBN 0-7923-3561-9
179. Fan Dainian and R.S. Cohen (eds.): *Chinese Studies in the History and Philosophy of Science and Technology.* 1996 ISBN 0-7923-3463-9
180. P. Forman and J.M. Sánchez-Ron (eds.): *National Military Establishments and the Advancement of Science and Technology.* Studies in 20th Century History. 1996 ISBN 0-7923-3541-4
181. E.J. Post: *Quantum Reprogramming.* Ensembles and Single Systems: A Two-Tier Approach to Quantum Mechanics. 1995 ISBN 0-7923-3565-1
182. A.I. Tauber (ed.): *The Elusive Synthesis: Aesthetics and Science.* 1996 ISBN 0-7923-3904-5
183. S. Sarkar (ed.): *The Philosophy and History of Molecular Biology: New Perspectives.* 1996 ISBN 0-7923-3947-9
184. J.T. Cushing, A. Fine and S. Goldstein (eds.): *Bohmian Mechanics and Quantum Theory: An Appraisal.* 1996 ISBN 0-7923-4028-0
185. K. Michalski: *Logic and Time.* An Essay on Husserl's Theory of Meaning. 1996 ISBN 0-7923-4082-5
186. G. Munévar (ed.): *Spanish Studies in the Philosophy of Science.* 1996 ISBN 0-7923-4147-3
187. G. Schubring (ed.): *Hermann Günther Graßmann (1809–1877): Visionary Mathematician, Scientist and Neohumanist Scholar.* Papers from a Sesquicentennial Conference. 1996 ISBN 0-7923-4261-5
188. M. Bitbol: *Schrödinger's Philosophy of Quantum Mechanics.* 1996 ISBN 0-7923-4266-6
189. J. Faye, U. Scheffler and M. Urchs (eds.): *Perspectives on Time.* 1997 ISBN 0-7923-4330-1
190. K. Lehrer and J.C. Marek (eds.): *Austrian Philosophy Past and Present.* Essays in Honor of Rudolf Haller. 1996 ISBN 0-7923-4347-6
191. J.L. Lagrange: *Analytical Mechanics.* Translated and edited by Auguste Boissonade and Victor N. Vagliente. Translated from the *Mécanique Analytique, novelle édition* of 1811. 1997 ISBN 0-7923-4349-2
192. D. Ginev and R.S. Cohen (eds.): *Issues and Images in the Philosophy of Science.* Scientific and Philosophical Essays in Honour of Azarya Polikarov. 1997 ISBN 0-7923-4444-8

Boston Studies in the Philosophy of Science

193. R.S. Cohen, M. Horne and J. Stachel (eds.): *Experimental Metaphysics.* Quantum Mechanical Studies for Abner Shimony, Volume One. 1997 ISBN 0-7923-4452-9
194. R.S. Cohen, M. Horne and J. Stachel (eds.): *Potentiality, Entanglement and Passion-at-a-Distance.* Quantum Mechanical Studies for Abner Shimony, Volume Two. 1997 ISBN 0-7923-4453-7; Set 0-7923-4454-5
195. R.S. Cohen and A.I. Tauber (eds.): *Philosophies of Nature: The Human Dimension.* 1997 ISBN 0-7923-4579-7
196. M. Otte and M. Panza (eds.): *Analysis and Synthesis in Mathematics.* History and Philosophy. 1997 ISBN 0-7923-4570-3
197. A. Denkel: *The Natural Background of Meaning.* 1999 ISBN 0-7923-5331-5
198. D. Baird, R.I.G. Hughes and A. Nordmann (eds.): *Heinrich Hertz: Classical Physicist, Modern Philosopher.* 1999 ISBN 0-7923-4653-X
199. A. Franklin: *Can That be Right?* Essays on Experiment, Evidence, and Science. 1999 ISBN 0-7923-5464-8
200. D. Raven, W. Krohn and R.S. Cohen (eds.): *The Social Origins of Modern Science.* 2000 ISBN 0-7923-6457-0
201. Reserved
202. Reserved
203. B. Babich and R.S. Cohen (eds.): *Nietzsche, Theories of Knowledge, and Critical Theory.* Nietzsche and the Sciences I. 1999 ISBN 0-7923-5742-6
204. B. Babich and R.S. Cohen (eds.): *Nietzsche, Epistemology, and Philosophy of Science.* Nietzsche and the Science II. 1999 ISBN 0-7923-5743-4
205. R. Hooykaas: *Fact, Faith and Fiction in the Development of Science.* The Gifford Lectures given in the University of St Andrews 1976. 1999 ISBN 0-7923-5774-4
206. M. Fehér, O. Kiss and L. Ropolyi (eds.): *Hermeneutics and Science.* 1999 ISBN 0-7923-5798-1
207. R.M. MacLeod (ed.): *Science and the Pacific War.* Science and Survival in the Pacific, 1939-1945. 1999 ISBN 0-7923-5851-1
208. I. Hanzel: *The Concept of Scientific Law in the Philosophy of Science and Epistemology.* A Study of Theoretical Reason. 1999 ISBN 0-7923-5852-X
209. G. Helm; R.J. Deltete (ed./transl.): *The Historical Development of Energetics.* 1999 ISBN 0-7923-5874-0
210. A. Orenstein and P. Kotatko (eds.): *Knowledge, Language and Logic.* Questions for Quine. 1999 ISBN 0-7923-5986-0
211. R.S. Cohen and H. Levine (eds.): *Maimonides and the Sciences.* 2000 ISBN 0-7923-6053-2
212. H. Gourko, D.I. Williamson and A.I. Tauber (eds.): *The Evolutionary Biology Papers of Elie Metchnikoff.* 2000 ISBN 0-7923-6067-2
213. S. D'Agostino: *A History of the Ideas of Theoretical Physics.* Essays on the Nineteenth and Twentieth Century Physics. 2000 ISBN 0-7923-6094-X
214. S. Lelas: *Science and Modernity.* Toward An Integral Theory of Science. 2000 ISBN 0-7923-6303-5
215. E. Agazzi and M. Pauri (eds.): *The Reality of the Unobservable.* Observability, Unobservability and Their Impact on the Issue of Scientific Realism. 2000 ISBN 0-7923-6311-6
216. P. Hoyningen-Huene and H. Sankey (eds.): *Incommensurability and Related Matters.* 2001 ISBN 0-7923-6989-0
217. A. Nieto-Galan: *Colouring Textiles.* A History of Natural Dyestuffs in Industrial Europe. 2001 ISBN 0-7923-7022-8

Boston Studies in the Philosophy of Science

218. J. Blackmore, R. Itagaki and S. Tanaka (eds.): *Ernst Mach's Vienna 1895–1930*. Or Phenomenalism as Philosophy of Science. 2001 ISBN 0-7923-7122-4
219. R. Vihalemm (ed.): *Estonian Studies in the History and Philosophy of Science*. 2001 ISBN 0-7923-7189-5
220. W. Lefèvre (ed.): *Between Leibniz, Newton, and Kant*. Philosophy and Science in the Eighteenth Century. 2001 ISBN 0-7923-7198-4
221. T.F. Glick, M.Á. Puig-Samper and R. Ruiz (eds.): *The Reception of Darwinism in the Iberian World*. Spain, Spanish America and Brazil. 2001 ISBN 1-4020-0082-0
222. U. Klein (ed.): *Tools and Modes of Representation in the Laboratory Sciences*. 2001 ISBN 1-4020-0100-2
223. P. Duhem: *Mixture and Chemical Combination*. And Related Essays. Edited and translated, with an introduction, by Paul Needham. 2002 ISBN 1-4020-0232-7
224. J.C. Boudri: *What was Mechanical about Mechanics*. The Concept of Force Betweem Metaphysics and Mechanics from Newton to Lagrange. 2002 ISBN 1-4020-0233-5
225. B.E. Babich (ed.): *Hermeneutic Philosophy of Science, Van Gogh's Eyes, and God*. Essays in Honor of Patrick A. Heelan, S.J. 2002 ISBN 1-4020-0234-3
226. D.D. Villemaire: *The Metaphysical Foundations of Modern Physical Science and E.A. Burtt: Historian and Philosopher*. 2002 ISBN 1-4020-0428-1
227. L.J. Cohen: *Knowledge and Language*. Selected Essays of L. Jonathan Cohen. Edited and with an introduction by James Logue. 2002 ISBN 1-4020-0474-5
228. G.E. Allen and R.M. MacLeod (eds.): *Science, History and Social Activism: A Tribute to Everett Mendelsohn*. 2002 ISBN 1-4020-0495-0

Also of interest:

R.S. Cohen and M.W. Wartofsky (eds.): *A Portrait of Twenty-Five Years Boston Colloquia for the Philosophy of Science, 1960-1985*. 1985 ISBN Pb 90-277-1971-3

Previous volumes are still available.

KLUWER ACADEMIC PUBLISHERS – DORDRECHT / BOSTON / LONDON

